Machine Learning for Computer Scientists and Data Analysts

Setareh Rafatirad • Houman Homayoun •
Zhiqian Chen • Sai Manoj Pudukotai Dinakarrao

Machine Learning for Computer Scientists and Data Analysts

From an Applied Perspective

Springer

Setareh Rafatirad
George Mason University
Fairfax, VA, USA

Houman Homayoun
University of California, Davis
Davis, CA, USA

Zhiqian Chen
Mississippi State University
Mississippi State, MS, USA

Sai Manoj Pudukotai Dinakarrao
George Mason University
Fairfax, VA, USA

ISBN 978-3-030-96755-0 ISBN 978-3-030-96756-7 (eBook)
https://doi.org/10.1007/978-3-030-96756-7

This Springer imprint is published by the registered company Springer Nature Switzerland AG
The registered company address is: Gewerbestrasse 11, 6330 Cham, Switzerland

Preface

The recent popularity gained by the field of machine learning (ML) has led to its adaptation into almost all the known applications. The applications range from smart homes, smart grids, and forex markets to military applications and autonomous drones. There exists a plethora of machine learning techniques that were introduced in the past few years, and each of these techniques fits greatly for a specific set of applications rather than a one-size-fits-all approach.

In order to better determine the application of ML for a given problem, it is non-trivial to understand the current state of the art of the existing ML techniques, pros and cons, their behavior, and existing applications that have already adopted them. This book thus aims at researchers and practitioners who are familiar with their application requirements, and are interested in the application of ML techniques in their applications not only for better performance but also for ensuring that the adopted ML technique is not an overkill to the considered application. We hope that this book will provide a structured introduction and relevant background to aspiring engineers who are new to the field, while also helping to revise the background for the researchers familiar with this field. This introduction will be further used to build and introduce current and emerging ML paradigms and their applications in multiple case studies.

Organization This book is organized into three parts that consist of multiple chapters. The first part introduces the relevant background information pertaining to ML, traditional learning approaches that are widely used.

- Chapter 1 introduces the concept of applied machine learning. The metrics used for evaluating the machine learning performance, data pre-processing, and techniques to visualize and analyze the outputs (classification or regression or other applications) are discussed.
- Chapter 2 presents a brief review of the probability theory and linear algebra that are essential for a better understanding of the ML techniques discussed in the later parts of the book.

- Chapter 3 introduces the machine learning techniques. Supervised learning is primarily discussed in this chapter. Multiple supervised learning techniques, learning techniques, and applications along with pros and cons for each of the techniques are discussed. A qualitative comparison of different supervised learning techniques is presented along with their suitability to different kinds of applications.
- Unsupervised learning is introduced in Chap. 4. The differences compared to the supervised learning and application scenarios are discussed first. Different supervised learning for different applications including classification and feature selection is discussed along with examples in this chapter.
- Reinforcement learning is a human learning-inspired technique, which can be laid between supervised and unsupervised learning techniques in the spectrum. Chapter 5 discusses the basics of reinforcement learning along with its variants together with a comparison among different techniques.

Building on top of the basic concepts of machine learning, advanced machine learning techniques used in real-world applications are discussed in the second part of this book.

- The majority of the supervised learning techniques and their learning mechanisms discussed in the first part of this book focus on offline or batch learning. However, the learning in real-world applications needs to happen in an online manner. As such, Chap. 6 introduces the online learning technique and different variants of online learning techniques.
- With a diverse spectrum of Web applications demonstrating the importance of learning from user behavior, recommender systems are widely used by the bulk of social media companies. Chapter 7 of this book discusses approaches for recommender learning.
- Chapter 8 offers approaches for graph learning. Graphs are used to depict things and their connections in a variety of real-world applications, including social networking, transportation, and disease spreading. Methods for learning graphs and the relationships between nodes are discussed.
- In addition to advancements in machine learning algorithms, researchers have also focused on exploiting the vulnerabilities in machine learning techniques. Chapter 9 introduces adversarial machine learning techniques that discuss techniques to inject the adversarial perturbations into the input samples to mislead the machine learning algorithms. In addition, the techniques to harden the machine learning techniques against these adversarial perturbations are discussed.

In addition to the advanced learning techniques, the application of machine learning algorithms with entire discussions dedicated to real-world applications is presented in the third part of this book.

- The application of machine learning techniques for health monitoring is one of the critical real-world applications, especially with the introduction of wearable devices including fitness trackers. Chapters 10 and 11 focus on the application of machine learning techniques for health applications, particularly in the context of wearable devices.
- Another pivotal application of machine learning is anomaly detection in the context of security. Here, security refers to the security of the computing systems including mobile devices. Chapter 12 focuses on the application of machine learning to detect malware applications in resource-constrained devices, where lightweight machine learning techniques are preferable compared to heavy deep learning techniques.
- In contrast to other applications discussed, the final chapter of this book discusses the application of machine learning for cloud resource management applications. In particular, memory management, and resource distribution according to the workload in a cognitive manner through machine learning techniques is discussed in this chapter.

What's New?
Numerous publications exist that give readers theoretical insights, and similarly, there are books that focus on practical implementation through programming exercises. However, our proposed book incorporates theoretical and practical perspectives, as well as real-world case studies, and covers advanced machine learning ideas. Additionally, this book contains various case studies, examples, and solutions covering topics ranging from simple forecasting to enormous network optimization and housing price prediction employing a massive database. Finally, this book includes real implementation examples and exercises that allow readers to practice and enhance their programming skills for machine learning applications.

Scope of Book
This book introduces the theoretical aspects of machine learning (ML) algorithms starting from simple neuron basics all the way to the complex neural networks including generative adversarial neural networks and graph convolutional networks. Most importantly, this book helps the readers in understanding the concepts of ML algorithms and provides the necessary skills for the reader to choose an apt ML algorithm for a problem that the reader wishes to solve.

Acknowledgements
The authors of this book would like to thank the colleagues at George Mason University, University of California Davis, and Mississippi State University, especially the members of the Hardware Architecture and Artificial Intelligence (HArt) lab. We would also like to express our deepest appreciation to the following faculty members and students for their support: Abhijit Dhavlle (GMU), Sanket Shukla (GMU), Sreenitha Kasarapu (GMU), Sathwika Bavikadi (GMU), Ali Mirzaein (GMU),

- "What Is Applied Machine Learning?": Mahdi Orooji (University of California Davis), Mitra Rezaei, Roya Paridar
- "Reinforcement Learning": Qi Zhang (University of South Carolina)

- "Online Learning": Shuo Lei (Virginia Tech), Yifeng Gao (University of Texas Rio Grande Valley), Xuchao Zhang (NEC Labs America)
- "Recommender Learning": Shanshan Feng (Harbin Institute of Technology, Shenzhen), Kaiqi Zhao (University of Auckland)
- "Graph Learning": Liang Zhao (Emory University)
- "SensorNet: An Educational Neural Network Framework for 167 Low-Power Multimodal Data Classification": Tinoosh Mohsenin (University of Maryland Baltimore County), Arnab Mazumder (University of Maryland Baltimore County), Hasib-Al- Rashid (University of Maryland Baltimore County)
- "Transfer Learning in Mobile Health": Hassan Ghasemzadeh (Arizona State University)
- "Applied Machine Learning for Computer Architecture Security": Hossein Sayadi (California State University, Long Beach)
- "Applied Machine Learning for Cloud Resource Management": Hossein Mohammadi Makrani (University of California Davis), Najme Nazari (University of California Davis)

Kaiqi Zhao (University of Auckland), Shanshan Feng (Harbin Institute of Technology, Shenzhen), Xuchao Zhang (NEC Labs America), Yifeng Gao (University of Texas Rio Grande Valley), Shuo Lei (Virginia Tech), Zonghan Zhang (Mississippi State University), and Qi Zhang (University of South Carolina).

Fairfax, VA, USA — Sai Manoj Pudukotai Dinakarrao
Fairfax, VA, USA — Setareh Rafatirad
Davis, CA, USA — Houman Homayoun
Mississippi State, MS, USA — Zhiqian Chen
November 2021

Contents

Part I
Basics of Machine Learning

Chapter 1
What Is Applied Machine Learning?

We begin this chapter by discussing the importance of understanding data in order to address various problems about the distribution of data, significant features, how to transform features, and how to construct models to perform a specific machine learning task in various problem domains. Let us begin our conversation with Artificial Intelligence (AI), a collection of concepts that enables computers to mimic human behavior. The primary objective of the field of artificial intelligence is to develop artificial algorithms that can be used to inform intelligent future judgments. Machine learning (ML) is an area of artificial intelligence that is concerned with instructing/training an algorithm to execute such tasks. It is a scientific technique for uncovering hidden patterns and conclusions in structured and unstructured data by building mathematical models using a sample dataset referred to as *training set*. Computing systems use machine learning models to transform data into actionable results and carry out specific tasks, such as detecting malicious activity in an IoT system, classifying an object in an autonomous driving application, or discovering interesting correlations between variables in a patient dataset in a health application domain. Machine learning algorithms include *regression, instance-based learning, regularization, decision tree, Bayesian, clustering, association-rule learning, reinforcement learning, support vector machines, ensemble learning, artificial neural network, deep learning, adversarial learning, federated learning, zero-shot learning*, and *explainable machine learning*.

Requirements of such techniques and applications will be discussed in the first part of this book.

1.1 Introduction

Machine learning can be approached in two distinct ways: theoretical machine learning and applied machine learning (Applied ML). Both the paths empower an individual to solve problems in disparate ways. Theoretical machine learning is con-

S. Rafatirad et al., *Machine Learning for Computer Scientists and Data Analysts*,
https://doi.org/10.1007/978-3-030-96756-7_1

cerned with an understanding of the fundamental concepts behind machine learning algorithms, mathematics, statistics, and probability theory. However, applied machine learning is about achieving the potential and impact of theoretical machine learning developments. Thus, the purpose of Applied Machine Learning is to get a sufficient understanding of fundamental machine learning principles and to address real-world issues utilizing tools and frameworks that incorporate machine learning algorithms. It is concerned with developing a workable learning system for a particular application. Indeed, skill in applied machine learning comes from solving numerous issues sequentially in multiple areas, which requires a grasp of the data and the challenges encountered. This is not an easy undertaking, as no dataset or algorithm exists that is optimal for all applications or situations.

Applied machine learning can be thought of as a search problem, where the objective is to find the optimal mapping of inputs to outputs given a set of data and a machine learning method. In other words, Applied Machine Learning illustrates how an algorithm learns and the justification for combining approaches and algorithms. The application of machine learning techniques has developed dramatically from a specialty to a mainstream practice. They have been applied to a variety of sectors to address specific issues, including autonomous driving, Internet of Things (IoT) security, computer system cybersecurity [1–3], multimedia computing, health [4, 5], and many more. Machine learning encompasses a broad variety of tasks, from data collection to pre-processing and imputation, from data exploration to feature selection, and finally, model construction and evaluation. At each stage of this pipeline, decisions are made based on two key factors: an awareness of the application's characteristics and the availability of the required data. The primary objective is to overcome machine learning issues posed by these factors. For instance, in autonomous driving, some of the machine learning problems include the size, completeness, and validity of the training set, as well as the safety of the deep neural networks utilized against adversarial perturbations that could force the system to misclassify an image [6]. Adversarial perturbations comprise minor image manipulations such as scaling, cropping, and changing the lighting conditions.

Another rapidly expanding application of machine learning is security for the Internet of Things (IoT). Due to advancements in the IoT technology stack, massive amounts of data are being generated in a variety of sectors with distinct characteristics. As a result, computing devices have become more connected than ever before, spanning the spectrum from standard computing devices (such as laptops) to resource-constrained embedded devices, servers, edge nodes, sensors, and actuators. The Internet of Things (IoT) network is a collection of internet-connected non-traditional computer devices that are often low power and have limited processing and storage capabilities. Parallel to the exponential growth of the Internet of Things, the number of IoT assaults has risen tremendously. Due to a lack of security protection and monitoring systems for IoT devices and networks, we are in desperate need of creating secure machine learning approaches for IoT device protection. As a result, such solutions must be sturdy, yet resource-constrained, making their development a difficult challenge. Thus, tasks such as developing safe and durable models and performing hardware analysis on trained models (in terms of hardware latency and area) are significant applied machine learning problems to address in this sector [1, 7, 8].

The majority of this book discusses the difficulties and best practices associated with constructing machine learning models, including understanding an application's properties and the underlying sample dataset.

1.2 The Machine Learning Pipeline

What is *Machine Learning Pipeline*? How do you describe the goal of machine learning? What are the main steps in the machine learning pipeline? We will answer these questions both through formal definitions and practical examples. Machine learning pipeline is meant to help with automating the machine learning workflow, in order to obtain actionable insights from big datasets. The goal of machine learning is to train an accurate model to solve an underlying problem. However, the term *pipeline* is misleading as many of the steps involved in the machine learning workflow may be repeated iteratively so to enhance and improve the accuracy of the model. The cyclical architecture of machine learning pipelines is demonstrated in Fig. 1.1.

Initially, the input (or collected) data is prepared before performing any analysis. This step includes tasks such as data cleaning, data imputation, feature engineering, data scaling/standardization, and data sampling for dealing with issues including noise, outliers, transforming categorical variables, normalizing/standardizing dataset features, and imbalanced (or biased) datasets.

In the *Exploratory Data Analysis* step (EDA), data is analyzed to understand its characteristics such as having a normal or skewed distribution (see Fig. 1.2). Skewedness in data affects a statistical model's performance, especially in the case of regression-based models. To prevent harming the results due to skewness, it is a common practice to apply a transformation over the whole set of values (such as log transformation) and use the transformed data for the statistical model.

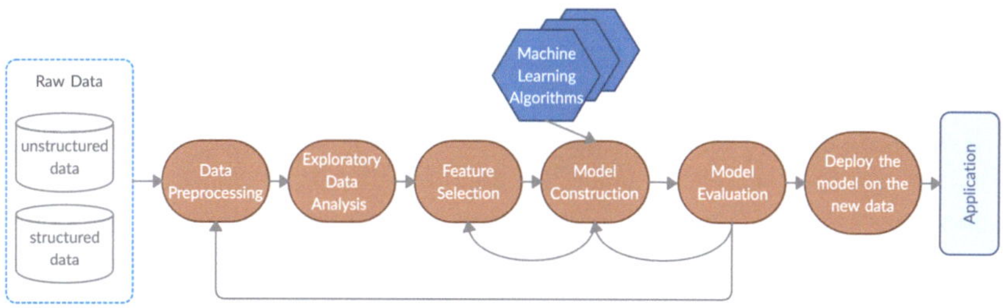

Fig. 1.1 The cyclical architecture of machine learning pipelines

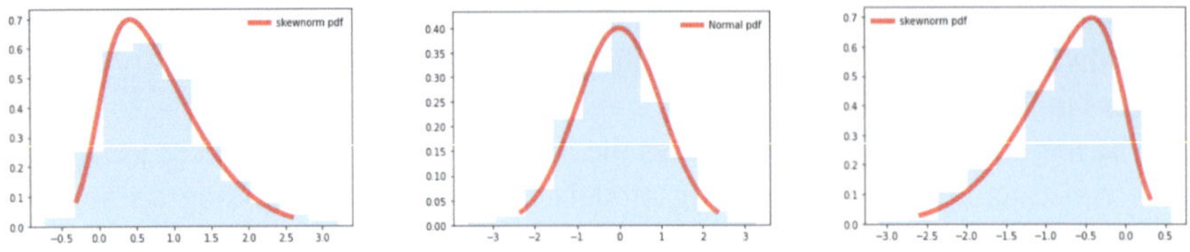

(a) (b) (c)

Fig. 1.2 Comparison of different data distributions. In Right-Skewed or positive distribution, most data falls to the positive, or the right side of the peak. In Left-Skewed or negative distribution, most data falls to the negative, or the left side the peak. (**a**) Right-skewed. (**b**) Normal distribution. (**c**) Left-skewed

Another prominent task performed during EDA is discovering the correlations between attributes of the dataset to identify the independent variables that are eventually used in the training process. For instance, if feature a_1 is highly correlated with feature a_2, then only one of those features should be considered for training a model. Furthermore, in datasets where there is a linear relationship between input and output variables, it is important to realize the relationships between the variables such as positive correlations (when an input variable increases/decreases as the target (i.e., output) variable increases/decreases) and negative correlations (when an input variable increases/decreases as the target (i.e., output) variable decreases/increases), or no correlation. Visualization techniques such as plotting the colinearity in the data using a correlation map, or a scatter plot matrix (also called pair-plot) can show a bi-variate or pairwise relationships between different combinations of variables in a dataset. An example of a correlation matrix is illustrated in Fig. 1.3.

Next, in the *Feature Selection* step, important features for training a machine learning model using a dataset are identified. Important benefits of *Feature Selection* include reducing over-fitting, improving the accuracy of the model, and reducing the training time. Attribute selection can be conducted in different ways. Leveraging known relationships between the variables can guide the selection of features. However, when the number of features grows, data-driven exploratory techniques come in handy. Some of the most common dimensionality reduction techniques include Principal Component Analysis (PCA), t-distributed Stochastic Neighbor Embedding (t-SNE), Independent Component Analysis (ICA), and clustering algorithms (e.g., Gaussian mixture model).

Real-world datasets contain many attributes, among which, just a subset of them help with the analysis. For instance, for lane detection in the autonomous driving applications, important features include edge, gradient, and intensity [9, 10] as they rely on the different intensity between the road surface and the lane markings. Once important features are identified to perform a particular machine learning task in an application, the prepared dataset is partitioned into a training and testing set; the training data is used to train a machine learning algorithm to construct a model, followed by the evaluation process, which is relied on the test data.

Fig. 1.3 A correlation matrix for an airline dataset

As illustrated in Fig. 1.1, training a model is an iterative process; depending on the performance factors and accuracy of the generated model, it is usually tuned iteratively to enhance the weights of its parameters until no further improvements are possible, or a satisfactory outcome is obtained.

1.3 Knowing the Application and Data

We live in the age of big data where data lives in various sources and repositories stored in different formats: structured and unstructured. Raw input data can contain *structured data* (e.g., numeric information, date), *unstructured data* (e.g., image, text), or a combination of both, which is called *semi-structured data*. Structured data is quantitative data that can fit nicely into a relational database such as the dataset in Table 1.2 where the information is stored in tabular form. Structured attributes can be transformed into quantitative values that can be processed by a machine. Unlike structured data, unstructured data needs to be further processed to extract structured information from it; such information is referred to as *data about data*, or what we call *metadata* in this book.

Table 1.1 demonstrates a dataset for a sample text corpus related to research publications in the public domain. This is an example of a semi-structured dataset, which includes both structured and unstructured attributes: the attributes of *year* and

Table 1.1 A semi-structured dataset collected from **GoogleScholar** data source

Year	Title	Citations	Authors	Conference	Abstract
1881	Surveylance: automatically detecting online survey scams	0	A Kharraz, W Robertson, E Kirda	39th S&P 2018: San Francisco, CA, USA	[… we present SURVEYLANCE, the first system that automatically identifies survey scams using machine learning techniques. Our evaluation demonstrates …]
1885	EyeTell: vide-assisted touchscreen keystroke inference from eye movements	2	Y Chen, T Li, R Zhang, Y Zhanga	39th S&P 2018: San Francisco, CA, USA	[… Keystroke inference attacks pose an increasing threat to ubiquitous mobile devices. This paper …]
1886	Understanding linux malware	4	E Cozzi, M Graziano, Y Fratantonioa	39th S&P 2018: San Francisco, CA, USA	[… For the past two decades, the security community has been fighting malicious programs for Windows-based operating …]
1989	SoK: keylogging side channels	1	J Monaco	39th S&P 2018: San Francisco, CA, USA	[… The first keylogging side channel attack was discovered over 50 years ago when Bell Laboratory researchers noticed an electro …]
1869	FuturesMEX: secure, distributed futures market exchange	2	F Massacci, CN Ngo, J Nie, D Venturia	39th S&P 2018: San Francisco, CA, USA	[… in a futures-exchange, such as the Chicago mercantile exchange, traders buy and sell contractual promises (futures) to acquire or deliver, at some future …]

citation are structured variables (categorical and numerical, respectively), whereas the *title*, *authors*, *conference*, and *abstract* contain unstructured data (i.e., raw text).

In applied machine learning problems, we begin with understanding the data behind an application in case there is no limited background knowledge available about an application. Knowing the application can help make accurate decisions about the important metadata to extract from the unstructured variables, which techniques to entertain for metadata extraction—e.g., in case of raw text, extract information such as the frequency of concepts using *bag-of-words* model, which metadata standard to use, how to encode features (e.g., one-hot encoding), as well as the selection of top features to help with generating the output model, which will be later deployed on unseen data (Table 1.2).

Understanding the data behind an application transpires through performing Exploratory Data Analysis (EDA), which is an approach used by the data analysts to use visual explorations to understand what is in the dataset and the data characteristics such as the relationships between the attributes and distribution of data. There are many visualization techniques to use for understanding the data within an application such as *correlation matrix*, *histogram*, *box plot*, and *scatter plot*.

Let us take a look at a sample *customer airline* dataset in Table 1.4, which contains 7 attributes including *INDEX*, *FARE*, *SEATCLASS*, *GUESTS*, *GENDER*, *AGE*, and the class variable *SUCCESS* for a fictitious airline *A*. The ultimate goal is to identify the factors that are helpful to understand why some customers are flying with the airline, and why others are canceling. Here is a brief description of the features:

- **CUSTOMERID**: A unique ID associated to a customer.
- **GUESTS**: Number of guests accompanying the customer.
- **SUCCESS**: Categorical variable that displays whether customer traveled or not.
- **SEATCLASS**: Categorical variable that displays the seat class of the customer.
- **AGE**: Numerical variable corresponding to the age of the customer.
- **GENDER**: Categorical variable describing the gender of the customer.
- **FARE**: Numeric variable for the total fare paid by the customer.
- **SUCCESS**: Categorical class variable indicating if the customer flies with the airline.

The correlation matrix illustrated in Fig. 1.3 is an example of a technique used to understand the relationships between the attributes of a dataset.

A correlation matrix is a tool to show the degree of association between a pair of variables in a dataset. It visually describes the direction and strength of a linear relationship between two variables. This correlation matrix visualizes the correlations between the variables of the airline dataset. According to this plot, *GENDER* and *SEATCLASS* have the highest correlations with the class variable; *GENDER* is positively correlated with *SUCCESS* (with the degree of +0.54), while *SEATCLASS* is negatively correlated (with the degree of −0.36).

Histogram is a graphical technique used to understand the distribution of data. Figure 1.4 illustrates the distribution of the airline dataset over its structured vari-

Table 1.2 Structured malware dataset, obtained from virusshare and virustotal, covering 5 different classes of malware

Bus-cycles	Branch-instructions	Cache-references	Node-loads	Node-stores	Cache-misses	Branch-loads	LLC-loads	L1-dcache-stores	Class
11463	37940	8057	1104	111	2419	37190	2360	38598	Backdoor
1551	5055	1096	165	17	333	4916	330	5003	Backdoor
29560	126030	20008	1769	146	4098	108108	5987	99237	Backdoor
26211	117761	14783	1666	48	4182	117250	4788	91070	Backdoor
30139	123550	20744	1800	158	4238	124724	6969	115862	Backdoor
12989	30012	9076	1252	136	5412	27909	2000	27170	Benign
6546	12767	4953	548	87	3683	13157	864	12361	Benign
8532	31803	7087	699	124	3240	34722	1970	34974	Benign
14350	27451	9157	1843	178	6611	28507	2411	24908	Benign
13837	25436	12235	1296	192	7148	24747	2533	23757	Benign
1068674	8211420	168839	42612	28574	73696	6298568	64166	6202146	Rootkit
1054761	8187337	162526	41245	28389	71576	6688738	67408	6655480	Rootkit
1046053	8196952	158955	40525	28113	70250	6981991	69597	6950106	Rootkit
1038524	8124926	157896	40207	28214	69910	7134795	71132	7148734	Rootkit
1030773	8069156	158085	39603	28265	69356	7230800	72226	7294250	Rootkit
999182	29000000	455	64	5	94	29000000	289	14000000	Trojan
999189	29000000	457	65	5	95	29000000	288	14000000	Trojan
999260	29000000	457	65	6	96	29000000	287	14000000	Trojan
999265	29000000	459	67	6	98	29000000	287	14000000	Trojan
999277	29000000	459	67	6	98	29000000	288	14000000	Trojan

989865	9128084	2549	169	37	268	9404871	923	9614242	Virus
989984	9130539	2529	168	37	266	9402351	920	9611680	Virus
990117	9132992	2510	167	36	264	9400377	914	9609689	Virus
990233	9135227	2491	165	36	262	9397484	909	9606756	Virus
990366	9137694	2473	164	36	260	9395002	903	9604237	Virus
760836	7851079	165236	8891	4047	13803	10530146	38930	4454651	Worm
765750	7957382	161998	8717	3967	13533	10573140	38205	4453953	Worm
770445	8059123	158884	8549	3891	13273	10606358	37508	4450452	Worm
774993	8157690	155888	8388	3818	13022	10660033	36824	4454598	Worm
779347	8251754	153008	8237	3747	12785	10693711	36171	4452344	Worm

Table 1.4 A sample dataset of an airline's customers

Index	Description	Success	Guests	Seat class	Customer ID	Fare	Age	Title	Gender
0	Braund, Mr. Owen Harris; 22	0	1	3	1	7.25	22	Mr	Male
1	Cumings, Mrs. John Bradley ...	1	1	1	2	71.3	38	Mrs	Female
2	Heikkinen, Miss. Laina; 26	1	0	3	3	7.92	26	Miss	Female
3	Futrelle, Mrs. Jacques Heath...	1	1	1	4	53.1	35	Mrs	Female
4	Allen, Mr. William Henry...	0	0	3	5	8.05	35	Mr	Male
5	Moran, Mr. James;	0	0	3	6	8.46	0	Mr	Male
6	McCarthy, Mr. Timothy J; 54	0	0	1	7	51.9	54	Mr	Male

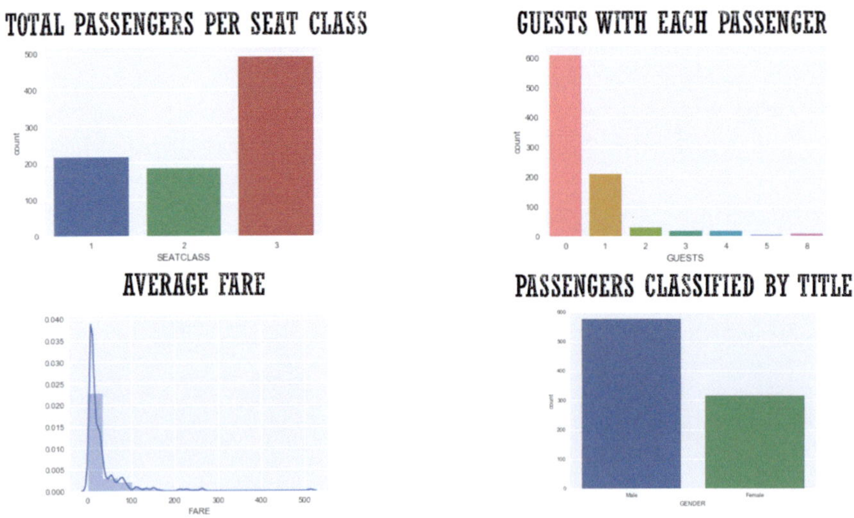

Fig. 1.4 Data distribution over the variables of an airline dataset

ables including SEAT CLASS, GUESTS, FARE, and customer TITLE. Histograms display a general distribution of a set of numeric values corresponding to a dataset variable over a range.

Plots are great means to help with understanding the data behind an application. Some example application of such plots is described in Table 1.5. It is important to note that every plot is deployed for a different purpose and applied to a particular type of data. Therefore, it is crucial to understand the need for such techniques used during the EDA step. Such graphical tools can help maximize insight, reveal underlying structure, check for outliers, test assumptions, and discover optimal factors.

As indicated in Table 1.5, several Python libraries offer very useful tools to plot your data. Python is a real generic programming language with a very large user community. It is purpose-built for large datasets and machine learning analysis. In this book, we focus on using Python language for various machine learning tasks and hands-on examples and exercises.

1.4 Getting Started Using Python

Before getting started using Python for applying machine learning techniques on a problem, you may want to find out which IDEs (Integrated Development Environment) and text editors are tailored for Python programming or looking at code samples that you may find helpful. IDE is a program dedicated to software development. A Python IDE usually includes an editor to write and handle Python code, build, execution, debugging tools, and some form of source control. Several Python programming environments exist depending on how advanced is a Python programmer to perform a machine learning task. For example, Jupyter Notebook is a very helpful environment for beginners who have just started with traditional machine learning or deep learning. Jupyter Notebook can be installed in a virtual environment using Anaconda-Navigator, which helps with creating virtual environments and installing packages needed for data science and deep learning. While Jupyter Notebook is more suitable for beginners, there are other machine learning frameworks such as TensorFlow that are mostly used for deep learning tasks. As such, depending on how advanced you are in Python programming, you may end up using a particular Python programming environment. In this book, we will begin with using Jupyter Notebook for programming examples and hands-on exercises. As we move toward more advanced machine learning tasks, we switch to TensorFlow. You can download and install Anaconda-Navigator on your machine using the following link by selecting Python 3.7 version: https://www.anaconda.com/distribution/.

Once it is installed, navigate to Jupyter Notebook and hit "Launch." You will then have to choose or create a workspace folder that you will use to store all your Python programs. Navigate to your workspace directory and hit the "New" button to create a new Python program and select Python 3. Use the following link to get familiar with the environment: https://docs.anaconda.com/anaconda/user-guide/getting-started/.

In the remaining part of this chapter, you will learn how to conduct preliminary machine learning tasks through multiple Python programming examples.

Table 1.5 Popular Python tools for understanding the data behind an application. https://github.com/dgrtwo/gleam

Plot type	Python library	Usage description	Example
Line plot	Plotly	Trends in data	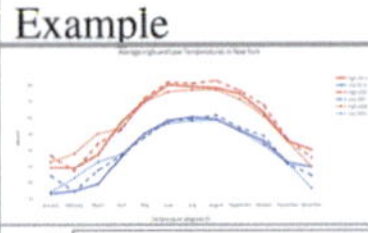
Scatter plot	Gleam	Multivariate data	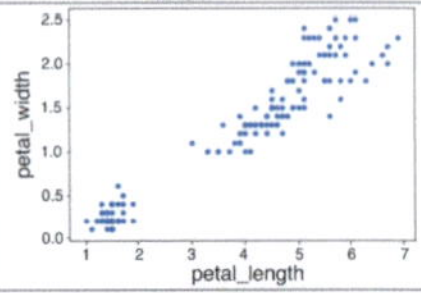
Layered area chart	ggplot	Compare trend over time	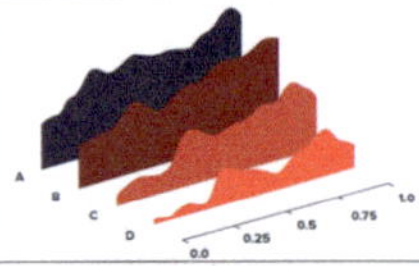
Nullity matrix	Missingno	Data sparsity	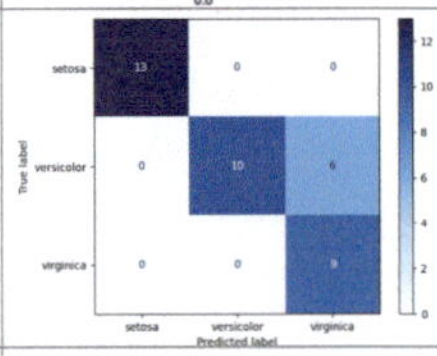
Bar plot	Bokeh	Streaming & real-time data	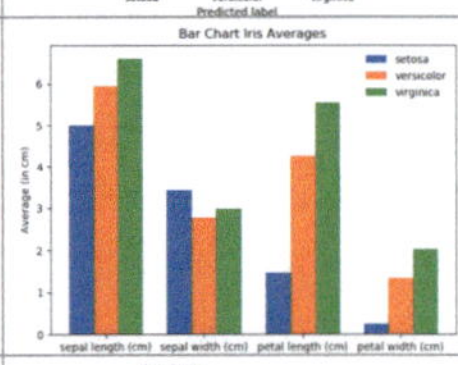
Scatter plot matrix	Seaborn	Bivariate data correlations	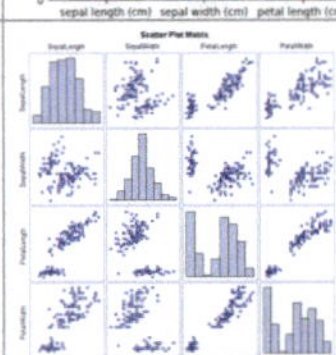
Box plot	Pygal, Seaborn	Outliers and data distribution	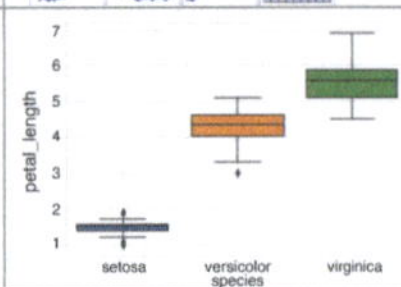
Histogram	Matplotlib	Outliers & data distribution	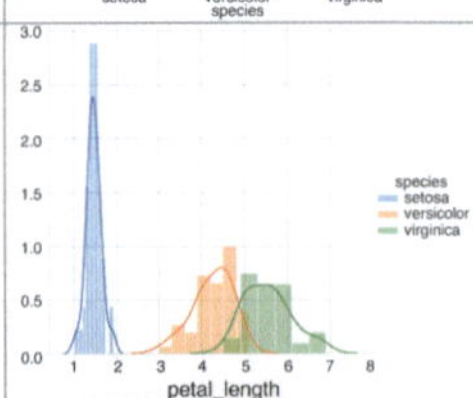
Heatmap, dot-density	Seaborn	Uses a system of color coding to represent different values	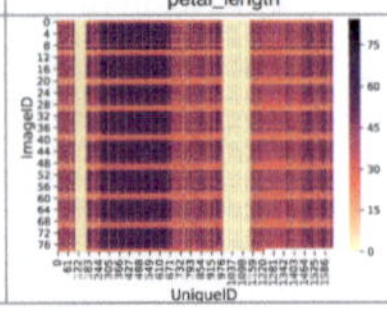

1.5 Metadata Extraction and Data Pre-processing

It is a no-brainer that data is a crucial aspect of machine learning. Data is used to train machine learning models and tune their parameters to improve their accuracy and performance. Data is available in various types, and different forms: structured and unstructured. Metadata contains information about a dataset. Such information describes the characteristics of data such as format, location, author, content, relationship, and data quality. It can also include information about features, models, and other artifacts from the machine learning pipeline. Metadata is highly structured and actionable information about a dataset.

The popularity of metadata grows due to the proliferation of devices that generate data and data integration, dealing with heterogeneity and diversity of data. Metadata extraction is the process of extracting salient features from a dataset. Depending on the type of data (e.g., text, image, so forth), its metadata is extracted and represented in different ways. For instance, weather information (metadata) can be generated using the timestamp and location information of the image provided in the image's EXIF tag that is used largely to encode contextual information related to image generation by digital cameras. Another example of metadata is Bag-of-Words (BOW) and its flavors such as Frequency Vectors, One Hot Encoding (OHE), and Term Frequency/Inverse Document Frequency (TF/IDF), which are used to generate metadata corresponding to a text document. Such representation encompasses words that stand for entities in a dataset, leading to the notion of *entity extraction*. Feature representation in a dataset is an important step. In some datasets, features are numeric, such as the attributes displayed in a tabular view in Table 1.2. However, there are many other datasets that contain categorical information, which would need feature engineering before performing any machine learning task. For instance, recording weather information in a dataset using categories such as *cloudy, windy, rainy*. Furthermore, applications such as *text classification, concept modeling, language modeling, image captioning, question answering, and speech recognition* are some examples where feature engineering is required to represent features numerically.

Let us consider a topic modeling application where the goal is to perform text classification. Table 1.6 illustrates a sample dataset that has *ID, content*, and *Topic* as the original attributes. However, the machine cannot use these attributes as is to perform mathematical computations as these features (except ID) are not numeric. Therefore, metadata extraction followed by feature engineering is required to transform these attributes into numeric values. Let us make this more concrete by focusing on the *content* attribute that contains the unstructured raw text. The data in this column cannot be directly used as features in its original form unless we extract metadata from it and provide a numeric encoding. For instance, one way is to extract N-grams, which is a contiguous sequence of n items (such as phonemes, syllables, letters, words, or base pairs) from a given sample of text or speech.

Example 1.1 (N-Gram Extraction)

Problem: Extract the N-grams from a given string of text and display all the extracted N-grams.

Solution: To extract N-grams as metadata, off-the-shelf Natural Language Processing (NLP) tools such as Natural Language Toolkit (NLTK) can be used, which is a leading platform for building Python programs to work with human language data. A widely used feature engineering technique in this situation is *One-Hot-Encoding*, which provides the mapping of categorical values into integer values. But first, we need to extract the categories. To perform this, one can use `re` and `nltk.util` packages to apply regular expression matching operation and finding n-grams to only retain useful content terms. The Python code below can be used to extract 2-grams and 4-grams. The result is displayed.

```
import re
from nltk.util import ngrams

#input text
text = """tighter time analysis for real-time traffic in on-chip \
networks with shared priorities"""
print('input text: ' + text)

tokens = [item for item in text.split(" ") if item != ""]

output2 = list(ngrams(tokens, 2)) #2-grams
output4 = list(ngrams(tokens, 4)) #4-grams

allOutput=[]
for bigram in output2:
    if bigram[0]!= "for" and bigram[1]!= "for" and bigram[0]!="in" and \
    bigram[1]!="in" and bigram[0]!="with" and bigram[1]!="with":
        allOutput.append(bigram)

print('\nall extracted bigrams are:')
print(allOutput)


allOutput=[]

for quadgram in output4:
    if quadgram[0]!= "for" and quadgram[1]!= "for" and quadgram[2]!= "for" \
    and quadgram[3]!= "for" and quadgram[0]!="in" and quadgram[1]!="in" \
    and quadgram[0]!="with" and quadgram[1]!="with":
        allOutput.append(quadgram)

print('\nall extracted quadgrams are:')
print(allOutput)

>input text: tighter time analysis for real-time traffic in on-chip \
>networks with shared priorities

>all extracted bigrams are:
>[('tighter', 'time'), ('time', 'analysis'),\
>('real-time', 'traffic'), ('on-chip', 'networks'), ('shared', 'priorities')]

```

Table 1.6 A sample dataset for text classification

ID	Content	Topic
1	Using benes networks at fault-tolerant and deflection routing based network-on-chips	Fault tolerant systems
2	Tighter time analysis for real-time traffic in on-chip networks with shared priorities	Network-on-chip analysis
3	Loosely coupled in situ visualization: a perspective on why it's here to stay	Scientific visualization
4	Lessons learned from building in situ coupling frameworks	In Situ visualization
5	An approach to lowering the in situ visualization barrier	In situ visualization
6	PROSA: protocol-driven NoC architecture	Computer architecture
7	Hybrid large-area systems and their interconnection backbone	Sensor phenomena and characterization
8	Bubble budgeting: throughput optimization for dynamic workloads by exploiting dark cores in many core systems	Resource management

```
>all extracted quadgrams are:
>[('real-time', 'traffic', 'in', 'on-chip'), \
>('on-chip', 'networks', 'with', 'shared')]
```

Metadata extraction is an important phase in machine learning. Once the features are extracted, the dataset should be pre-processed to get prepared for training. Preprocessing includes data cleaning and data imputation, outlier detection, and data exploration. Outliers in a dataset are those samples that show abnormal distance from the other samples in the dataset. There are various methods to detect outliers. One simple technique is to visually identify irregular sample using a scatter plot, or a histogram when the problem is not very complex. For more complex problems, techniques such as one-class SVM, Local Outlier Factor, and Isolation Forest. In outlier detection, it is important to include the output variable as the outliers form around the clusters related to the output variable.

1.6 Data Exploration

Data Exploration or Exploratory Data Analysis (EDA) is an important part of data analysis. It is a systematic process to understand the data, maximize insight, discover latent correlations between variables, identify important variables, outliers, and anomalies, and perform dimensionality reduction using various data visualization and statistical techniques. Data exploration or data understanding is where an analyst takes a general view of the data to make some sense of it.

Exploratory Data Analysis (EDA) is understanding the datasets by summarizing their main characteristics often plotting them visually. This step is very important especially when we arrive at modeling the data in order to apply machine learning. Plotting in EDA consists of histograms, box plot, scatter plot, and many more. It often takes much time to explore the data. Through the process of EDA, we can ask to define the problem statement or definition on our dataset, which is very important. Some of the common questions one can ask during EDA are:

- What kind of variations exist in data?
- What type of knowledge is discovered from the covariance matrix of data in terms of the correlations between the variables?
- How are the variables distributed?
- What kind of strategy to follow with regard to the outliers detected in a dataset?

Some typical graphical techniques widely used during EDA include histogram, confusion matrix, box plot, scatter plot, principal component analysis (PCA), and so forth. Some of the available popular Python libraries used for EDA include seaborn, pandas, matplotlib, and NumPy. In this section, we will illustrate multiple examples showing how EDA is conducted on a sample dataset.

1.7 A Practice for Performing Exploratory Data Analysis

The selection of techniques for performing Exploratory Data Analysis (EDA) depends on the dataset. There is no single method or common methods in order to perform EDA. Based on this section, you can practice some common methods and plots that would be used in the EDA process.

We will perform the EDA for Fisher's Iris dataset to illustrate different EDA techniques. The Iris dataset contains 3 classes of 50 instances each, where each class refers to a type of iris flower. The features in the dataset are sepal length, sepal width, petal length, and petal width (Fig. 1.5). One class is linearly separable from the other two; the latter are NOT linearly separable from each other. The predicted attribute is the class of the Iris flower. The objective is to classify flowers into one of the categories. In this section, we will perform the EDA on the Iris dataset and observe the trend.

Fig. 1.5 Flower attributes "Sepal and Petal"

1.7.1 Importing the Required Libraries for EDA

Let us begin the EDA by importing some libraries required to perform EDA.

```
import pandas as pd
import seaborn as sns      #visualization
import matplotlib.pyplot as plt #visualization
import numpy as np
```

1.7.2 Loading the Data Into Dataframe

The first step to performing EDA is to represent the data in a Dataframe form, which provides one with extensive usage for data analysis and data manipulation. Loading the data into the Pandas dataframe is certainly one of the most preliminary steps in EDA, as we can see that the value from the dataset is comma separated. So all we have to do is to just read the CSV file into a dataframe and pandas dataframe does the job for us. First, download *iris.csv* from https://raw.githubusercontent.com/uiuc-cse/data-fa14/gh-pages/data/iris.csv. Loading the data and determining its statistics can be done using the following command:

```
import pandas as pd
import matplotlib.pyplot as plt

data = pd.read_csv('iris.csv')
print('size of the dataset and the number of features are:')
print(data.shape)
print('\ncolumn names in the dataset:')
print(data.columns)
print('\nnumber of samples for each flower species:')
print(data["species"].value_counts())

data.plot(kind='scatter', x='petal_length', y='petal_width')
plt.show()

> # size of the dataset and the number of features are:
>(150, 5)

># column names in the dataset:
>Index(['sepal_length', 'sepal_width', 'petal_length', 'petal_width','species
    '], dtype='object')

># number of samples for each flower species:
>virginica      50
>setosa         50
>versicolor     50
>Name: species, dtype: int64
```

The value_counts() method helps to understand whether the dataset is balanced or imbalanced. Based on the output of this method, Iris dataset is a balanced dataset with 50 samples/data points per species. Now let us use some visualization to better understand data including distribution of observations, classes, correlation of attributes, and identifying potential outliers.

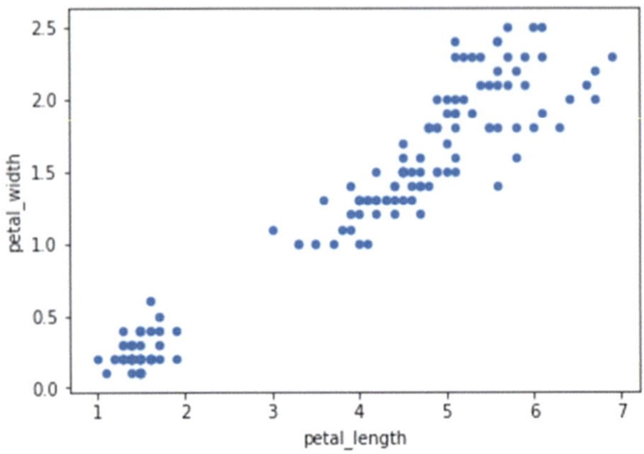

Fig. 1.6 2D scatter plot for iris dataset, based on two attributes "petal-length" and "petal-width"

1.7.3 Data Visualization

2D Scatter Plot

A scatter plot can display the distribution of data. Figure 1.6 shows a 2D scatter plot for visualizing the iris data (the command is included in the previous code snippet). The plot observed is a 2D scatter plot with petal_length on x-axis and petal_width on y-axis. However, with this plot, it is difficult to understand per class distribution of data. Using a color-coded plot can help plot the color coding for each flower/species/type of class. This can be done using seaborn(sns) library by executing the following commands:

```
import seaborn as sns
sns.set_style("whitegrid")
sns.FacetGrid(data, hue="species", height=4) \
    .map(plt.scatter, "petal_length", "petal_width") \
    .add_legend()
plt.show()
```

Looking at this scatter plot in Fig. 1.6, it is a bit difficult to make sense of the data since all data points are displayed with the same color regardless of their label (i.e., category). However, apply color coding to the plot and we can say a lot about the data by using a different color for each label. Figure 1.7 shows the color-coded scatter plot coloring *setosa* with blue, *versicolor* with orange, and *virginica* with green. One can understand how data is distributed across the two axes of *petal-width* and *petal-length* based on the flower species. The plot clearly shows the distribution across three clusters (blue, orange, and green), two of which are non-overlapping (blue and orange), and two overlapping ones (i.e., orange and green).

One important observation that can be realized from this plot is that petal-width and petal-length attributes can distinguish between *setosaa* and *versicolor* and between *setosa* and *versicolor*. However, the same attributes cannot distinguish

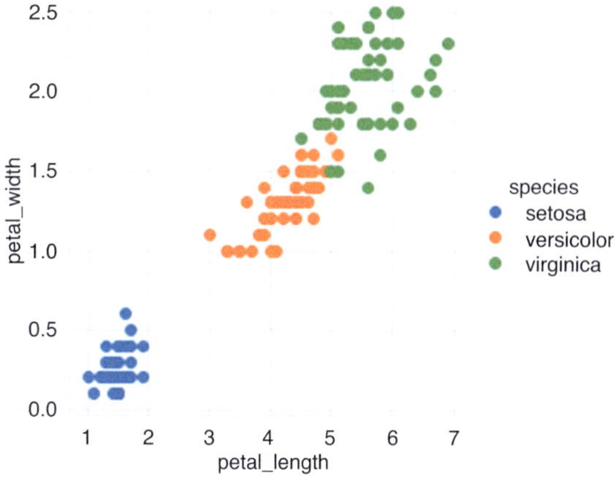

Fig. 1.7 2D color-coded scatter plot for iris dataset to visualize the distribution of the iris dataset

versicolor from *virginica* due to their overlapping clusters. This implies that the analyst should explore other attributes to train an accurate classifier and perform a reliable classification. So here is the summary of our observations:

- Using *petal-length* and *petal-width* features, we can distinguish *setosa* flowers from others. How about using all the attributes?
- Separating *versicolor* from *viginica* is much harder as they have considerable overlap using *petal-width* and *petal-length* attributes. Would one obtain the same observation if instead *sepal-width* and *sepal-length* attributes were used?

We have also included the 3D scatter plot in the Jupyter notebook for this tutorial. A sample tutorial for 3D scatter plot with *Plotly Express* can be found here, which needs a lot of mouse interaction to interpret data. https://plot.ly/pandas/3d-scatter-plots/ (What about 4D, 5D, or n-D scatter plot?)

Pair-Plot

When the number of features in a dataset is high, pair-plot can be used to clearly visualize the correlations between the dataset variables. The pair-plot visualization helps to view 2D patterns (Fig. 1.8) but fails to visualize higher dimension patterns in 3D and 4D. Datasets under real-time study contain many features. The relation between all possible variables should be analyzed. The pair plot gives a scatter plot between all combinations of variables that you want to analyze and explains the relationship between the variables (Fig. 1.8).

To plot multiple pairwise bivariate distributions in a dataset, you can use the *pairplot()* function in seaborn. This shows the relationship for (n, 2) combination of variables in a Dataframe as a matrix of plots and the diagonal plots are the univariate plots. Figure 1.8 illustrates the pair-plot for iris dataset, which lead to the following observations:

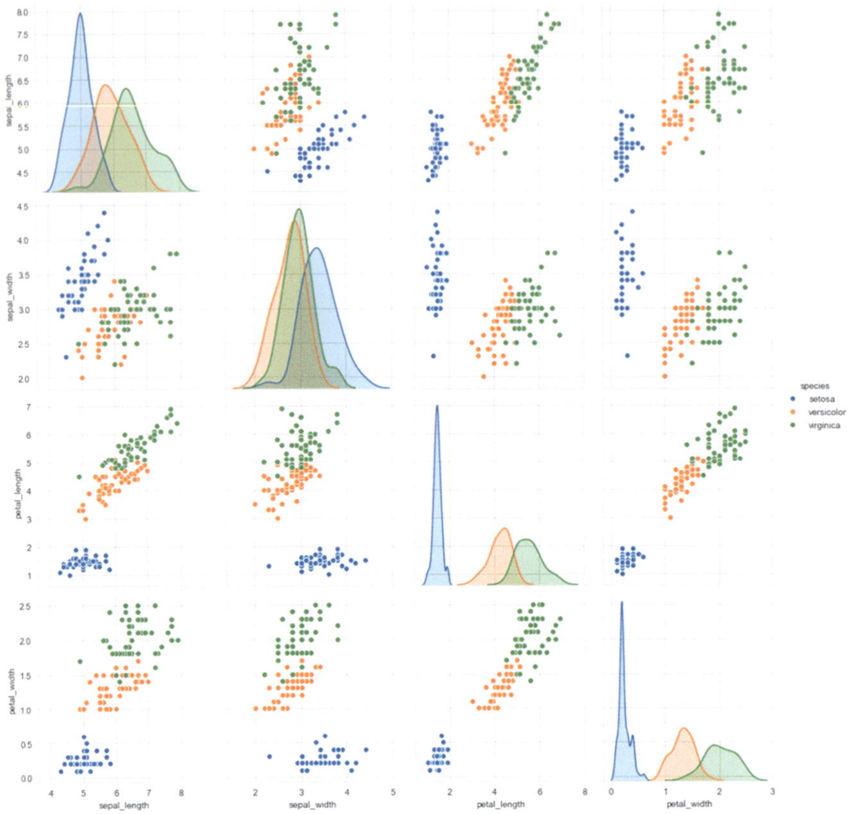

Fig. 1.8 Pair-plot over the variables of iris dataset

- Petal-length and petal-width are the most useful features to identify various flower types.
- While Setosa can be easily identified (linearly separable), Virnica and Versicolor have some overlap (almost linearly separable).

With the help of pair-plot, we can find "lines" and "if-else" conditions to build a simple model to classify the flower types.

```
plt.close();
sns.set_style("whitegrid");
sns.pairplot(iris, hue="species", height=3);
plt.show()
```

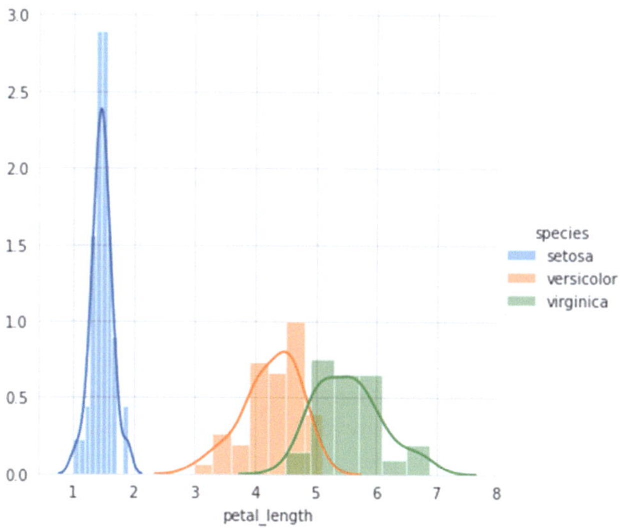

Fig. 1.9 Histogram plot showing frequency distribution for variable "petal_length"

Histogram Plot

A histogram plot is a diagram, which shows the underlying frequency plot/distribution of different variables in a dataset. The plot will allow us to inspect the data for its underlying distribution (e.g., normal distribution), outliers, skewness, and many more Fig. 1.9. We can view a histogram plot by using seaborn library with the help of following commands:

```
sns.FacetGrid(iris, hue="species", height=5) \
   .map(sns.distplot, "petal$\_$length") \
   .add_legend();
plt.show()
```

Probability Distribution Function

A probability distribution function (PDF) is a statistical function that describes all the possible values, likelihoods that a random variable is possible within a given range. The range is bounded between the minimum and maximum possible values, but where the possible value is likely to be plotted on the probability distribution depends on several factors that include the distribution's mean (average), standard deviation, skewness, and kurtosis.

Cumulative Distribution Function

The cumulative distribution function (CDF) of a random variable is another method to describe the distribution of random variables. The advantage of the CDF is that it can be defined for any kind of random variable (discrete, continuous, and mixed). The cumulative distribution function is applicable for describing the distribution of random variables that is either continuous or discrete.

Box Plot

A box and whisker plot also called a box plot displays the five-number summary of a set of data. The five-number summary is the minimum, first quartile, median, third quartile, and maximum. In a box plot, we draw a box from the first quartile to the third quartile. A vertical line goes through the box at the median. The whiskers go from each quartile to the minimum or maximum. A box and whisker plot is a way of summarizing a set of data measured on an interval scale. It is often used in explanatory data analysis. This type of graph is used to show the shape of the distribution, its central value, and its variability. In a box and whisker plot:

- The ends of the box are the upper and lower quartiles, so the box spans the interquartile range.
- The median is marked by a vertical line inside the box.
- The whiskers are the two lines outside the box that extend to the highest and lowest observations.

The following code snippet shows how a box plot is used to visualize the distribution of the iris dataset. Figure 1.10 shows the box plot visualization across the iris dataset "species" output variable.

```
sns.boxplot(x='species',y='petal_length', data=data)
plt.show()
```

Violin Plots

Violin plots are a method of plotting numeric data and can be considered a combination of the box plot with a kernel density plot. In the violin plot (Fig. 1.11), we can find the same information as in box plots:

- Median.
- Interquartile range.
- The lower/upper adjacent values are defined as first quartile-1.5 IQR and third quartile + 1.5 IQR, respectively. These values can be used in a simple outlier detection (Turkey's fence) techniques, where observations lying outside of these "fences" can be considered outliers.

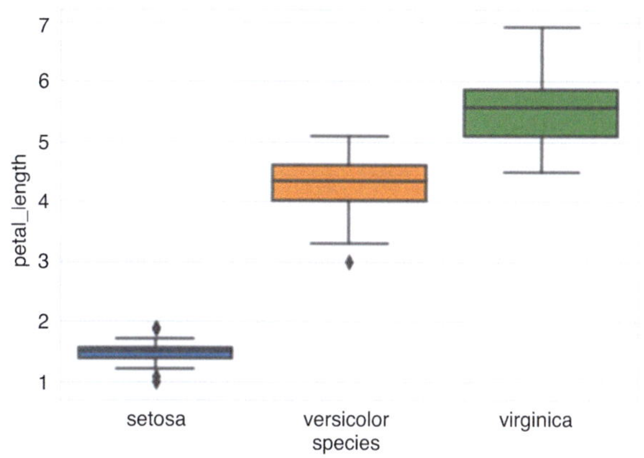

Fig. 1.10 Box plot for Iris dataset over "species" variable

Fig. 1.11 Violin plot over the variable "petal_length" of the iris dataset

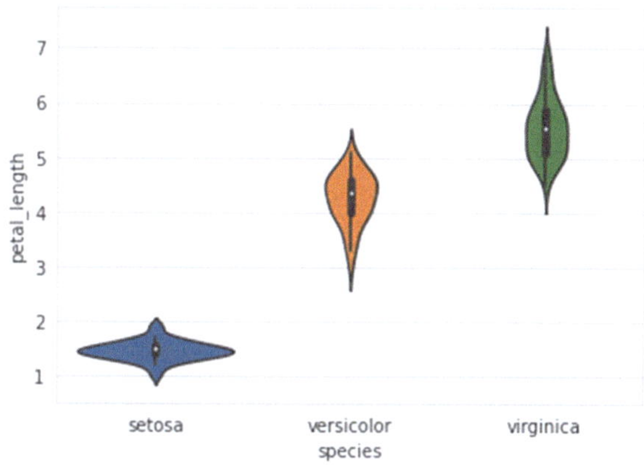

Violin plots can be easily visualized using seaborn library as follows:

```
sns.violinplot(x="species", y="petal$\_$length", data=iris, size=8)
plt.show()
```

Univariate, Bivariate, and Multivariate Analysis

Univariate is a term commonly used in statistics to describe a type of data that consists of observations on only a single characteristic or attribute. A simple example of univariate data would be the salaries of workers in the industry. Like all the other data, univariate data can be visualized using graphs, images, or other analysis tools after the data is measured, collected, reported, and analyzed.

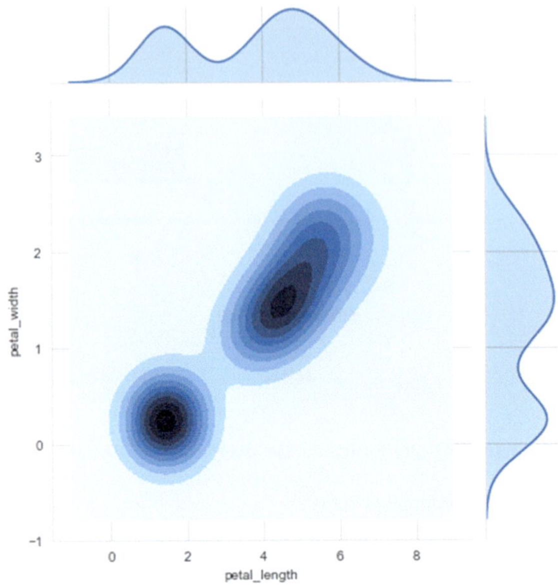

Fig. 1.12 Bivariate relationship of two attributes in Iris dataset. The univariate profiles are plotted in the margin

Data in statistics are sometimes classified according to how many variables are in a study. For example, "height" might be one variable and "weight" might be another variable. Depending on the number of variables being looked at, the data is univariate, or it is bivariate.

Multivariate data analysis is a set of statistical models that examine patterns in multi-dimensional data by considering at once with several data variables. It is an expansion of bivariate data analysis, which considers only two variables in its models. As multivariate models consider more variables, they can examine more complex analyses/phenomena and find the data patterns that can more accurately represent the real world. These three analyses can be done by using seaborn library in the following manner, depicted in Fig. 1.12, showing the bivariate distribution of "petal-length" and "petal-width," as well as the univariate profile of each attribute in the margin.

```
sns.jointplot(x="petal_length", y="petal_width", data=data, kind="kde")
plt.show()
```

Visualization techniques are very effective, helping the analyst understand the trends in data.

1.7.4 Data Analysis

In addition to data visualization, extracting the information related to data is non-trivial. Here, we discuss different kinds of information that can be extracted related to the data.

Standard Deviation

The standard deviation is a statistic that measures the dispersion of a dataset relative to its mean and is calculated as the square root of the variance. The standard deviation is calculated as the square root of variance by finding each data point's deviation in the dataset relative to the mean. If the data points are far from the mean, there is a higher deviation within the dataset. The more dispersed the data, the larger the standard deviation; conversely, the more dispersed the data, the smaller the standard deviation.

```
print("\n Std-dev:");
print(np.std(iris_setosa["petal_length"]))
print(np.std(iris_virginica["petal_length"]))
print(np.std(iris_versicolor["petal_length"]))

>Std-dev:
>0.17191858538273286
>0.5463478745268441
>0.4651881339845204
```

Mean/Average

The mean/average is the most popular and well-known measure of central tendency. It can be used with both discrete and continuous data, although its use is most often with continuous data. The mean is the sum of all values in the dataset divided by all the values in the dataset. So, if we have n data points in a dataset and they have values $x_1, x_2, \cdots, x_n$, the sample mean, usually denoted by x, is $x = (x_1+x_2+\cdots+x_n)/n$

```
print("Means:")
print(np.mean(iris_setosa["petal_length"]))
# Mean with an outlier.
print(np.mean(np.append(iris_setosa["petal_length"],50)))
print(np.mean(iris_versicolor["petal_length"]))

>Means:
>1.4620000000000002
>2.4137254901960787
>4.26
```

Run the above commands to see the output.

Variance

Variance in statistical context is a measurement of the spread between numbers in a dataset. That is, it measures how far each number in the set is from the mean and therefore from every other number in the set. Variance is calculated by taking the differences between each number in the dataset and the mean, then squaring the differences to make them positive, and finally dividing the sum of the squares by the number of values in the dataset.

```
print("Variance:")
print(np.var(iris_setosa["petal_length"]))
# Variance with an outlier.
print(np.var(np.append(iris_setosa["petal_length"],50)))
print(np.var(iris_versicolor["petal_length"]))

>Variance:
>0.029556000000000001
>45.31804690503652
>0.21640000000000012
```

Median

The median is the central/middle number in a set of sorted ascending or descending list of numbers and can be more descriptive of that dataset than the average. If there is an odd amount of numbers, the median value is the number that is in the middle, with the same amount of numbers below and above. If there is an even amount of numbers in the list, the middle pair must be determined, added together, and divided by two to find the median value.

```
print("\n Medians:")
print(np.median(iris_setosa["petal_length"]))
# Median with an outlier
print(np.median(np.append(iris_setosa["petal_length"],50)))
print(np.median(iris_virginica["petal_length"]))
print(np.median(iris_versicolor["petal_length"]))

>Medians:
>1.5
>1.5
>5.55
>4.35
```

Percentile

Percentiles are used to understand and interpret data. The nth percentile of a set of data is the value at which n percent of the data is below it. They indicate the values below which a certain percentage of the data in a dataset is found. Percentiles can be calculated using the formula $n = (P/100) \times N$, where P = percentile, N = number of values in a dataset (sorted from smallest to largest), and n = ordinal rank of a given value.

```
print("\n 90th Percentiles:")
print(np.percentile(iris_setosa["petal_length"],90))
print(np.percentile(iris_virginica["petal_length"],90))
print(np.percentile(iris_versicolor["petal_length"], 90))

>90th Percentiles:
>1.7
>6.3100000000000005
>4.8
```

Quantile

A quantile is a statistical term describing a division of observations into four defined intervals based upon the values of the data and how they compare to the entire set of observations. The median is an estimator but says nothing about how the data on either side of its value is spread or dispersed. The quantile measures the spread of values above and below the mean by dividing the distribution into four groups. We can map the four groups formed from the quantiles. The first group of values contains the smallest number up to Q1; the second group includes Q1 to the median; the third set is the median to Q3; and the fourth category comprises Q3 to the highest data point of the entire set. Each quantile contains 25% of the total observations. Generally, the data is arranged from smallest to largest: 1. First quantile: the lowest 25% of numbers 2. Second quantile: between 25.1 and 50% (up to the median) 3. Third quantile: 51–75% (above the median) 4. Fourth quantile: the highest 25% of numbers

```
print("\n Quantiles:")
print(np.percentile(iris_setosa["petal_length"],np.arange(0, 100, 25)))
print(np.percentile(iris_virginica["petal_length"],np.arange(0, 100, 25)))
print(np.percentile(iris_versicolor["petal_length"], np.arange(0, 100, 25)))

>Quantiles:
>[1.    1.4    1.5    1.575]
>[4.5   5.1    5.55   5.875]
>[3.    4.    4.35 4.6  ]
```

Interquartile Range

The IQR describes the middle 50% of values when ordered from lowest to highest. To find the interquartile range (IQR), initially, find the median (middle value) of the lower and upper half of the data. These values are quartile 1 (Q1) and quartile 3 (Q3). The IQR is the difference between Q3 and Q1.

Mean Absolute Deviation

The mean absolute deviation of a dataset is the average distance between each data point and the mean. It gives us an idea about the variability in a dataset. The idea is to calculate the mean, calculate how far away each data point is from the mean using positive distances, which are also called absolute deviations, add those deviations together, and divide the sum by the number of data points.

```
from statsmodels import robust

print ("\n Median Absolute Deviation")
print(robust.mad(iris_setosa["petal_length"]))
print(robust.mad(iris_virginica["petal_length"]))
print(robust.mad(iris_versicolor["petal_length"]))

>Median Absolute Deviation
>0.14826022185056031
>0.6671709983275211
>0.5189107764769602
```

1.7.5 Performance Evaluation Metrics

Evaluating the performance of machine learning classifiers is an important step in implementing effective ML-based countermeasure techniques. In machine learning and statistics, there are a variety of measures that can be deployed to evaluate the performance of a detection method in order to show its detection accuracy. Table 1.7 lists the standard evaluation metrics used for performance analysis of malware and side-channel attacks detection and classification. For analyzing the detection rate of ML-based security countermeasures, malicious applications' samples are often considered as positive instances. As a result, the True Positive Rate (TPR) metric, or the hit rate, represents sensitivity that stands for the proportion of correctly identified positives. It is basically the rate of malware samples (i.e., positive instances) correctly classified by the classification model. The True Negative Rate (TNR) also represents specificity that measures the proportion of correctly identified negatives. In addition, the False Positive Rate (FPR) is the rate of benign files (i.e., negative instances) wrongly classified (i.e., misclassified as malware samples).

The F-measure (F-score) in ML is interpreted as a weighted average of the precision (p) and recall (r). The precision is the proportion of the sum of true positives versus the sum of positive instances and the recall is the proportion of instances that are predicted positive of all the instances that are positive. F-measure is a more comprehensive evaluation metric over accuracy (percentage of correctly classified samples) since it takes both the precision and the recall into consideration. More importantly, F-measure is also resilient to the class imbalance in the dataset, which is the case in our experiments. The Detection Accuracy (ACC) measures the rate of the correctly classified positive and negative samples, which evaluates the correct classification rate across all tested samples.

Table 1.7 Evaluation metrics for performance of ML security countermeasures

Evaluation metric	Description
True positive (TP)	Correct positive prediction
False positive (FP)	Incorrect positive prediction
True negative (TN)	Correct negative prediction
False negative (FN)	Incorrect negative prediction
Specificity: true negative rate	$TNR = TN/(TN + FP)$
False positive rate	$FPR = FP/(FP + TN)$
Precision	$P = TP/(FP + TP)$
Recall: true positive rate	$TPR = TP/(TP + FN)$
F-measure (F-score)	$Fmeasure = 2 \times (P \times R)/(P + R)$
Detection accuracy	$ACC = (TP + TN)/(TP + FP + TN + FN)$
Error rate	$ERR = (FP + FN)/(P + N)$
Area under the curve	$AUC = \int_0^1 TPR(x)dx = \int_0^1 P(A > \tau(x))dx$

Precision and recall are not adequate for showing the performance of detection even contradictory to each other because they do not include all the results and samples in their formula. F-score (i.e., F-measure) is then calculated based on precision and recall to compensate for this disadvantage. Receiver Operating Characteristic (ROC) is a statistical plot that depicts a binary detection performance while its discrimination threshold setting is changeable. The ROC space is supposed by FPR and TPR as x and y axes, respectively. It helps the detector to determine trade-offs between TP and FP, in other words, the benefits and costs. Since TPR and FPR are equivalent to sensitivity and (1-specificity), respectively, each prediction result represents one point in the ROC space in which the point in the upper left corner or coordinate (0, 1) of the ROC curve stands for the best detection result, representing 100% sensitivity and 100% specificity (perfect detection point).

Example 1.2 (Performance Evaluation of ML-Based Malware Detectors)
Problem: A neural network ML classifier is applied on various HPC samples for hardware-assisted malware detection. Assuming that the FN=2, FP=1, TP=8, and TN=6, evaluate the performance of the neural network ML in classifying malware from benign samples by calculating Accuracy, Precision, Recall, and F-measure metrics.
Solution: As mentioned before, the detection accuracy calculates the rate of the correctly classified positive and negative samples:

$$ACC = \frac{TP + TN}{TP + FP + TN + FN} = \frac{8 + 6}{8 + 1 + 6 + 2} = 0.82. \qquad (1.1)$$

(continued)

Example 1.2 (continued)

Precision measures the percentage of malware (positive) samples that are correctly classified as malware:

$$P = \frac{TP}{FP + TP} = \frac{8}{1 + 8} = 0.89. \tag{1.2}$$

Recall measures the percentage of actual malware samples that were correctly classified by the ML-based detector:

$$R = \frac{TP}{TP + FN} = \frac{8}{8 + 2} = 0.8. \tag{1.3}$$

Now, we can calculate F-measure that is interpreted as a weighted average of the precision and recall:

$$F - Measure = \frac{2 \times (P \times R)}{P + R} = \frac{2 \times (0.89 \times 0.8)}{0.89 + 0.8} = 0.84. \tag{1.4}$$

1.8 Putting It All Together

Applied machine learning, is a rapidly growing field due to its interdisciplinary nature. It is considered a search problem for finding the optimal mapping of inputs and outputs given data and a machine learning method. We provided an introduction to the machine learning pipeline and described data metadata extraction, feature engineering, and preprocessing, data exploration and visualization, and data standardization and analysis. Each of these important tasks was described using a hands-on example. We deferred the training of machine learning models to the next chapters. We also covered several performance evaluation metrics such as TPR, TNR, precision and recall, F-measure, detection accuracy (ACC), and ROC. We also covered the rationale for the application of these metrics.

1.9 Exercise Problems

Problem 1.1 Describe the Machine learning Pipeline.

Problem 1.2 Download Haberman Cancer Survival dataset from Kaggle. You may have to create a Kaggle account to download data (https://www.kaggle.com/gilsousa/habermans-survival-data-set). Then provide a comprehensive description of the dataset including dataset size, the number of features (dimensions), type of features (numeric, nominal, discrete, continuous, binary, so forth), and class attribute (dependent variable).

Problem 1.3 Plot the distribution of data to show the number of data points per class and describe if the dataset is balanced or not. If the dataset is imbalanced or skewed, what solution do you propose as a remedy?

Problem 1.4 Identify outliers (if any) in the dataset and propose a solution to deal with the outliers and explain why it is a suitable approach to be applied to this dataset. You can use a visualization technique such as a box plot or a scatter plot to identify outliers.

Problem 1.5 Perform a high-level statistical analysis of the dataset in terms of reporting the mean, median, mean absolute deviation, and quantile before dealing with potential outliers.

Problem 1.6 Perform Bi-variate analysis (correlation matrix, pair-plots) to find a combination of useful features (i.e., independent variables) for classification.

Problem 1.7 Download the Airline .json file from https://github.com/sathwikabavikadi/Machine-Learning-for-Computer-Scientists-and-Data-Analysts and convert to .csv file and import into a dataframe.

Problem 1.8 Write your Python code to extract gender, age, and tile (such as "Mr") attributes from the "Description" field. Use *pandas* library.

Problem 1.9 Using the output of question 1.8, write a Python code to perform data imputation on age and gender attributes. Explain your approach. You can use *numpy* library.

Problem 1.10 Write a Python code to plot the distribution of Gender attributes after imputation using a histogram plot.

Problem 1.11 Write a Python code to plot the distribution of Age attribute and plot the box plots.

Problem 1.12 Write a Python code to plot the correlations between the dataset attributes. You can use seaborn and matplotlib libraries. In case of finding correlations between independent variables report them.

Problem 1.13 Outline the EDA techniques discussed in this chapter and the significance of these techniques.

Problem 1.14 Discuss the prominence of data pre-processing.

Chapter 2
A Brief Review of Probability Theory and Linear Algebra

2.1 Introduction

In daily life, we encounter various series of events and experiments that are based on probability and have no certainty about the outcome. Probability theory is an advantageous tool for quantitatively describing and forecasting the outcomes of probability-based investigations. By applying probability theory to a problem, one can simplify its understanding, evaluate it using the relevant mathematical model, and forecast probable outcomes based on the probability. Two examples are provided here to help you gain a better understanding of probability theory's applicability.

Consider rolling a fair dice as a simple example. When we are rolling a fair dice, there is no certainty in the output to be achieved. It can be said that the output of this experiment is based on probability. In more detail, in rolling a fair dice, the outcome would be "1" with the probability of 1/6. Also, the outcome would be "2" with the probability of 1/6. Similarly, each of the numbers of the dice would occur with the probability of 1/6. In other words, if we repeat this experiment too many times, the outcome "1" would be achieved in 16.66% of the time. A similar interpretation is also applied to other possible outcomes. It can be seen that the possible outcome is based on probability. This analysis and interpretation are possible using the concept of the probability theory according to the definition of the probability theory.

Another example in this field is the entering and existing rate of the customers in a restaurant. Using the probability theory, the entry rate of the costumers, the time duration each customer spends in the restaurant, and their existing rate can be easily modeled and analyzed mathematically. In particular, the average income of the restaurant can be estimated. In fact, according to these predictions and analyzes, one can take action to improve the performance of the restaurant.

Probability theory, in general, covers a broad range of applications. In any subject where complete information is unavailable and hence no certainty about the outcome, the issue can be controlled through the use of probability theory. Other

S. Rafatirad et al., *Machine Learning for Computer Scientists and Data Analysts*,
https://doi.org/10.1007/978-3-030-96756-7_2

applications of probability theory include weather forecasting, victory or defeat in a contest, wireless communication, machine learning, and even the drug distribution problem in the medical area. This section is considered to be a discussion of the fundamental notions of probability theory. Additionally, the topic of matrix algebra is given and studied in relation to machine learning, which is one of the applications of probability theory.

2.2 Fundamental of the Probability

Consider a probability-based experiment where there is no certainty in the outcome. Each of the possible outcomes is known as an event. As an example, consider we are tossing a coin. In this case, two possible outcomes could be achieved: "Head" and "Tail," each of them is called the event. Each of the events is specified with a probability. In this example, the probability of achieving "Head" is 1/2, or equivalently, we have:

$$\mathrm{P(Head)} = \frac{1}{2}.$$

Similarly, the probability of achieving "Tail" would be 1/2. Generally, the probability of an event is shown as $P(X = x_i)$. In this example, x_i could be "Head" or "Tail." Note that the probability of an event is always a non-negative, less than or equal to one value:

$$0 \leq \mathrm{P}(X = x_i) \leq 1. \tag{2.1}$$

The coin-tossing experiment is a simple example in which two possible outcomes would be achieved. In order to express the concept of probability in a more complex form, consider two random variables X and Y. Each of the random variables can take value from their corresponding dictionaries. More precisely, if the random variable X takes the value x_i, and the random variable Y takes the value y_j, then we have:

$$i \in D_X,$$

$$j \in D_Y,$$

where D_X and D_X are the dictionaries corresponding to random variables X and Y, respectively. Consider the case in which N possible outcomes could be achieved from the combination of these two random variables. The probability of the event x_i corresponding to the random variable X is denoted by $P(X = x_i)$. Similarly, the probability of the event y_j corresponding to the random variable Y is denoted by $P(Y = y_j)$. Now, consider that we are interested in finding the probability of

$X = x_i$ and $Y = y_j$ jointly. This probability is known as the joint probability and is denoted by $P(X = x_i, Y = y_j)$. The joint probability of X and Y is written as below:

$$P(X = x_i, Y = y_j) = \frac{n_{ij}}{N}, \tag{2.2}$$

where n_{ij} denotes the number of events in which the probabilities $X = x_i$ and $Y = y_j$ occurred jointly. The schematic of the joint probability is depicted in Fig. 2.1. In this figure, $P(X = x_i)$ and $P(Y = y_j)$ are denoted with the red and blue colors, respectively. Note that the region where the events x_i and y_j are met simultaneously is denoted by n_{ij}, as mentioned above.

As shown in Fig. 2.1, the number of events where $X = x_i$ is denoted by c_i. Also, the number of events where $Y = y_i$ is denoted by r_j. Therefore, $P(X = x_i)$ and $P(Y = y_j)$ can be written as follows:

$$\begin{aligned} p(X = x_i) &= \frac{c_i}{N}, \\ p(Y = y_j) &= \frac{r_j}{N}. \end{aligned} \tag{2.3}$$

In the above equation, c_i and r_j are achieved as below:

$$\begin{aligned} c_i &= \sum_{j \in D_Y} n_{ij}, \\ r_j &= \sum_{i \in D_X} n_{ij}. \end{aligned} \tag{2.4}$$

According to (2.3) and (2.4), the probability of $X = x_i$ and $Y = y_j$ can be rewritten as below:

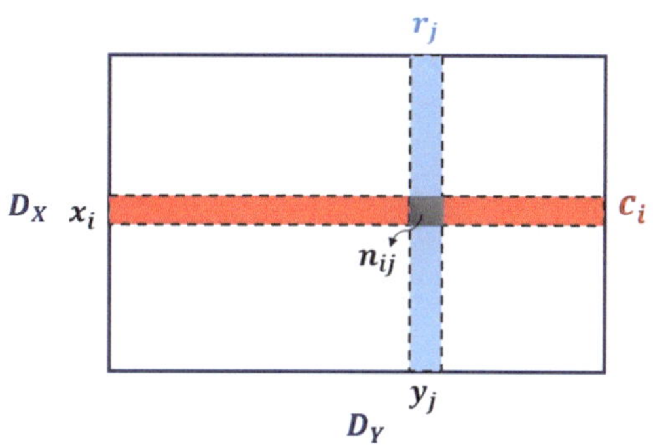

Fig. 2.1 Illustration of the probability theory

$$P(X = x_i) = \sum_{j \in D_Y} \frac{n_{ij}}{N},$$
$$P(Y = y_j) = \sum_{j \in D_X} \frac{n_{ij}}{N}. \tag{2.5}$$

In particular, consider $P(X = x_i)$. It can be seen from the above formula that the probability of $X = x_i$ is independent of the random variable Y by performing a summation over $j \in D_Y$. This is called the "marginal probability" of X which can be rewritten as below:

$$P(X = x_i) = \sum_{j \in D_Y} P(X = x_i, Y = y_j). \tag{2.6}$$

Similarly, the marginal probability of Y (or equivalently, $P(Y = y_j)$) could be found by performing a summation over $i \in D_X$. Note that (2.6) is obtained referring to (2.3) and (2.5). Also, note that (2.6) is known as the "sum rule" of the probability.

As an example, consider we have two random variables X and Y, where X corresponds to a coin-tossing experiment, and Y corresponds to rolling a fair dice experiment. The dictionaries of X and Y are specified as below:

$$D_X = \{1, 2, \cdots, 6\},$$
$$D_Y = \{\text{Head, Tail}\}.$$

Now, consider that we are interested in finding the probability of X ="Head" and $Y = 1$. Here, $n_{ij} = 1$ where i corresponds to the event "Head" in random variable X, and j corresponds to the event "1" corresponds to random variable Y. Moreover, the total number of events obtained from the combination of X and Y is $N = 12$. According to (2.2), the joint probability of X and Y would be obtained as below:

$$P(X = \text{Head}, Y = 1) = \frac{n_{ij}}{N} = \frac{1}{12}. \tag{2.7}$$

Also, inspired by (2.6), the marginal probabilities of X and Y would be obtained as below:

$$P(X = \text{Head}) = \sum_{j \in D_Y} P(X = \text{Head}, Y = y_j) = \frac{6}{12},$$
$$P(Y = 1) = \sum_{i \in D_X} P(X = x_i, Y = 1) = \frac{2}{12}. \tag{2.8}$$

Consider the case in which the event $X = x_i$ occurred given the knowledge that the event $Y = y_j$ has already occurred. The probability of $X = x_i$ given $Y = y_j$

is known as the conditional probability and is denoted by $p(X = x_i|Y = y_j)$. The conditional probability is formulated as below:

$$P(X = x_i|Y = y_j) = \frac{n_{ij}}{r_j}. \tag{2.9}$$

It can be seen from the above equation that the given information $(Y = y_j)$ limits the denominator of (2.2) compared to the non-conditional case. Referring to (2.9), the equation (2.2) can be reformulated as below:

$$P(X = x_i, Y = y_j) = \frac{n_{ij}}{N} = \frac{n_{ij}}{r_j} \times \frac{r_j}{N}, \tag{2.10}$$

which shows the relation between the joint probability and the conditional probability. Using (2.3) and (2.9), the above equation is rewritten as below:

$$P(X = x_i, Y = y_j) = P(X = x_i|Y = y_j)P(Y = y_j), \tag{2.11}$$

which is known as the product rule of probability. Generally, we have:

$$\textbf{sum rule}: \quad p(X) = \sum_Y p(X, Y),$$

$$\textbf{product rule}: \quad P(X, Y) = P(X|Y)P(Y), \tag{2.12}$$

where $P(X = x_i)$ and $P(Y = y_j)$ are written as $P(X)$ and $P(Y)$ for simplicity. Again, consider the above example where the two experiments, coin-tossing and rolling a fair dice, are considered. According to (2.4) and (2.9), the conditional probability of X="Head," given the knowledge that $Y = 1$, is obtained as below:

$$P(X = \text{Head}|Y = 1) = \frac{1}{2}. \tag{2.13}$$

Note that in the probability theory, we have symmetry property as below:

$$\textbf{symmetry property}: \quad p(X, Y) = p(Y, X).$$

According to the product rule and the symmetry property, we have:

$$P(Y|X)P(X) = P(X|Y)P(Y). \tag{2.14}$$

Dividing both sides of the above equation by $P(X)$, we obtain:

$$P(Y|X) = \frac{P(X|Y)P(Y)}{P(X)}. \tag{2.15}$$

This equation is known as the "Bayes rule." This theorem is used in the cases in which the conditional probability $P(X|Y)$ is known, but we are interested in the conditional probability $P(Y|X)$. Using the sum rule and the product rule presented in (2.12), the Bayes theorem can be rewritten as follows:

$$P(Y|X) = \frac{P(X|Y)P(Y)}{\sum_y p(X|Y)P(Y)}. \tag{2.16}$$

Independence

The random variables X and Y are said to be independent if the probability of the events corresponding to random variable X does not affect the probability of the events corresponding to random variable Y. For two independent random variables X and Y we have:

$$P(X|Y) = P(X) \quad \text{or} \quad P(Y|X) = P(Y), \tag{2.17}$$

which indicates that knowing that the event $Y = y_j$ occurred does not change the value of the probability of $X = x_i$. In the example discussed above where the coin-tossing and rolling dice are considered as X and Y, it can be concluded from (2.8) and (2.13) that X and Y are independent; knowing the probability of $Y = 1$ does not affect the probability of X="Head." If two random variables X and Y are independent, the joint probability would be formulated as below:

$$P(X, Y) = P(X|Y)P(Y) = P(X)P(Y), \tag{2.18}$$

which is concluded from (2.12) and (2.17). It can be seen from (2.18) that the joint probability of two independent random variables equals the production of the probability of each random variable. Note that X and Y are said to be "unconditionally independent" if their joint distribution can be represented as the product of their marginal probabilities.

2.3 Discrete Random Variable

In probability theory, a random variable is assigned to a variable that its value is obtained from a random process or an experiment. It can be said that a random variable is a function that maps a sample space into real numbers. For instance, tossing a coin for three times can be considered as a random experiment. The possible outcomes obtained from this experiment constitute the sample space denoted by S:

$$S = \{HHH, HHT, HTH, THH, HTT, THT, TTH, TTT\}, \tag{2.19}$$

where H and T stand for "head" and "tail," respectively. It can be seen that the sample space consists of 2^3 components. Now, consider that we want to find out the number of obtained Hs. Our goal in this example, or equivalently, the number of heads obtained in this experiment, is defined as a random variable, denoted by X. We assign a real number to each of the outcomes. In this example, we assign the number "0" to the last case in which H occurred zero times. Also, the number "1" would be assigned to the cases where one H occurred (e.g., THT), and so on. This is the meaning of mapping the sample space into real numbers. It can be concluded that depending on the outcome of the experiment, the value of the corresponding random variable would be 0, 1, 2, or 3. This set of possible values that can be assigned to the random variable is considered as the range of the random variable X, denoted by R_X:

$$R_X = \{0, 1, 2, 3\}. \tag{2.20}$$

Note that the above example is a discrete random variable. Generally, random variables can be categorized into two main parts: discrete random variables and continuous random variables. In this section, discrete random variables are discussed. Continuous random variable would be described and explained in the next section. Discrete or continuous nature of a random variable can be identified by its range; discrete random variables are assigned to the random variables that their ranges are countable, as it can be seen from the above example.

2.3.1 Probability Mass Function

Consider the range of the random variable X as follows:

$$R_X = \{x_1, x_2, x_3, \cdots\}, \tag{2.21}$$

where x_is are the possible values that can be assigned to the random variable X. Note that the random variables are usually denoted by the capital letters. Also, to show the numbers in the range, the lowercase letters are usually used. Here, we want to find out the probabilities of each event. The probability of occurring x_i, or equivalently, $P(X = x_i)$, is known as the probability mass function (PMF) of the random variable X, which is denoted by $P_X(x_i)$:

$$P_X(x_i) = P(X = x_i), \quad i = 1, 2, 3, \cdots. \tag{2.22}$$

It can be concluded that for a random variable, the probability of occurring an event is identified by its PMF. In other words, the distribution of a discrete random variable is described using the PMF. Note that the PMF of a random variable should satisfy the following two properties:

$$0 \leq p_X(x_i) \leq 1, \quad \text{for all } x_i, \tag{2.23}$$

$$\sum_{x_i \in S} p_X(x_i) = 1. \tag{2.24}$$

From the first property, it can be seen that the PMF is a value between 0 and 1 that shows how likely the event is. If $P_X(x_i)$ is close to 0, it is very unlikely that event x_i occurs. Obviously, if $P_X(x_i)$ is close to 1, the event a_i is very likely to occur. Consider the experiment where a coin is tossed for three times, similar to the previous example, and we are interested in the number of Hs. The sample space and the range of the random variable X are shown in (2.19) and (2.20), respectively. The PMF of this random variable is obtained according to (2.22). By expanding this equation, we have:

$$\begin{aligned} p_X(0) &= P(\{TTT\}) = \frac{1}{8}, \\ p_X(1) &= P(\{HTT, THT, TTH\}) = \frac{3}{8}, \\ p_X(2) &= P(\{HHT, HTH, THH\}) = \frac{3}{8}, \\ p_X(3) &= P(\{HHH\}) = \frac{1}{8}. \end{aligned} \tag{2.25}$$

In order to make sure that the PMF is written correctly, the properties of (2.24) should be evaluated. It can be seen that the first property is satisfied. Moreover, the summation of the probabilities equals 1, which satisfies the second property. Therefore, it can be concluded that the PMF of this example is written correctly. Figure 2.2 shows the PMF plot of this example.

Example 2.1 (Probability Mass Function (PMF))
Problem: Consider a contest where the probability of victory is p. The contest is repeated continuously until the first victory is achieved. If the contest is repeated for X times, the range of X is written as below:

$$R_X = \{1, 2, 3, \cdots\}.$$

Find the PMF.

(continued)

Example 2.1 (continued)
Solution: To find the PMF of X, we have:

$$
\begin{aligned}
P(X = 1) &= p \\
P(X = 2) &= (1 - p)p \\
&\vdots \\
P(X = n) &= (1 - p)^{(n-1)}p \\
\Rightarrow P_X(x) &= (1 - p)^{(n-1)}p \quad \text{for } n = 1, 2, \cdots .
\end{aligned}
$$

The above distribution is called the Geometric distribution. In order to check if the result is correct or not, the property presented in (2.24) is used:

$$
\begin{aligned}
\sum_{n=1}^{\infty}(1 - p)^{(n-1)}p = p\sum_{j=0}^{\infty}(1 - p)^j \\
= p \times \frac{1}{1 - (1 - p)} = 1,
\end{aligned}
$$

which guarantees the correctness of the obtained PMF.

Fig. 2.2 The PMF plot corresponding to (2.25)

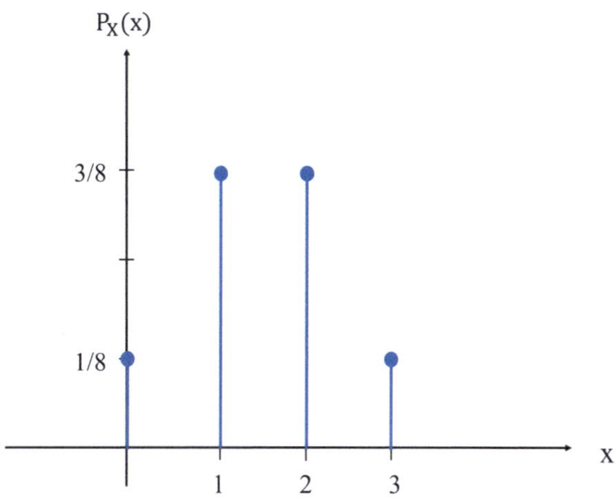

2.3.2 Cumulative Distribution Function

Another way to describe the distribution of a discrete random variable is possible using the cumulative distribution function (CDF). Note that CDF can be defined for the continuous random variables too. The CDF of the random variable X is defined as below:

$$F_X(x) = P(X \leq x), \quad \text{for } x \in R. \tag{2.26}$$

From the above equation, it can be seen that the CDF of the random variable X is obtained by accumulating the probability of occurring the events from $-\infty$ until the present x. Therefore, it can be concluded that the CDF of a random variable is always a non-decreasing function. Note that the following property could be easily obtained from the definition of the CDF:

$$P(a < x \leq b) = F_X(b) - F_X(a), \quad \text{for } a \leq b. \tag{2.27}$$

Again, consider the coin-tossing example mentioned above. The CDF of the corresponding random variable could be obtained as below:

$$F_X(x) = \begin{cases} 0 & \text{for } x < 0 \\ \frac{1}{8} & \text{for } 0 \leq x < 1 \\ \frac{4}{8} & \text{for } 1 \leq x < 2 \\ \frac{7}{8} & \text{for } 2 \leq x < 3 \\ 1 & \text{for } x \geq 3 \end{cases}. \tag{2.28}$$

Figure 2.3 shows the plot of the above CDF. It can be seen from the figure that the CDF is a non-decreasing function, as mentioned before.

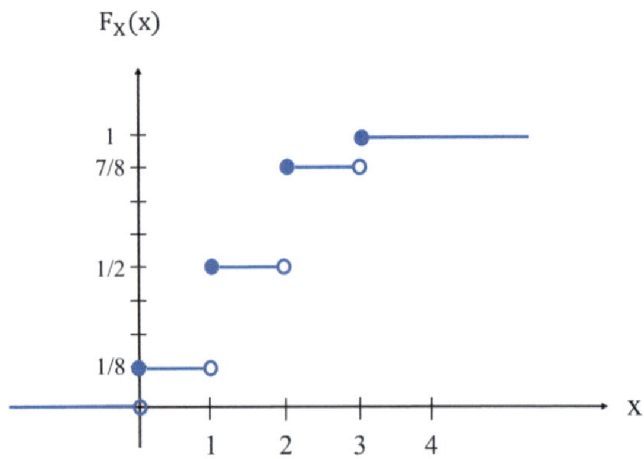

Fig. 2.3 The CDF plot corresponding to (2.28)

Example 2.2 (Cumulative Distribution Function (CDF))
Problem: Consider a discrete random variable with the following PMF:

$$P_N(n) = \frac{e^{-\lambda}\lambda^n}{n!},$$

which is called the Poisson distribution. Obtain the CDF.
Solution: According to the definition, the CDF of the above random variable is obtained as below:

$$F_N(n) = \begin{cases} 0 & \text{for } n < 0 \\ \sum_{k=0}^{n} \frac{e^{-\lambda}\lambda^k}{k!} & \text{for } n \geq 0 \end{cases}.$$

Note that $F_N(n) = 1$ for $n = \infty$. Also, consider we are interested in finding $P(2 < n \leq 5)$. According to (2.27) we have:

$$\begin{aligned} P(2 < N \leq 5) &= \sum_{k=0}^{5} \frac{e^{-\lambda}\lambda^k}{k!} - \sum_{k=0}^{2} \frac{e^{-\lambda}\lambda^k}{k!} \\ &= \sum_{k=3}^{5} \frac{e^{-\lambda}\lambda^k}{k!} \\ &= e^{-\lambda}\left(\frac{\lambda^3}{3!} + \frac{\lambda^4}{4!} + \frac{\lambda^5}{5!}\right). \end{aligned}$$

2.3.3 Expectation and Variance

The probability distribution can be described by its moments. However, in some of the common probability distributions such as Poisson or Normal distributions, only the first and the second moments are sufficient for the distributions to be described. Therefore, it can be concluded that among the moments, the first and the second moment are the most common parameters employed to describe and even compare the distributions. The first-moment metric is known as the expectation. Moreover, using the expectation and the second-moment metrics, another metric named variance is defined. In this section, it is considered to introduce and explain the expectation and the variance, as two common parameters, in more detail. Note that some of the most common probability distributions would be introduced later in Sect. 2.5 where the expectation and the variance parameters are sufficient to describe them.

Expectation

In probability theory, the expectation or the mean value of a random variable is defined as the value that is expected, on average, to be obtained from a random process after infinite repetition. This is one of the most common parameters which is performed to describe the distribution of a random variable, predict the behavior of the process, or compare the results of two or more random variables.

The expected value of the random variable X, denoted by $\mathbb{E}[X]$, is defined as below:

$$\mathbb{E}[X] \triangleq \sum_{x_i \in R_X} x_i P_X(x_i), \tag{2.29}$$

which is formulated as the sum of the value of each event multiplied to its corresponding probability. It can be seen from the formula that as the probability of an event increases, the contribution of that event would be greater in the expectation calculation. Consider $g(X)$ as a function of the random variable X. In order to calculate the expectation of $g(X)$, we have:

$$\mathbb{E}[g(X)] = \sum_{x \in R_X} g(x) P_X(x). \tag{2.30}$$

The above formula is known as the law of the unconscious statistician (LOTUS) which is applicable in some cases where calculating the expectation value using the direct formula is hard to solve.

Again, consider the previous example in which a coin is tossed three times. Referring to (2.29), the expectation of the random variable in this example would be obtained as below:

$$\begin{aligned}\mathbb{E}[X] &= \left(0 \times \frac{1}{8}\right) + \left(1 \times \frac{3}{8}\right) + \left(2 \times \frac{3}{8}\right) + \left(3 \times \frac{1}{8}\right) \\ &= \frac{3}{2}\end{aligned}$$

which means that the mean value of the process would be convergent or converges only $\frac{3}{2}$ if the experiment is repeated infinite times.

Example 2.3 (Expectation)

Problem: Consider the Geometric distribution with the following PMF:

$$P_X(x) = (1-p)^{(n-1)} p \quad \text{for } n = 1, 2, 3, \cdots.$$

Find the expectation for the aforementioned Geometric distribution.

(continued)

Example 2.3 (continued)
Solution: According to (2.29), the expectation of the above distribution is obtained as below:

$$\mathbb{E}[X] = \sum_{n=1}^{\infty} n(1-p)^{(n-1)} p.$$
$$= p \sum_{j=0}^{\infty} (j+1)(1-p)^{j}. \tag{2.31}$$

In order to obtain the solution of the above formula, consider the following relation:

$$\sum_{j=0}^{\infty} (1-p)^{(j+1)} = \frac{1-p}{p}.$$

Taking the derivation from both sides of the above equation with respect to p, we have:

$$-\sum_{j=0}^{\infty} (j+1)(1-p)^{j} = \frac{-1}{p^2}. \tag{2.32}$$

Referring to (2.31) and (2.32), the expectation would be obtained as below:

$$\mathbb{E}[X] = p \times \frac{1}{p^2} = \frac{1}{p}.$$

Variance

In addition to the expectation (or equivalently, the mean) parameter, the variance of a random variable is another important parameter that can be used to identify the distribution of that random variable. The variance of a random variable indicates how the samples are spread over the mean distribution of that random variable. Consider the expectation value of a random variable X to be $\mathbb{E}[X] = \mu_X$. The variance of the random variable would be defined as follows:

$$\begin{aligned} Var[X] &\triangleq \mathbb{E}[(X-\mu_X)^2] \\ &= \mathbb{E}[X^2 - 2X\mu_X + \mu_X^2] \\ &= \mathbb{E}[X^2] - 2\mu_X\mathbb{E}[X] + \mu_X^2 \\ &= \mathbb{E}[X^2] - \mu_X^2 \end{aligned} \tag{2.33}$$

or equivalently:

$$Var[X] = \mathbb{E}[X^2] - \mathbb{E}^2[X], \tag{2.34}$$

where $\mathbb{E}[X^2]$ is the second-moment parameter of the distribution of the random variable X. Note that the variance of a random variable X is also denoted by σ_X^2. Also, note that the variance is always a non-negative parameter ($\sigma_X^2 \geq 0$). Comparing the expectation and the variance, it can be concluded that the expectation or the mean value of a random variable shows the average position of the distribution. However, the variance metric shows how the samples are distributed around the mean value. A lower value of the variance metric indicates that by selecting a sample from the distribution, it is expected that the selected sample is close to its mean value. It is worth to note that for a random variable X, we have:

$$Var[aX+b] = a^2 Var[X], \tag{2.35}$$

where a and b are constant. The proof of the above relation is left as an exercise. Referring to (2.34), the variance of the coin-tossing example mentioned before would be obtained as below:

$$\mathbb{E}[X^2] = \left(0^2 \times \frac{1}{8}\right) + \left(1^2 \times \frac{3}{8}\right) + \left(2^2 \times \frac{3}{8}\right) + \left(3^2 \times \frac{1}{8}\right) = 3.$$

Therefore, we have:

$$Var[X] = \mathbb{E}[X^2] - \mathbb{E}^2[X] = 3 - \left(\frac{3}{2}\right)^2 = \frac{3}{4}.$$

Another measure, called the standard deviation, also could be defined which is simply the square root of the variance. It is defined as:

$$std(X) \triangleq \sigma_X = \sqrt{Var[X]}. \tag{2.36}$$

The concept of the standard deviation and the variance are similar, except that their units are different. It is easy to find out that the standard deviation of the above coin-tossing example would be $\frac{\sqrt{3}}{2}$.

Example 2.4 (Variance)
Problem: Consider the Geometric distribution with the following PMF:

$$P_X(x) = (1-p)^{(n-1)}p \quad \text{for } n = 1, 2, 3, \cdots.$$

Find the variance for the above distribution.
Solution: In order to obtain the variance of the above distribution, one can use LOTUS presented in (2.30). Accordingly, we have:

$$\begin{aligned}
\mathbb{E}[X(X-1)] &= \sum_{n=1}^{\infty} n(n-1)(1-p)^{(n-1)}p. \\
&= p\sum_{j=0}^{\infty}(j+1)j(1-p)^{j} \\
&= -p\frac{\partial}{\partial p}\sum_{j=0}^{\infty} j(1-p)^{j+1} \\
&= -p\frac{\partial}{\partial p}\left[(1-p)\sum_{j=0}^{\infty} j(1-p)^{j} \times \frac{(1-p)^{-1}}{(1-p)^{-1}}\right] \\
&= -p\frac{\partial}{\partial p}\left[(1-p)^{2}\sum_{j=0}^{\infty} j(1-p)^{j-1}\right] \\
&= p\frac{\partial}{\partial p}\left[(1-p)^{2}\frac{\partial}{\partial p}\sum_{j=0}^{\infty}(1-p)^{j}\right]. \qquad (2.37)
\end{aligned}$$

Also, we have:

$$\sum_{j=0}^{\infty}(1-p)^{j} = \frac{1}{p}.$$

Therefore, (2.37) can be rewritten as below:

$$\begin{aligned}
\mathbb{E}[X(X-1)] &= \mathbb{E}[X^2] - \mathbb{E}[X] \\
&= p\frac{\partial}{\partial p}\left[(1-p)^{2}\frac{\partial}{\partial p}\left(\frac{1}{p}\right)\right]
\end{aligned}$$

(continued)

Example 2.4 (continued)

$$= \frac{2(1-p)}{p^2}. \tag{2.38}$$

The expectation of the Geometric distribution is $1/p$. Therefore, according to (2.38) we have:

$$\mathbb{E}[X^2] = \frac{2(1-p)}{p^2} + \mathbb{E}[X]$$
$$= \frac{2(1-p)}{p^2} + \frac{1}{p} = \frac{2-p}{p^2}.$$

Finally, the variance of the Geometric distribution is obtained as below:

$$Var[X] = \mathbb{E}[X^2] - \mathbb{E}^2[X]$$
$$= \frac{2-p}{p^2} - \frac{1}{p^2} = \frac{1-p}{p^2}.$$

2.4 Continuous Random Variable

In discrete random variables, the range of the random variable is countable, as mentioned before. However, in a continuous random variable, the range is defined as an interval (or the summation of intervals), and therefore, it would not be countable anymore. Moreover, for a continuous random variable, we have $P(X = x) = 0$ for all $x \in \mathbb{R}$. In the following, the distribution of the continuous random variables will be discussed in more detail.

2.4.1 Probability Density Function

In order to describe the distribution of a continuous random variable, the probability density function (PDF) is defined (instead of the PMF defined for discrete random variables). The PDF of the random variable X is denoted by $f_X(x)$ and its concept is similar to the PMF defined for discrete random variables. The PDF of the continuous random variable X would be defined as below:

$$f_X(x) = \frac{dF_X(x)}{dx}, \tag{2.39}$$

where $F_X(x)$ is the CDF of the random variable X, as described in Sect. 2.3.2. Note that $F_X(x)$ is differentiable at x. The above equation declares that the PDF of a continuous random variable equals the derivative of its corresponding CDF. The relation between the PDF and the CDF of a continuous random variable X is obtained as follows:

$$F_X(x) = P(X \leq x) = \int_{-\infty}^{x} f_X(u)du, \quad -\infty < x < \infty. \tag{2.40}$$

Similar to the result obtained from the CDF of a discrete random variable, for the continuous random variable we can also conclude that the CDF is a non-decreasing function. Moreover, we have

$$P(a < X \leq b) = F_X(b) - F_X(a) = \int_{a}^{b} f_X(u)du. \tag{2.41}$$

Note that the above equation is similar to (2.27) described in the discrete random variable section. Generally, it can be said that the formulas of the continuous random variable are similar to the discrete random variables, except that the integral operator is replaced with the summation operator. This issue will be more clearly seen in the following.

Let us investigate the PDF and the CDF of the continuous random variable with an example; consider an experiment in which a number is randomly chosen from the interval [0, 10]. This can be considered as a continuous random variable according to the definition of the continuous random variable. This example is known as the Uniform distribution which will be discussed in more detail later in Sect. 2.5. The PDF of this example would be written as below:

$$f_X(x) = \begin{cases} \frac{1}{10} & \text{for } 0 \leq x \leq 10 \\ 0 & \text{otherwise} \end{cases}. \tag{2.42}$$

Figure 2.4 shows the PDF plot of this example. It can be seen from the figure that the Uniform distribution is a continuous random variable. According to (2.40), the CDF would be obtained as below:

$$\begin{aligned} F_X(x) &= \int_{-\infty}^{x} f_X(u)du \\ &= \int_{0}^{x} \frac{1}{10}du = \frac{1}{10}x, \end{aligned}$$

for $0 \leq x \leq 10$. Also, for $x < 0$ and $x > 10$ intervals, the CDF would be 0 and 1, respectively. In other words, we have:

Fig. 2.4 The PDF plot corresponding to (2.42)

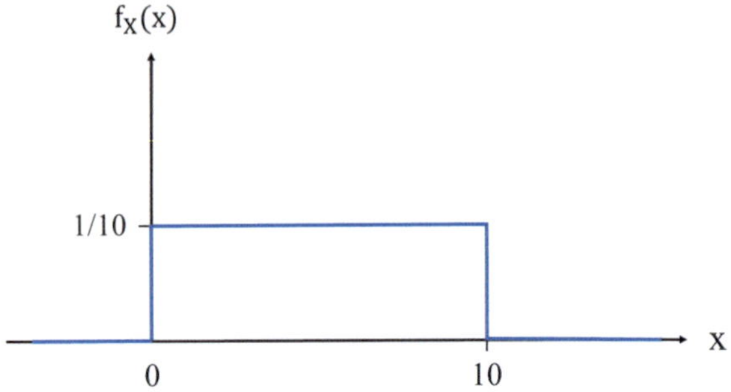

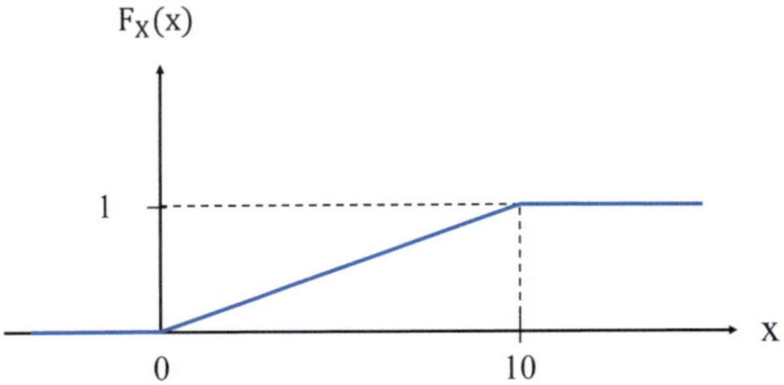

Fig. 2.5 The CDF plot corresponding to (2.43)

$$F_X(x) = \begin{cases} 0 & \text{for } x < 0 \\ \frac{1}{10}x & \text{for } 0 \leq x \leq 10 \\ 1 & \text{for } x > 10 \end{cases} . \tag{2.43}$$

Figure 2.5 depicts the CDF plot corresponding to (2.43). It can be clearly seen that the CDF plot is a non-decreasing function, as mentioned before.

Example 2.5 (Continuous Random Variable)

Problem: Consider a random variable with the following PDF:

$$f_X(x) = \alpha^2 e^{-\frac{\alpha}{5}x} \quad \text{for } x \geq 0,$$

Determine the α value and CDF.

Solution: The properties of the discrete random variable are valid for continuous random variables, except that the integral operator is replaced with the summation operator. Therefore, according to (2.24), we have:

(continued)

Example 2.5 (continued)

$$\int_{-\infty}^{\infty} f_X(x)dx = 1$$

$$\int_{0}^{\infty} \alpha^2 e^{-\frac{\alpha}{5}x}dx = 1 \rightarrow \alpha = \frac{1}{5}.$$

Also, the CDF of the above random variable is obtained as below:

$$\begin{aligned} F_X(x) &= \int_0^{\infty} \frac{1}{25} e^{-\frac{1}{25}x} dx \\ &= 1 - e^{-\frac{1}{25}x}. \end{aligned}$$

2.4.2 Expectation and Variance

Similar to the discrete random variables, the expectation and variance are also defined in continuous random variables. These two common parameters are applicable in describing the probability distribution of the continuous random variables such as Exponential distribution. In the following, these parameters are discussed separately. As will be seen, the definition of the expectation and variance in continuous random variables is similar to the ones in discrete random variables. The difference is that the integral operator is replaced with the summation operator.

Expectation

The expectation value of a continuous random variable X would be obtained from the following formula:

$$\mathbb{E}[X] = \int_x x f_X(x) dx. \tag{2.44}$$

It can be seen that the above relation is similar to the expectation formula of the discrete random variable, defined in (2.29), with this difference that the integral operator is replaced with the summation operator. Considering the uniform distribution example where its PDF is presented in (2.42), the corresponding expectation value would be obtained as below:

$$\mathbb{E}[X] = \int_0^{10} \frac{x}{10} dx = 5. \tag{2.45}$$

Example 2.6 (Continuous Random Variable Expectation)
Problem: Consider a continuous random variable with the following PDF:

$$f_X(x) = \frac{1}{25} e^{-\frac{1}{25}x} \quad \text{for } x \geq 0.$$

Determine the expectation.
Solution: The expectation of the random variable is calculated as below:

$$\begin{aligned}\mathbb{E}[X] &= \int_0^{\infty} \frac{x}{25} e^{-\frac{1}{25}x} dx \\ &= \left(-xe^{-\frac{x}{25}} - 25e^{-\frac{x}{25}}\right)_0^{\infty} = 25.\end{aligned}$$

Variance

Before computing the variance of the continuous random variable, LOTUS should be formulated. For continuous random variables, LOTUS is formulated as below:

$$\mathbb{E}[g(X)] = \int_{-\infty}^{\infty} g(x) f_X(x) dx. \tag{2.46}$$

Comparing (2.30) and (2.46), it can be seen that the definition of LOTUS is similar for discrete and continuous random variables. The difference is that for the continuous random variable, the integral operator is replaced with the summation operator, as shown in the above equation. Referring to the formulation of the variance obtained in (2.34), the variance of the continuous random variable equals $\mathbb{E}[(X - \mu_X)^2]$. According to (2.46), it can be concluded that $g(X) = (X - \mu_X)^2$, and therefore, the variance of the continuous random variable would be written as follows:

$$\begin{aligned}Var[X] &\triangleq \mathbb{E}[(X - \mu_X)^2] \\ &= \int_{-\infty}^{\infty} (x - \mu_X)^2 f_X(x) dx \\ &= \int_{-\infty}^{\infty} x^2 f_X(x) dx - 2\mu_X \int_{-\infty}^{\infty} x f_X(x) dx + \mu_X^2 \int_{-\infty}^{\infty} f_X(x) dx \\ &= \int_{-\infty}^{\infty} x^2 f_X(x) dx - \mu_X^2.\end{aligned} \tag{2.47}$$

As an example, consider the Uniform distribution in the interval [0, 10]. According to the above formula, the variance of the distribution would be obtained as follows:

$$\int_{-\infty}^{\infty} x^2 f_X(x)dx = \int_{0}^{10} x^2 \frac{1}{10} dx = \frac{100}{3}.$$

Note that the expectation of the considered distribution is presented in (2.45). Therefore, we have:

$$\begin{aligned} Var[X] &= \int_{-\infty}^{\infty} x^2 f_X(x)dx - \mu_X^2 \\ &= \frac{100}{3} - (5)^2 = \frac{25}{3}. \end{aligned}$$

Example 2.7 (Continuous Random Variable Variance)
Problem: Consider the following PDF:

$$f_X(x) = \frac{1}{25} e^{-\frac{1}{25}x} \quad \text{for } x \geq 0.$$

Determine the variance.
Solution: In order to obtain the variance of the above distribution, we have:

$$\mathbb{E}[X^2] = \int_{0}^{\infty} \frac{x^2}{25} e^{-\frac{1}{25}x} dx = 2(25)^2.$$

Also, it is obtained that the expectation of the above distribution is 25. Therefore, the variance would be obtained as below:

$$Var[X] = \mathbb{E}[X^2] - \mathbb{E}^2[X] = 2(25)^2 - (25)^2 = 625.$$

2.5 Common Distributions

There exist some special distributions that have a wide range of applications in the machine learning and are commonly used in modeling the experiments. These common distributions are identified and separated with special names. Due to the applicability of these distributions, it is worth investigating their properties and memorize them. In the following, we will review some of the commonly used probability distributions. The content is divided into two categories. In the first category, the discrete distributions are discussed. In the second category, some continuous distributions are introduced and investigated.

2.5.1 Discrete Distributions

Bernoulli Distribution

The Bernoulli distribution is one of the most simple discrete distributions in which two outcomes can occur with the probability of p and $q = p - 1$, respectively. As an example, consider tossing a coin. If the outcome is H, the value "1" would be assigned to the outcome. Also, If the outcome is T, the corresponding value would be "0". Each of the events occurs with a certain probability. It means we have:

$$P\{\text{Head}\} = p$$
$$P\{\text{Tail}\} = 1 - p.$$

The PMF of X can be written as

$$P_X(x) = \begin{cases} p & \text{for } x = 1 \\ 1 - p & \text{for } x = 0 \\ 0 & \text{otherwise} \end{cases}. \tag{2.48}$$

According to (2.29), the expected value of the Bernoulli distribution is:

$$\mathbb{E}[X] = (1)(p) + (0)(1 - p) = p. \tag{2.49}$$

In order to compute the variance, we have:

$$\mathbb{E}[X^2] = (1^2)(p) + (0^2)(1 - p) = p, \tag{2.50}$$

and therefore, the variance of the Bernoulli distribution would be obtained as below:

$$Var[X] = \mathbb{E}[X^2] - \mathbb{E}^2[X] = p - p^2 = pq. \tag{2.51}$$

Note that the Bernoulli distribution is expressed as $X \sim Bernoulli(p)$.

Binomial Distribution

The Binomial distribution is also a discrete distribution. Consider we toss a coin n times independently. Similar to the Bernoulli distribution, the probability of H is p, and the probability of T is $1 - p$ for each trial. Consider that we are interested in finding the total number of obtained Hs after n trials. The PMF of this distribution is written as below:

$$P_X(k) = \binom{n}{k} p^k (1-p)^{n-k}, \quad x = 0, 1, \ldots, n. \tag{2.52}$$

It can be seen that the Binomial distribution is identified by two parameters: n and p. The Binomial distribution is expressed as $X \sim Binomial(n, p)$. It is worth noting that the Binomial distribution can be considered as the sum of n independent Bernoulli distributions. By taking this issue into account, the expectation of this distribution would be easily calculated as below:

$$\mathbb{E}[X] = np. \tag{2.53}$$

Similarly, the variance of the Binomial distribution would be obtained from summing the variance of n independent Bernoulli distributions:

$$Var[X] = npq. \tag{2.54}$$

Figure 2.6 shows three Binomial distributions with different parameters N and p.

The Multinomial Distributions

As it was described, The binomial distribution is defined when there are only two possible outcomes, i.e., true/false or heads/tails. When the binomial case

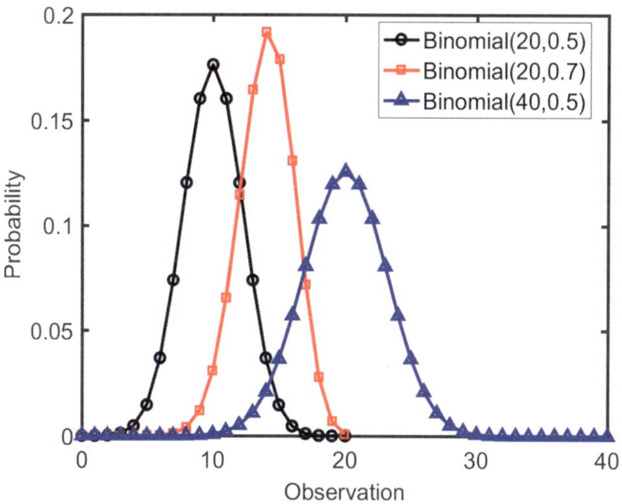

Fig. 2.6 Illustration of the binomial distribution with different parameters N and p

is extended to the multi-dimensional case, the distribution is called multinomial distribution. In the multinomial distribution, there are a set of k possible outcomes $(X_1, X_2, X_3, \cdots, X_k)$ with associated probabilities $(p_1, p_2, p_3, \cdots, p_k)$. Note that ,the sum of probabilities must equal 1 because one of the results is sure to occur. Let us assume n trials and let the number of times that X_i occurs be denote by x_i. With this notation, the PDF of concurrently observing $\{x_1, \cdots, x_k\}$ is given by

$$f(x_1, \cdots, x_k; n, p_1, \ldots, p_k) = \binom{n}{x_1, \ldots, x_k} p_1^{x_1} \ldots p_k^{x_k}, \tag{2.55}$$

where $\binom{n}{x_1, \ldots, x_k} = \frac{n!}{x_1!\, x_2! \cdots x_k!}$ is the multinomial coefficient.

The Poisson Distribution

The Poisson distribution is one of the most common discrete distributions that has a wide range of applications in different scenarios. It can be used for modeling a series of discrete events in which the average time between the events is known, but the exact timing of events is random. For instance, consider that we are interested in observing the number of cars that visit a certain street at a given time interval (e.g., 1 to 2 pm). Also, suppose that we know the number of cars visiting the considered street is $\lambda = 20$, on average, according to the information obtained from the previous days. We can model this scenario using the Poisson distribution with the parameter $\lambda = 20$, and its PMF is of the form:

$$P_X(k) = \frac{\lambda^k e^{-\lambda}}{k!}, \quad k = 0, 1, 2, \ldots. \tag{2.56}$$

The Poisson distribution is expressed as $X \sim Poisson(\lambda)$. Note that in Poisson distribution, the arrival of an event is independent of its previous events. According to (2.29), the expected value of the Poisson distribution is calculated as below:

$$\begin{aligned} \mathbb{E}[X] &= \sum_{k=0}^{\infty} k \frac{\lambda^k e^{-\lambda}}{k!} \\ &= e^{-\lambda} \sum_{k=1}^{\infty} \frac{\lambda^k}{(k-1)!} = \lambda. \end{aligned} \tag{2.57}$$

After some manipulation, one can show that the variance of Poisson distribution is also given by λ. This is an interesting feature for this distribution where its expectation and its variance are equal. Figure 2.7 shows the Poisson distribution for different values of λ.

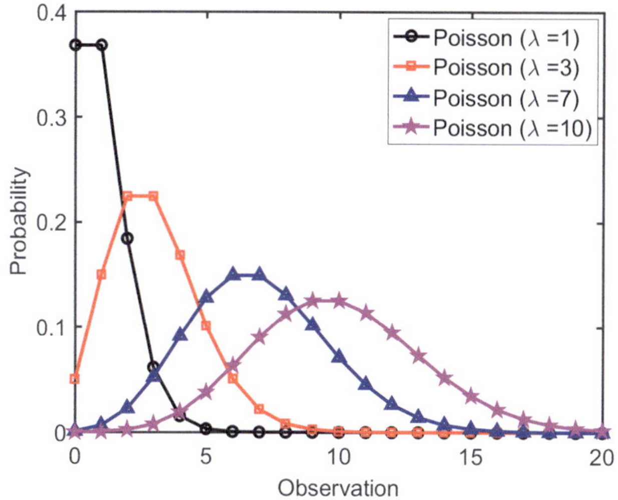

Fig. 2.7 Illustration of the Poisson distribution with different values of λ

2.5.2 *Continuous Distributions*

In this section we review some commonly used continuous probability distributions.

Gaussian (Normal) Distribution

One of the most important probability distributions for continuous random variables is the Normal, or the Gaussian distribution. This distribution plays an important role in the probability theory and machine learning. The Normal distribution is a symmetric distribution where most of the observations cluster around its central peak (mean). Extreme values in both tails of the distribution are similarly unlikely. The random variable X has a Normal distribution if its PDF is defined as follows:

$$f_X(x) = \frac{1}{\sigma_X\sqrt{2\pi}} \exp\left[-\frac{(x-\mu_X)^2}{2\sigma_X^2}\right], \quad -\infty < x < \infty, \tag{2.58}$$

where μ_X and σ_X^2 are the mean value and the variance of the distribution, respectively. The Normal distribution is identified by these two parameters and expressed as $X \sim \mathcal{N}(\mu_X, \sigma_X^2)$. The shape of the distribution changes based on the values of these parameters. Changing the mean value shifts the entire curve on the X-axis. The variance defines the width of the Normal distribution. Changing the variance either tightens or spreads out the width of the distribution along the X-axis. Figure 2.8 shows the effect of changing the parameters on the shape of the distribution.

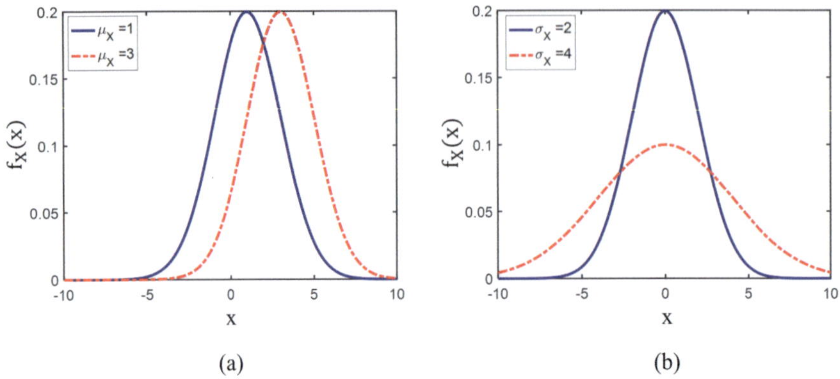

Fig. 2.8 Normal distribution with (**a**) $\sigma_X^2 = 2$ and $\mu_X \in \{1, 3\}$, (**b**) $\mu_X = 0$ and $\sigma_X^2 \in \{2, 4\}$

A continuous random variable X is said to be a standard Normal (or standard Gaussian) random variable, $X \sim \mathcal{N}(0, 1)$, if its mean value and its variance equal to "0" and "1", respectively. According to (2.58), the PDF of the standard Normal distribution is written as:

$$f_X(x) = \frac{1}{\sqrt{2\pi}} e^{-(x^2/2)}. \tag{2.59}$$

According to (2.40), it can be concluded that in order to obtain the CDF of the Gaussian distribution, we can integrate the PDF function as follows:

$$F_X(x) = Q(\frac{X-\mu}{\sigma}) = \int_{-\infty}^{x} \frac{1}{\sqrt{2\pi\sigma_X^2}} \exp\left[-\frac{(u-\mu_X)^2}{2\sigma_X^2}\right] du, \tag{2.60}$$

where $Q(.)$ is given by,

$$Q(x) = \int_{-\infty}^{x} \frac{1}{\sqrt{2\pi}} e^{u^2/2} du. \tag{2.61}$$

Q-function does not have a closed-form solution and is available in tabulated form. Another important function in the context of Normal distribution is "Error function," which is given by

$$\text{erf}(x) = \frac{2}{\sqrt{\pi}} \int_0^x e^{-t^2} dt. \tag{2.62}$$

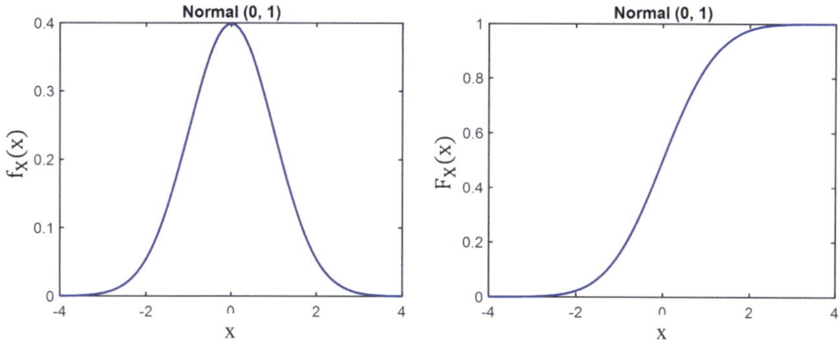

(a) (b)

Fig. 2.9 The graph of the (**a**) PDF and (**b**) CDF of the standard Normal distribution

After some algebra, one can show that $Q(x) = \frac{1}{2} - \frac{1}{2}\text{erf}(\frac{x}{\sqrt{2}})$. Figure 2.9 shows the PDF and the CDF plot of the standard Normal distribution.

Exponential Distribution

When events occur independently over non-overlapping intervals, the time interval between the occurrence of two consecutive events can be modeled by the exponential distribution. The PDF of the exponential distribution is written as

$$f_X(x) = \begin{cases} \lambda e^{-\lambda x} & \text{for} \quad x \geq 0, \\ 0 & \text{otherwise} \end{cases} \tag{2.63}$$

where λ is denoted as the rate parameter of the exponential distribution. According to (2.40), the CDF of the exponential distribution would be obtained as below:

$$F_X(x) = \int_0^x \lambda e^{-\lambda u} du = 1 - e^{\lambda x}. \tag{2.64}$$

Figure 2.10 shows the PDF and the CDF of the exponential distribution for three different values of λ. Following (2.44) and (2.47), the expectation and the variance of the exponential distribution are obtained as

$$\mathbb{E}[X] = \int_0^\infty x\lambda e^{-\lambda x} dx = \frac{1}{\lambda} \tag{2.65}$$

$$Var[x] = \int_0^\infty x^2 \lambda e^{-\lambda x} dx - \frac{1}{\lambda^2} = \frac{1}{\lambda^2}. \tag{2.66}$$

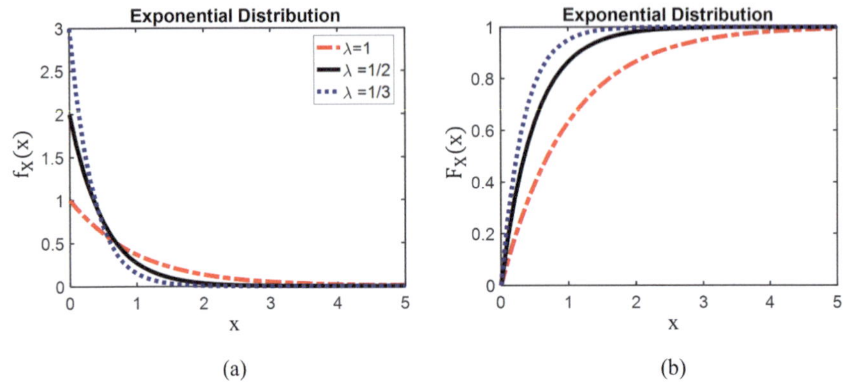

Fig. 2.10 The plot of the (**a**) PDF and (**b**) CDF of the Exponential distribution for three different values of λ

Memory-Less Property of Exponential Distribution

Assume two positive real number, $\alpha, \beta \geq 0$. For two event of $\{X > \alpha + \beta\}$ and $\{X > \alpha\}$, we can write

$$\begin{aligned} P\{X > \alpha + \beta \mid X > \alpha\} &= \frac{e^{-(\alpha+\beta)}}{e^{-\alpha}} \\ &= P\{X > \beta\}. \end{aligned} \tag{2.67}$$

For example, if X is the waiting time, then (2.67) implies that if an event does not arrive at time α, and has to wait additional time β to arrive, the probability of waiting for additional time β depends only on β (not on α), and this probability is identical to the probability of waiting for time β. This fact is called "Memory-less" property of exponential distribution.

Uniform Distribution

In continuous distributions, the uniform distribution is one of the most simple probability distributions which plays an important role in machine learning. It is concerned with events that are equally likely to occur. The continuous random variable X is said to be uniformly distributed on the interval $[a, b]$, and expressed as $X \sim U(a, b)$, when its PDF on the given interval is written as:

$$f_X(x) = \begin{cases} \frac{1}{b-a} & \text{for} \quad a \leq x \leq b \\ 0 & \text{otherwise.} \end{cases} \tag{2.68}$$

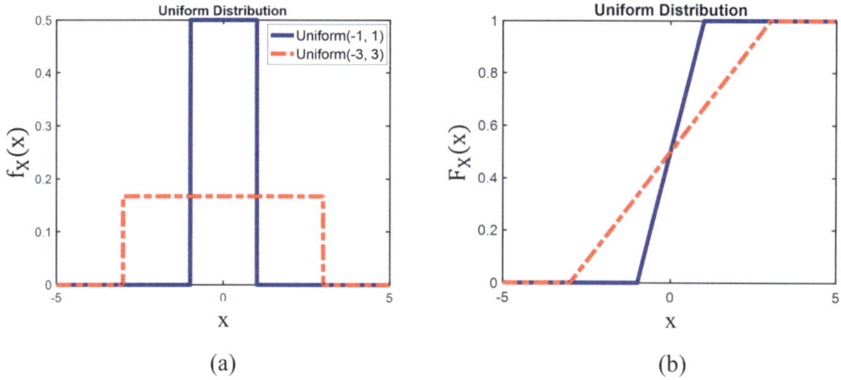

Fig. 2.11 The plot of the (**a**) PDF and (**b**) CDF of the Uniform distribution for two different intervals

Consequently, the CDF of the Uniform distribution would be:

$$F_X(x) = \begin{cases} 0 & \text{for } x < a \\ \frac{x-a}{b-a} & \text{for } a \leq x \leq b \\ 1 & \text{for } x > b \end{cases} \tag{2.69}$$

Figure 2.11 depicts the PDF and the CDF of uniform distribution for two different intervals. According to (2.44), the expected value of the Uniform distribution would be obtained as:

$$\mathbb{E}[X] = \int_a^b x f_X(x) dx = \int_a^b \frac{x}{b-a} da = \frac{a+b}{2}. \tag{2.70}$$

In order to calculate the variance of the uniform distribution, we have:

$$\mathbb{E}[X^2] = \int_a^b \frac{x^2}{b-a} dx = \frac{a^2+ab+b^2}{3}, \tag{2.71}$$

and therefore, the variance would be obtained as below:

$$Var[X] = \mathbb{E}[X^2] - \mathbb{E}^2[X] = \frac{(b-a)^2}{12}. \tag{2.72}$$

2.6 Joint Probability Distributions

The probability distribution of a random variable is discussed in both cases of discrete and continuous forms in the previous sections. In this section, the joint probability distribution will be discussed in which the distribution of two or more random variables has to be considered jointly. If two random variables are considered, the joint probability distribution is called the bivariate distribution. It can be generalized to multivariate probability distribution as the number of random variables increases. The joint probability distribution can be categorized into two parts: the joint distribution of discrete and continuous random variables. Generally, it can be said that the formulations of the joint distribution of discrete and continuous random variables are similar, with this difference that in the joint distribution of discrete random variables, the summation operator is used. However, in the joint distribution of continuous random variables, the integral operator is replaced with the summation operator. In the following, each of the categories will be discussed separately in more detail.

2.6.1 *Joint Distribution: Discrete Random Variables*

From Sect. 2.3 we know that for a discrete random variable, the PMF can describe the distribution of the random variable. Here, we are interested in investigating the distribution of two (or more) discrete random variables at the same time. Therefore, joint PMF should be defined as below:

$$P_{XY}(x, y) = P(X = x, Y = y). \tag{2.73}$$

Consider the range of the random variables X and Y as R_X and R_Y, respectively. The range of the joint problem would be written as below:

$$R_{XY}\{(x, y)|x \in R_X, y \in R_Y\}. \tag{2.74}$$

Similar to the properties obtained in (2.24), for the joint PMF we have:

$$\begin{aligned} 0 \leq P_{XY}(x_i, y_i) &\leq 1, \quad \text{for all } x_i, y_i, \\ \sum_{x_i, y_i \in R_{XY}} P_{XY}(x_i, y_i) &= 1. \end{aligned} \tag{2.75}$$

We will be focused on the joint PMF of two random variables. However, it should be noted that the joint PMF can be defined for more random variables (e.g., $P_{XYZ}(x, y, z)$) similar to this manner.

As an example, consider the following joint PMF ($P_{XY}(x, y)$) consists of two random variables X and Y:

	$Y = 1$	$Y = 2$
$X = 1$	$\frac{1}{3}$	$\frac{1}{12}$
$X = 2$	$\frac{1}{3}$	$\frac{1}{4}$

(2.76)

The range of the joint PMF is $R_{XY} = \{(x, y)|x, y \in \{1, 2\}\}$. Note that the properties mentioned in (2.75) are also met. It is worth noting that if two random variables X and Y are independent, the joint distribution will be the multiplication of the distribution of each random variable:

$$P_{XY}(x, y) = P_X(x)P_Y(y), \tag{2.77}$$

Joint CDF

For two random variables X and Y, the joint CDF can be defined as:

$$F_{XY}(x, y) = P(X \leq x, Y \leq y). \tag{2.78}$$

Consider the above example, the joint CDF would be obtained as follows:

$$F_{XY}(x, y) = \begin{cases} 0 & \text{for } x < 1 \text{ or } y < 1 \\ P(X \leq 1, Y \leq 1) = \frac{1}{3} & \text{for } 1 \leq x, y < 2 \\ P(X = 1, Y \leq 2) = \frac{5}{12} & \text{for } 1 \leq x < 2, y \geq 2 \\ P(X \leq 2, Y = 1) = \frac{2}{3} & \text{for } 1 \leq y < 2, x \geq 2 \\ 1 & \text{for } x, y > 2 \end{cases}$$

Marginal PMF

In order to obtain the PMF of one random variable from the joint PMF, a summation procedure should be applied over the range of the other random variable. This is called the marginal PMF which is defined as below:

$$\begin{aligned} P_X(x) &= \sum_{y_i \in R_Y} P_{XY}(x, y_i), \\ P_Y(y) &= \sum_{x_i \in R_X} P_{XY}(x_i, y). \end{aligned} \tag{2.79}$$

Considering the previous example where its joint PMF is presented in (2.76), the marginal PMF of X can be obtained as below:

$$P_X(1) = P_{XY}(1,1) + P_{XY}(1,2) = \frac{1}{3} + \frac{1}{12} = \frac{5}{12},$$

$$P_X(2) = P_{XY}(2,1) + P_{XY}(2,2) = \frac{1}{3} + \frac{1}{4} = \frac{7}{12}.$$

Therefore, we have:

$$P_X(x) = \begin{cases} \frac{5}{12} & \text{for} \quad x = 1 \\ \frac{7}{12} & \text{for} \quad x = 2 \\ 0 & \text{otherwise} \end{cases},$$

Similar to this manner, the marginal PMF of Y is written as below:

$$P_Y(y) = \begin{cases} \frac{2}{3} & \text{for} \quad y = 1 \\ \frac{1}{3} & \text{for} \quad y = 2 \\ 0 & \text{otherwise} \end{cases},$$

2.6.2 Joint Distribution: Continuous Random Variables

Similar to the discrete random variables, for continuous random variables the joint PDF could be defined as $f_{XY}(x, y)$. Also, we have:

$$\int_{-\infty}^{\infty} \int_{-\infty}^{\infty} f_{XY}(x, y) dx dy = 1, \tag{2.80}$$

which is similar to the property of the discrete joint PDF mentioned in (2.75). For instance, consider the following joint PDF:

$$f_{XY}(x, y) = \alpha \left(\frac{x}{2} + y^3 \right), \quad 0 \le x, y \le 1, \tag{2.81}$$

where α is a constant. According to (2.80), the constant α would be obtained as below:

$$\int_0^1 \int_0^1 \alpha \left(\frac{x}{2} + y^3\right) dxdy = 1$$
$$\int_0^1 \int_0^1 \left(\frac{x}{2} + y^3\right) dxdy = \frac{1}{\alpha} \tag{2.82}$$
$$\int_0^1 \left(\frac{1}{4} + y^3\right) dy = \frac{1}{\alpha} \Rightarrow \alpha = 2.$$

Note that if two random variables X and Y are independent, the joint distribution would be $f_{XY}(x, y) = f_X(x) f_Y(y)$.

Joint CDF

The joint CDF of two continuous random variables X and Y can be defined as:

$$F_{XY}(x, y) = \int_{-\infty}^{y} \int_{-\infty}^{x} f_{XY}(x, y) dxdy. \tag{2.83}$$

Consider the joint PDF expressed in (2.81). For $0 \leq x, y \leq 1$, the joint CDF would be obtained as below:

$$\begin{aligned} F_{XY}(x, y) &= \int_0^y \int_0^x \left(x + 2y^3\right) dxdy \\ &= \int_0^y \left(\frac{x^2}{2} + 2xy^3\right) dy \\ &= \frac{1}{2}\left(x^2 y + xy^4\right). \end{aligned}$$

Therefore, we have:

$$F_{XY}(x, y) = \begin{cases} 0 & \text{for} \quad x, y < 0 \\ \frac{1}{2}\left(x^2 y + xy^4\right) & \text{for} \quad 0 \leq x, y \leq 1 \\ 1 & \text{for } x, y > 1 \end{cases},$$

Marginal PDF

The definition of the marginal PDF in continuous random variables is similar to the discrete random variables, except that the integral operator is replaced by the summation operator in (2.79). Therefore, the marginal PDF of the random variables X and Y would be obtained as below:

$$f_X(x) = \int_{y \in R_Y} f_{XY}(x, y) dy,$$
$$f_Y(y) = \int_{x \in R_X} f_{XY}(x, y) dx. \tag{2.84}$$

Again, consider the joint PDF in (2.81). The marginal PDF of X would be obtained as follows:

$$\begin{aligned} f_X(x) &= \int_0^1 \left(x + 2y^3\right) dy \\ &= x + \frac{1}{2}, \quad 0 \leq x \leq 1. \end{aligned} \tag{2.85}$$

Similarly, the marginal PDF of Y is:

$$\begin{aligned} f_Y(y) &= \int_0^1 \left(x + 2y^3\right) dx \\ &= \frac{1}{2} + 2y^3, \quad 0 \leq y \leq 1. \end{aligned} \tag{2.86}$$

Note that is this example, X and Y are not independent, since $f_{XY}(x, y) \neq f_X(x) f_Y(y)$.

2.6.3 *Covariance and Correlation*

The covariance and the correlation metrics are two parameters that are defined between the random variables in order to evaluate the relation between them. In the following, these two parameters are introduced and discussed.

Covariance

One parameter could be defined between two random variables, is the covariance metric. This parameter gives some information about the relation between the random variables. The covariance between two random variables X and Y is defined as below:

$$\begin{aligned} cov[X, Y] &= \mathbb{E}_{XY}\left[(X - \mathbb{E}[X])\,(Y - \mathbb{E}[Y])\right] \\ &= \mathbb{E}_{XY}[XY] - \mathbb{E}[X]\mathbb{E}[Y]. \end{aligned} \tag{2.87}$$

It is obvious from the above equation that $cov[X, Y] = cov[Y, X]$. If two random variables are independent, we have $\mathbb{E}_{XY}[XY] = \mathbb{E}[X]\mathbb{E}[Y]$, and therefore, the covariance metric will be "0". Also, for two random variables X and Y, we have:

$$cov\,[(aX + b), (cY + d)] = ac \times cov\,[X, Y]\,, \tag{2.88}$$

where a, b, c, and d are constant. For instance, consider the joint PDF in (2.81), where the marginal PDFs of the random variables X and Y are calculated in (2.85) and (2.86), respectively. In order to obtain the covariance between X and Y, we have:

$$E\,[XY] = \int_0^1 \int_0^1 xy\left(x + 2y^3\right)dxdy = \frac{11}{30}.$$

Also, according to (2.84), the expectation of X and Y would be obtained as follows:

$$\begin{aligned} E\,[X] &= \int_0^1 x\left(x + \frac{1}{2}\right)dx = \frac{7}{12}, \\ E\,[Y] &= \int_0^1 y\left(\frac{1}{2} + 2y^3\right)dy = \frac{13}{20}. \end{aligned} \tag{2.89}$$

Therefore, the covariance between X and Y is achieved as below:

$$cov[X, Y] = \frac{11}{30} - \left(\frac{7}{12}\right)\left(\frac{13}{20}\right) = -\frac{1}{80}. \tag{2.90}$$

Example 2.8 (Covariance)

Problem: Consider two independent random variables $X \sim \mathcal{N}(0, 1)$ and $Y \sim \mathcal{N}(1, 4)$. Two random variables V and W are defined as below:

$$\begin{aligned} V &= 2X + XY \\ W &= XY - Y. \end{aligned}$$

Find the covariance between V and W.

Solution: In order to find $cov\,[V, W]$, we have:

$$\begin{aligned} cov\,[V, W] &= cov\,[(2X + XY), (XY - Y)] \\ &= 2cov\,[X, XY] - 2cov\,[X, Y] + cov\,[XY, XY] - cov\,[XY, Y]. \end{aligned}$$

(continued)

Example 2.8 (continued)
Since X and Y are independent, we conclude that $cov\,[X, Y] = 0$. Also, the above relation can be rewritten as below:

$$\begin{aligned} cov\,[V, W] &= 2\left(\mathbb{E}[X^2Y] - \mathbb{E}[X]\mathbb{E}[XY]\right) + \mathbb{E}[X^2Y^2] - \mathbb{E}^2[XY] \\ &\quad - \mathbb{E}[XY^2] + \mathbb{E}[XY]\mathbb{E}[Y] \\ &= 2\mathbb{E}[X^2]\mathbb{E}[Y] - 2\mathbb{E}^2[X]\mathbb{E}[Y] + \mathbb{E}[X^2]\mathbb{E}[Y^2] - \mathbb{E}^2[X]\mathbb{E}^2[Y] \\ &\quad - \mathbb{E}[X]\mathbb{E}[Y^2] - \mathbb{E}[X]\mathbb{E}^2[Y]. \end{aligned}$$

We know that $\mathbb{E}[X] = 0$, $\mathbb{E}^2[X] = 1$, $\mathbb{E}[Y] = 1$ and $\mathbb{E}^2[Y] = 5$. Replacing these values into the above formula, the covariance of the random variables V and W would be obtained:

$$cov\,[V, W] = 2(1)(1) + (1)(5) = 7.$$

Correlation

The correlation coefficient is another metric which is defined as a normalized version of the covariance. The correlation coefficient between X and Y is denoted by ρ_{XY} and is defined as below:

$$\rho_{XY} = \frac{cov[X, Y]}{\sigma_X \sigma_Y}. \tag{2.91}$$

This parameter can be changed between $-1 \leq \rho \leq 1$. If two random variables X and Y are independent, the covariance between these two random variables, and consequently, the correlation coefficient would be zero. However, the converse is not necessarily true; if two random variables X and Y are uncorrelated, then X and Y may or may not be independent.
As a practice, it is considered to calculate the correlation coefficient of two random variables X and Y that the corresponding joint PDF is presented in (2.81). Referring to (2.85) and (2.86), we have

$$\begin{aligned} \mathbb{E}[X^2] &= \int_0^1 x^2\left(x + \frac{1}{2}\right)dx = \frac{5}{12}, \\ \mathbb{E}[Y^2] &= \int_0^1 y^2\left(\frac{1}{2} + 2y^3\right)dy = \frac{1}{2}. \end{aligned}$$

Therefore, the variance of the random variables X and Y can be obtained as below:

$$\sigma_X^2 = \frac{5}{12} - \left(\frac{7}{12}\right)^2 = \frac{11}{144},$$

$$\sigma_Y^2 = \frac{1}{2} - \left(\frac{13}{20}\right)^2 = \frac{31}{400}.$$

Note that the mean values of X and Y are calculated in (2.89). Also, the covariance is presented in (2.90). The correlation coefficient can be easily obtained as follows:

$$\rho_{XY} = \frac{-\frac{1}{80}}{\sqrt{\frac{11}{144} \times \frac{31}{400}}} \simeq -0.16.$$

It can be concluded from the obtained value that the random variables X and Y are negatively correlated; the random variable X would be decreased as the random variable Y increases. Similarly, if the correlation between two random variables X and Y is positive (i.e., $\rho_{XY} > 0$), it can be concluded that the random variables are positively correlated in the way that one random variable is increased as the other one increases.

Example 2.9 (Correlation)
Problem: Consider two independent random variables $X \sim \mathcal{N}(0, 1)$ and $Y \sim \mathcal{N}(1, 4)$. Two random variables V and W are defined as below:

$$V = 2X + Y$$
$$W = X + 1.$$

Determine the correlation between the variables.
Solution: In order to see if the random variables V and W are correlated or not, the corresponding correlation coefficient should be obtained. At the first step, the covariance of V and W is obtained as below:

$$cov\,[V, W] = cov\,[(2X + Y), (X + 1)]$$
$$= 2Var\,[X] + cov\,[X, Y] = 2.$$

Then, the variance of V and W are calculated separately as below:

$$\sigma_V^2 = Var\,[2X + Y] = 4Var\,[X] + Var\,[Y] = 8,$$

(continued)

Example 2.9 (continued)

$$\sigma_W^2 = Var\,[X+1] = Var\,[X] = 1.$$

Finally, the correlation coefficient would be obtained as follows:

$$\rho_{VW} = \frac{2}{\sqrt{8 \times 1}} = \frac{1}{\sqrt{2}},$$

which indicates that the random variables V and W are correlated.

2.6.4 Multivariate Gaussian Distribution

The one-dimensional normal random variable is introduced and expressed as one of the most commonly used distributions in Sect. 2.5. In this section, we are focused on the joint distribution of two or more normal random variables. Consider we have N random variables $X_1, X_2, \ldots, X_N$. We can express them as a column vector X as follows:

$$\mathbf{X} = \begin{bmatrix} X_1 \\ X_2 \\ \vdots \\ X_N \end{bmatrix}, \tag{2.92}$$

where $\mathbf{X}$ is an N-dimensional vector since it consists of N random variables. The expected value of the random vector $\mathbf{X}$ is defined as below:

$$\mu = \mathbb{E}[\mathbf{X}] = \begin{bmatrix} \bar{X}_1 \\ \bar{X}_2 \\ \vdots \\ \bar{X}_N \\ , \end{bmatrix} \tag{2.93}$$

The covariance matrix of the random vector $\mathbf{X}$ is denoted by Σ, which is a $N \times N$ matrix and is obtained from the following formula:

$$\Sigma = \mathbb{E}\left[(\mathbf{X} - \mathbb{E}[\mathbf{X}])(\mathbf{X} - \mathbb{E}[\mathbf{X}])^T\right]$$

$$= \begin{bmatrix} var[X_1] & cov[X_1, X_2] & \cdots & cov[X_1, X_N] \\ cov[X_2, X_1] & var[X_2] & \cdots & cov[X_2, X_N] \\ \vdots & \vdots & \ddots & \vdots \\ cov[X_N, X_1] & cov[X_N, X_2] & \cdots & var[X_N] \end{bmatrix}. \tag{2.94}$$

Note that the diagonal elements of the above matrix equal the variance of ith random variable, or equivalently, $\sigma^2_{X_i}$, for $i = 1, 2, \cdots, N$. This is because of this fact that $cov[X, X] = \sigma^2_X$. Also, note that the covariance matrix is symmetric with respect to the diagonal elements of the matrix, since $cov[X, Y] = cov[Y, X]$, as mentioned before. Now, considering the $N \times 1$ mean vector μ, and the $N \times N$ covariance matrix Σ, the multivariate Gaussian distribution is defined as follows:

$$\mathcal{N}(\mathbf{X}|\mu, \Sigma) = \frac{1}{(2\pi)^{N/2}|\Sigma|^{1/2}} \exp\left\{-\frac{1}{2}(\mathbf{X} - \mu)^T \Sigma^{-1}(\mathbf{X} - \mu)\right\}, \tag{2.95}$$

where $|.|$ indicates the determinant operator. For instance, consider we have two normal distributions as $X \sim N(0, 1)$ and $Y \sim N(1, 4)$ with the correlation coefficient $\rho_{XY} = -1/2$. We are interested in finding the joint PDF, or equivalently, the bivariate normal distribution of X and Y. The mean vector can be constructed using (2.93):

$$\mu = \begin{bmatrix} \mu_X \\ \mu_Y \end{bmatrix} = \begin{bmatrix} 0 \\ 1 \end{bmatrix}.$$

Also, according to (2.91), we obtain the covariance between X and Y:

$$\begin{aligned} cov[X, Y] &= \rho_{XY}\,(\sigma_X \sigma_Y) \\ &= -\frac{1}{2}\,(1 \times 2) = -1. \end{aligned}$$

Therefore, the covariance matrix can be constructed according to (2.94):

$$\Sigma = \begin{bmatrix} 1 & -1 \\ -1 & 4 \end{bmatrix}, \quad \text{and} \quad \Sigma^{-1} = \frac{1}{3}\begin{bmatrix} 4 & 1 \\ 1 & 1 \end{bmatrix}.$$

Note that $|\Sigma| = 3$ and $N = 2$. By replacing the obtained mean vector and the covariance matrix in (2.95), the bivariate normal distribution would be obtained as below:

$$\mathcal{N}(\mathbf{X}|\mu, \Sigma) = \frac{1}{(2\pi)\sqrt{3}} \exp\left\{-\frac{1}{6}\left(4x^2 + (y - 1)^2 + 2\,(xy - x)\right)\right\}.$$

2.7 Matrix Decomposition

Many of the observations and realizations in machine learning are presented in the form of vectors. It is mainly due to the pre-classification step, called "feature extraction." When a subject (image, text, voice, etc.) is going to be classified via the machine learning algorithm, its features are extracted and a single observation will be extended into a vector of features corresponding to the observation. As a result, most of the calculations were manipulated in the vector and matrix domain. So, matrix algebra plays a crucial role in machine learning. In this section, we are going to review the matrix decomposition theory, which is widely used in different aspects of machine learning, including but not limited to feature selection, principal components analysis, and feature generation.

Just as integers can be decomposed into prime factors, matrices also can be decomposed into factors. The various matrix decomposition techniques have different properties. However, one of the most important computational methods in machine learning is the Singular Value Decomposition(SVD). In this section, we are going to discuss what the singular value decomposition is. But before that, we should learn about Eigen value, Eigen vector, and Eigen decomposition.

2.7.1 *Eigenvalue Decomposition*

Let us assume $A \in \mathbb{R}^{n \times n}$ be a square matrix. The eigenvector of this matrix is a non-zero vector v which satisfies the linear equation

$$Av = \lambda v, \tag{2.96}$$

where scalar λ is an eigenvalue of A, corresponding to eigenvector v. If we move λv to the left side in equation 2.96, we can rewrite it as $(A - \lambda I) = 0$. Where I is the identity matrix of dimension n. Hence, we can find the eigenvector v by solving this equation, providing the eigenvalue λ is known. The eigenvalues of A are non-zero solutions of the following "characteristic equation":

$$P(\lambda) = \det(A - \lambda I) = 0, \tag{2.97}$$

where $\det(\cdot)$ is the determinant of a matrix. $P(\lambda)$ is a polynomial equation of degree n, called the "characteristic polynomial" of A, and its roots are the eigenvalues of A as:

$$P(\lambda) = \det(A - \lambda I) = (\lambda_1 - \lambda)(\lambda_2 - \lambda)\ldots(\lambda_n - \lambda), \tag{2.98}$$

where the $\lambda_1, \lambda_2, \ldots, \lambda_n$ are eigenvalues of A. Let us have an example. Suppose we have the following matrix and we want to find its λ's and v's.

$$A = \begin{bmatrix} 2 & 1 \\ 1 & 2 \end{bmatrix}. \tag{2.99}$$

First, we should subtract λ from the diagonal of A to find $A - \lambda I$. Having the determinant of $A - \lambda I$, the characteristic polynomial is written as

$$P(\lambda) = \det \begin{bmatrix} 2-\lambda & 1 \\ 1 & 2-\lambda \end{bmatrix} = \lambda^2 - 4\lambda + 3. \tag{2.100}$$

By solving the characteristic polynomial, its roots are given by $\lambda_1 = 1$ and $\lambda_2 = 3$ which are two eigenvalues of A. Now, by solving equation $(A - \lambda I)v = 0$ for each eigenvalue, we have the corresponding eigenvectors as follows:

$$v_{\lambda=1} = \begin{bmatrix} 1 \\ -1 \end{bmatrix}, \qquad v_{\lambda=3} = \begin{bmatrix} 1 \\ 1 \end{bmatrix} \tag{2.101}$$

Eigendecomposition

Based on the presented eigenvalue and eigenvector of a matrix, now we can introduce the "Eigendecomposition" of a matrix. Suppose matrix A be a square real-valued $n \times n$ matrix with n linearly independent eigenvectors $\mathbf{q}_i \in \mathbb{R}^n$. For each eigenvector, we have

$$A\mathbf{q}_i = \lambda_i \mathbf{q}_i, \qquad \text{for } i \in \{1, 2, \cdots, n\}. \tag{2.102}$$

We can define a square matrix Q with $\mathbf{q}_i$ in each column as follows,

$$Q = [\mathbf{q}_1 \mid \mathbf{q}_2 \mid \cdots \mid \mathbf{q}_n]. \tag{2.103}$$

We define a diagonal matrix Λ whose diagonal elements are the eigenvalues of A as

$$\Lambda = \text{diag}(\lambda_1, \lambda_2, \ldots, \lambda_n) = \begin{bmatrix} \lambda_1 & 0 & \cdots & 0 \\ 0 & \lambda_2 & & \vdots \\ \vdots & & \ddots & 0 \\ 0 & \cdots & 0 & \lambda_n \end{bmatrix}. \tag{2.104}$$

We can rewrite Eq. (2.96) with all eigenvalues and eigenvectors of A as follows,

$$AQ = Q\Lambda. \tag{2.105}$$

Due to the linear independency of eigenvectors, we can write the eigendecomposition as

$$A = Q\Lambda Q^{-1}. \tag{2.106}$$

Following the previous example, we can write Q and its inverse as

$$Q = \begin{bmatrix} 1 & 1 \\ -1 & 1 \end{bmatrix}, \quad Q^{-1} = \begin{bmatrix} 1/2 & -1/2 \\ 1/2 & 1/2 \end{bmatrix} \tag{2.107}$$

Then we can decompose A to three matrix as,

$$A = \begin{bmatrix} 1 & 1 \\ -1 & 1 \end{bmatrix} \begin{bmatrix} 1 & 0 \\ 0 & 3 \end{bmatrix} \begin{bmatrix} 1/2 & -1/2 \\ 1/2 & 1/2 \end{bmatrix} \tag{2.108}$$

2.7.2 *Singular Value Decomposition*

The eigendecomposition mentioned in Sect. 2.7.1 is applicable only to square matrices. If A is not square, the eigendecomposition is undefined. Singular Value Decomposition (SVD) is a generalized extension of the eigendecomposition, when A is $m \times n$ matrix, for an arbitrary m and n. Using SVD, matrix $A_{m\times n}$ can be decomposed into the product of three simpler matrices called U,Σ, and V as

$$A = U\Sigma V^T, \tag{2.109}$$

where $U \in \mathbb{R}^{n\times n}$ and $V \in \mathbb{R}^{m\times m}$ are unitary and orthogonal matrices, and $\Sigma \in \mathbb{R}^{n\times m}$ is a non-square diagonal matrix. the diagonal elements of Σ are called singular values of A and they are sorted in descending order. columns of U are called left singular vectors and columns of V are called right singular vectors.

If we want to express the relationship between the two methods of decomposition and interpret SVD in terms of eigendecomposition, then we have the following relation. Using SVD, we have:

$$\begin{aligned} A^T A &= (U\Sigma V^T)^T U\Sigma V^T \\ &= V\Sigma^T U^T U\Sigma V^T. \end{aligned} \tag{2.110}$$

Since U and V are unitary matrices, we have $U^T U = V^T V = I$, so we have $U^T = U^{-1}$ and $V^T = V^{-1}$. We can simplify equation (2.110) as follows,

$$\begin{aligned} A^T A &= V\Sigma^T \Sigma V^T = V\Sigma^2 V^T \\ &= V\Sigma^2 V^{-1}. \end{aligned} \tag{2.111}$$

Similarly we have,

$$AA^T = U\Sigma^2 U^T = U\Sigma^2 U^{-1}. \tag{2.112}$$

These two equations have the same format as eigendecomposition in (2.106). The comparison of two decompositions reveals

- The columns of U are the left singular vectors of A and eigenvectors of AA^T
- The columns of V are the right singular vectors of A and eigenvectors of $A^T A$
- Non-zero singular values of A are non-zero elements of Σ and square roots of eigenvalues of $A^T A$ or AA^T

2.8 Putting It All Together

In applied machine learning, many real-world applications have a non-deterministic nature and varying behavior, since there is no guarantee to always get the same output for the same input. Since uncertainties in machine learning stem from various factors such as several parameters and complex environments, probability theory has a key role in machine learning.

We described the fundamentals of probability including joint probability and conditional probability for dependent and independent events, discrete and continuous random variables, types of probability distributions, and matrix decomposition.

Probability is the basis of machine learning in several algorithms such as Naive Bayes and Bayesian Networks. It is used in classification algorithms to predict a probability of class membership. Algorithms such as decision trees make decisions based on probability. Probability frameworks such as Maximum Likelihood are used to train machine learning algorithms such as logistic regression, neural networks, so forth. Fundamentals of probability described in this chapter are used in the following chapters to capture the uncertainty in non-deterministic application problems and domains.

2.9 Exercise Problems

Problem 2.1 If $X \sim Poisson(\lambda)$, proof:

(a) $\mathbb{E}[X] = \lambda$.
(b) $var[X] = \lambda$.

Problem 2.2 If A is an invertible square matrix

(a) Proof that the eigenvalues of A and A^T are the same.
(b) Find the singular value decomposition of the inverse of A.

Problem 2.3 (Maximum Likelihood Classifier) Assume there are two classes of ω_0 and ω_1. The distribution of a random variable, X, given a class of ω_i for $i \in \{0, 1\}$ is given by

$$P(X|\omega_i) = \mathcal{N}(\mu_i, \sigma^2), \quad \text{for } \mu_0 < \mu_1.$$

A classifier is going to determine the corresponding class of the observed random variable $X = x$, based on the following rule:

$$x \in \begin{cases} \omega_1 & P(x|\omega_1) > P(x|\omega_0) \\ \omega_0 & P(x|\omega_1) < P(x|\omega_0) \end{cases}$$

The error occurs when $x \in \omega_0$ but it is classified as ω_1, and vice versa. Find the probability of error for this classifier.

Problem 2.4 Suppose $A = U\Sigma V^T$ is a square matrix. Prove that the absolute value of the determinant of A is obtained by the product of diagonal elements of Σ.

Problem 2.5 Assume $X_1, X_2, \ldots . X_n$ are independent and identically distributed (iid) random variables with common PDF and CDF of f_X and F_X, respectively. Express the PDF and CDF of each of the following random variables in terms of the f_X and F_X.

(a) $Y_1 = \mathbf{max}\,\{X_1, X_2, \ldots X_n\}$
(b) $Y_2 = \mathbf{min}\,\{X_1, X_2, \ldots X_n\}$

Problem 2.6 Assume continuous real random variable X defined in $(-\infty, +\infty)$ with PDF and CDF of $f_X(x)$ and $F_X(x)$, respectively. Find PDF and CDF of conditional random variable $\{X|a \leq X \leq b\}$.

Problem 2.7 Assume X is a continuous random variable with the probability density function as,

$$f_X(x) = \begin{cases} x^2(\alpha x + \frac{1}{3}) & 0 \leq x \leq 1 \\ 0 & \text{Otherwise} \end{cases}.$$

(a) Find α such that f_X be a PDF
(b) For a new random variable Y, defined as $Y = \frac{3}{X} + 6$, find $E(Y)$ and $Var(Y)$.

Problem 2.8 Assume X is a continuous random variable with the probability density function as,

$f_X(x)$= $1/10e^{-}x$ for x$\geq$ 0 Determine the variance.

Problem 2.9 Suppose that we have a continuous random variable X with the uniform distribution over the interval $[0, 1]$. Let us define $Y = -ln(1 - X)$. Find the distribution of Y.

Problem 2.10 Consider two independent random variables $X \sim \mathcal{N}(0, 1)$ and $Y \sim \mathcal{N}(1, 8)$. Two random variables V and W are defined as below:

$$V = X + 2Y$$
$$W = X + 10.$$

Determine the correlation between the variables.

Problem 2.11 Consider two independent random variables $X \sim \mathcal{N}(0, 1)$ and $Y \sim \mathcal{N}(1, 5)$. Two random variables V and W are defined as below:

$$V = X + Y$$
$$W = XY - 6.$$

Determine the correlation between the variables.

Problem 2.12 Suppose that we have two independent random variables X and Y. If we know that $Var(X + Y) = 3$ and $Var(X - 2Y) = 6$. Find $Var(X)$ and $Var(Y)$.

Problem 2.13 Assume that X is a multivariate Gaussian random variable with mean μ and covariance matrix Σ defined as below:

$$\mu_X = [0, 1]^T, \qquad \Sigma_X = \begin{bmatrix} 1 & 0.5 \\ 0.5 & 2 \end{bmatrix}.$$

(a) Use Python to generate and plot $N = 100$ observations of random variable X.
(b) Use Python to find the experimental mean and covariance matrix of the multivariate Gaussian random variable, X, based on the observed data in the previous part.

Problem 2.14 Assume that the random variable X is distributed with Poisson distribution with parameter $\lambda = 3$.

(a) Use Python to generate $N = 10$ observations of random variable X.
(b) Use Python to estimate the parameter λ, based on the observations in part (a).
(c) Repeat parts (a) and (b) for $N = 1000$ trials, and plot the distribution of the estimated parameter.
(d) Find the mean and the variance of the estimated parameter λ.
(e) Repeat part (c) for $N = 1$, $N = 10$, $N = 100$ $N = 1000$, and compare the mean and the variance of the estimated parameter λ for $N = 10$ and $N = 1000$.
(f) Use Python to plot the Gaussian with calculated mean and variance for $N = 10$ and $N = 1000$, and compare it with the plotted distribution in part (c).

Chapter 3
Supervised Learning

3.1 Introduction

This chapter introduces multiple supervised machine learning techniques that are vastly popular in the modern day applications. Supervised learning refers to learning the data that is annotated with the labels and extracting the relationship between input data and the labels. The supervised learning technique explores the relation between the input X and output Y to learn the function f that maps the X and Y, i.e., $f : X \implies Y$. In supervised learning, for the purpose of training, i.e., extracting the relationship between input and out to build a model, the input data X is annotated with corresponding labels Y and fed to the machine learning algorithm, i.e., $(x_1, y_1), (x_2, y_2), \cdots, (x_n, y_n)$ are fed to train the learning algorithm to build the model $f(\cdot)$.

One of the main advantages of supervised learning compared to other learning techniques (Unsupervised and Reinforcement learning techniques) is the low complexity and forming a better model. Furthermore, for the applications whose objective is defined or determined during the design time, supervised learning techniques perform efficiently compared to unsupervised techniques or reinforcement learning techniques and can be optimized for performance and other relevant constraints.

Example 3.1 (Supervised Classification)
Problem: Explain supervised learning for classification.
Solution: Consider a problem of classification of different animals and Supervised learning technique is constructed to address this. Figure shown below is an illustrative example of supervised learning classification problem. The leftmost part of the image shows the data input to the supervised learning.

(continued)

S. Rafatirad et al., *Machine Learning for Computer Scientists and Data Analysts*,
https://doi.org/10.1007/978-3-030-96756-7_3

Example 3.1 (continued)
The input data comprises the input images (X) and the corresponding labels (Y) for the classification task in the provided example. In this example, the supervised learning technique is a fully connected neural network (details presented in Sect. 3.4.3) with hidden layers.

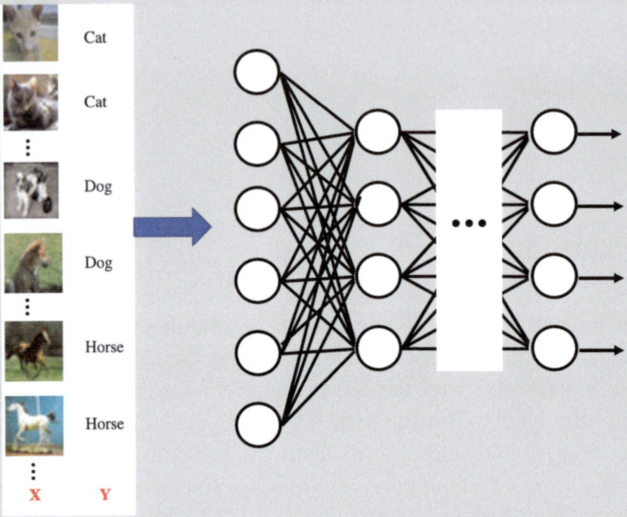

Example 3.2 (Supervised Regression)
Problem: Explain supervised regression with a prediction example.
Solution: In addition to classification, supervised learning can also be utilized for different applications including regression. Consider a simple case of house prediction. The input data X comprises information such as # bedrooms and the output comprises the house price (Y), as shown in Fig. 3.1. The supervised learning technique aims to determine the curve that fits the aforementioned X and Y, denoted by $f(X, Y)$. This is a simple case comprising single variable, but, in reality, the input could comprise multiple features such as area of the house, distance to the city center, and access to the public transportation.

It needs to be noted that the input data can also be a sequence of symbols or numbers or images or text, and the output can be labels (for classification) or the prediction for the series, depending on the application. The above examples are

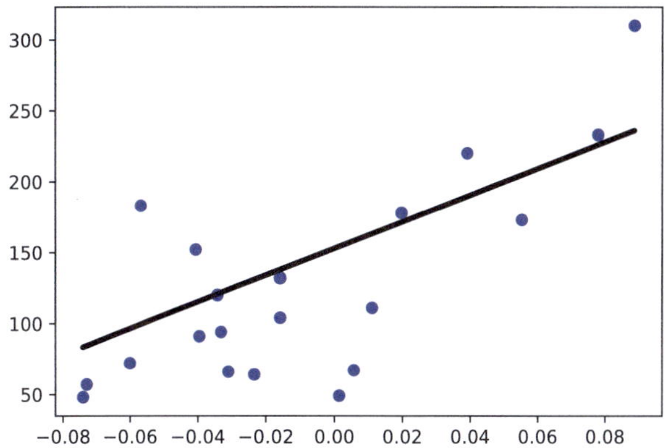

Fig. 3.1 Regression showing # bedrooms/area vs price (in K$)

mere instances of supervised classifiers and the input could comprise plethora of features.

3.2 Preparing Data

One of the main challenges in utilizing the machine learning techniques, especially the supervised learning, is the reliability and nature of the data. Often the obtained data from real-world scenarios such as autonomous driving or navigation systems are embedded with noise and could be missing some features and can be of large dimension for processing. While this chapter does not focus on the data cleaning or preprocessing techniques, preprocessing the data is a pivotal challenge in the machine learning applications. We discuss some of the most common problems and solutions for those challenges.

3.2.1 Data Abstraction

In machine learning problems, it is indispensable to identify important subset of features to train a model. Reducing the dimensionality of data by projecting it onto a subspace improves the efficiency of the model and the accuracy of the predictions. Data abstraction in machine learning is about applying techniques to simplify the representation of a machine learning problem and enable the problem to use less memory and computational power. Some of the popular techniques used in machine learning to reduce the dimensionality of data include but not limited to *Linear Discriminant Analysis (LDA), Neural autoencoder, t-distributed stochastic neighbor embedding (t-SNE), Principal component analysis (PCA), Pearson Cor-*

relation Coefficient, Recursive Feature Elimination (RFE), Self Organizing Maps (SOM), Spearman Correlation Coefficient, Chi-Squared Test, Kendall Tau Test. The choice of a dimensionality reduction method depends on the features and the class label in a dataset, being numeric, or categorical feature(s). For instance, in filter feature selection methods (such as Chi-Squared Test), the input X and output Y variables should have one of the following traits: a) X and Y are numerical, b) X is numerical and Y is categorical, c) X is categorical and Y is numerical, d) X and Y are categorical.

An example scenario where Filter Methods are applied is when input variable `Genre` in a music dataset can have four different categories (Rock, Jazz, Pop, and Hip hop) and the output variable Y can represent the likelihood of popularity (high, medium, low).

Unlike Filter Methods that are supervised, Principal component analysis (PCA) is an unsupervised feature selection technique since it does not employ labels in the computation. Principal Component Analysis is a matrix factorization technique used widely in reducing the dimensionality of a dataset through maximizing the variance in the dataset. Prior to applying PCA technique on a dataset D of $n = N$ rows and M features, first, the dataset needs to be normalized using a feature scaling technique such as *zero mean and unit variance*. *StandardScaler* from *sklearn* can be used to standardize unscaled data prior to PCA.

Example 3.3 (Data Abstraction)

Problem: Consider a dataset that has five attributes (Gender, Age, Fare, Seatclass, and Guests) and the dependent variable "Success," shown below. Perform the data abstraction using Principal Component Analysis (PCA) and display the principal components and plot the principal components and the correlation matrix.

	GENDER	AGE	FARE	SEATCLASS	GUESTS	SUCCESS
0	0	22.0	7.2500	3	1	0
1	1	38.0	71.2833	1	1	1
2	1	26.0	7.9250	3	0	1
3	1	35.0	53.1000	1	1	1
4	0	35.0	8.0500	3	0	0

Solution: Figure in Example 3.3 displays the top 5 rows of an example dataset with five attributes (Gender, Age, Fare, Seatclass, and Guests) and the dependent variable "Success." The following code snippet shows how this dataset is scaled using *StandardScaler* from *sklearn* Python machine learning library. Once a dataset is standardized, PCA is applied to the data. Principal

(continued)

Example 3.3 (continued)
Component Analysis is a robust technique that uses the dependencies between the features to reduce the number of attributes without losing too much information. It identifies attributes with higher variations in the dataset and discards those that have less to no variations. In other words, it discovers the projections that maximize the variance in the data.
The following code snippet shows how PCA technique is applied on the sample dataset in Figure of Example 3.3, and the principal components are reported. Assuming there are two top attributes to identify, Fig. 3.2 shows the top two principal components (i.e., the transformed features), and Fig. 3.3 visualizes the distribution of the principal components that shows that the samples are separable based on the projected features.

```
from sklearn.preprocessing import StandardScaler

features = ['GENDER', 'AGE', 'FARE' , 'SEATCLASS', 'GUESTS']
# Separating out the features
x = df.loc[:, features].values
# Separating out the target
y = df.loc[:,['SUCCESS']].values
# Standardizing the features
x = StandardScaler().fit_transform(x)

from sklearn.decomposition import PCA
pca = PCA(n_components=2)
principalComponents = pca.fit_transform(x)
principalDf = pd.DataFrame(data = principalComponents
             , columns = ['principal component 1', 'principal component 2'])

print('Principal Components')
print(principalComponents)

# finalDf is the final DataFrame before plotting the data.
finalDf = pd.concat([principalDf, df[['SUCCESS']]], axis = 1) #Concatenating DataFrame along axis = 1

>#  Principal Components
>[[-1.28414106  0.20415353]
 >[ 2.05402655  0.61501247]
 >[-0.59366741  0.29796843]
 >...
 >[-0.55850761  1.15368635]
 >[ 0.73639025 -0.57593183]
 >[-0.96230157 -0.81250573]]


# finalDf is the final DataFrame before plotting the data.
finalDf = pd.concat([principalDf, df[['SUCCESS']]], axis = 1) #Concatenating DataFrame along axis = 1

#showing the top five rows of 'final Df', showing features of dataset D projected on a #subspace with two principal components.
finalDf.head()
```

	principal component 1	principal component 2	SUCCESS
0	-1.284141	0.204154	0
1	2.054027	0.615012	1
2	-0.593667	0.297968	1
3	1.754380	0.638726	1
4	-0.875667	-0.936015	0

Fig. 3.2 Principal components for the dataset in Example 3.3

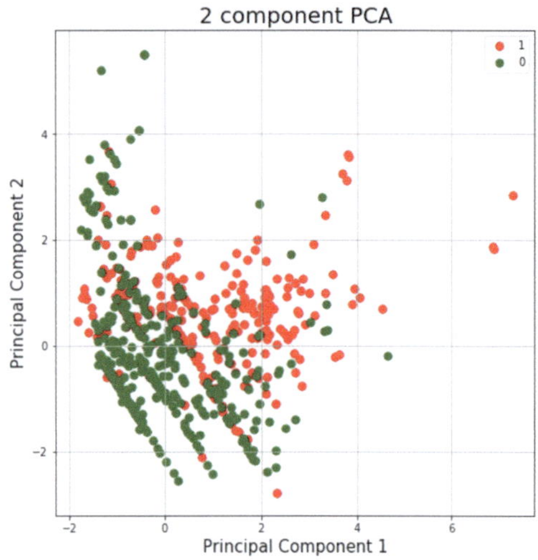

Fig. 3.3 2-D distribution plot of principal components for the dataset in Fig. 3.2

To learn which two features in the dataset correspond to the principal components, the following code snippet is used to perform cross-correlation between the features and principal components. The plot is displayed in Fig. 3.4. The vertical axis shows the principal components, and the horizontal axis indicates the features of the dataset. The color map shows that the principal components are constructed based on the combination of the dataset features.

```
#visualizing the principal components:
fig = plt.figure(figsize = (8,8))
ax = fig.add_subplot(1,1,1)
ax.set_xlabel('Principal Component 1', fontsize = 15)
ax.set_ylabel('Principal Component 2', fontsize = 15)
ax.set_title('2 component PCA', fontsize = 20)
targets = [1,0]
colors = ['r', 'g']
for target, color in zip(targets,colors):
    indicesToKeep = finalDf['SUCCESS'] == target
    ax.scatter(finalDf.loc[indicesToKeep, 'principal component 1']
               , finalDf.loc[indicesToKeep, 'principal component 2']
               , c = color
               , s = 50)
ax.legend(targets)
```

```
<matplotlib.axes._subplots.AxesSubplot at 0x11cf97e10>
```

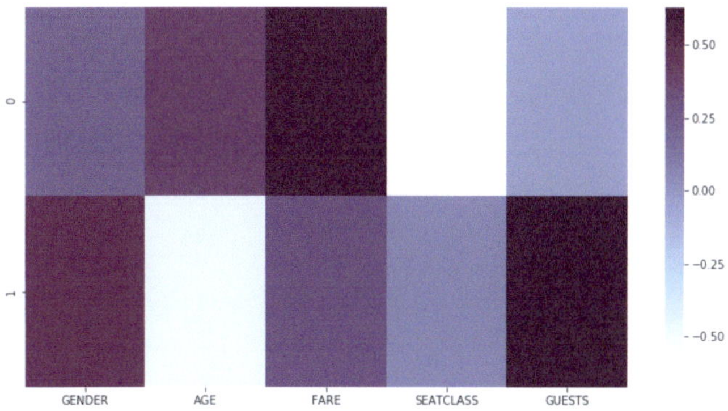

Fig. 3.4 Correlation of principal components to dataset features. The principal components (PC) correspond to combinations of the dataset features, and the PCs are captured as an attribute of the fitted PCA object

```
ax.grid()

print(pca.explained_variance_ratio_)
percent_variance = np.round(pca.explained_variance_ratio_* 100, decimals =2)
print(percent_variance)

>[0.34471984 0.2789678 ]
>[34.47 27.9 ]
```

In PCA, the variance measures the spread of each feature from its average value. In a dataset D with M number of attributes, for a feature X_i (a one dimensional array), the variance σ^2 is calculated as follows:

$$\sigma_i^2 = \frac{\sum_{j=1}^{N} X_i^{(j)} - \overline{X_i}}{N - 1} \tag{3.1}$$

such that $\overline{X_i}$ represents the average value of X_i feature.

3.2.2 *Dealing with Missing Data*

In addition to cleaning the data, one of the main challenges faced by the machine learning community is the lack of attributes[1] in the input data. An example of missing attributes is shown in Table 3.1. A sample population statistics for different zip codes are collected, shown in Table 3.1. It can be observed for data sample 100, 196, the population information is missing and for the sample 350, the zip code

[1] Attributes refer to the input data or the features in the input data.

Table 3.1 Sample population statistics (missing attributes)

Sample #	Zipcode	Population
1	22,043	5000
2	95,616	10,000
⋮	⋮	⋮
100	20,001	—
⋮	⋮	⋮
196	43,604	—
⋮	⋮	⋮
350	—	675
⋮	⋮	⋮

Table 3.2 Population Statistics with modified attributes (replaced with out of range values)

Sample #	Zipcode	Population
1	22,043	5000
2	95,616	10,000
⋮	⋮	⋮
100	20,001	-1
⋮	⋮	⋮
196	43,604	-1
⋮	⋮	⋮
350	-1	675
⋮	⋮	⋮

is missing. Such scenarios are encountered in real-world applications either due to improper storage of data or deliberately missing the data or inability to collect the data properly. Naïve technique to handle missing data include discarding the samples with missing attributes when the number of data samples with missing attributes is minimal. However, adopting such a technique can lead to information loss and is inefficient.

To address such challenge efficiently, two classes of techniques are widely utilized. One class of techniques proposes to replace the missing features with values that are out of the range as shown in Table 3.2. The population and zip codes cannot be negative numbers, i.e., negative numbers are out of range, thus the missing numbers are replaced with -1 in the data. Utilizing the modified data, the machine learning models will be trained. In some cases, the missing features will be replaced by "*unknown*" or "$\pm\infty$", depending on the data type (textual or numeric), as they are outliers in all the cases.

On the other hand, the missing features are replaced with the average of the other values, i.e., a smoothing is performed to replace the missing features. For instance, in Table 3.3, the missing features are replaced with the averages for samples 100,

Table 3.3 Population Statistics with modified attributes (replaced with mean)

Sample #	Zipcode	Population
1	22,043	5000
2	95,616	10,000
⋮	⋮	⋮
100	20,001	7632
⋮	⋮	⋮
196	43,604	7632
⋮	⋮	⋮
350	34,231	675
⋮	⋮	⋮

196, and 350. In addition to replacing the missing attributes with the mean, as in Table 3.3, interpolation techniques such as Inverse Distance Weighted (IDW) interpolation, Natural Neighbor Inverse Distance Weighted (NNIDW) interpolation, Linear interpolation, and Spline interpolation are adopted. Furthermore, other central tendency measures such as Median are also used instead of the mean. To make the missing feature estimation further realistic, the central tendency metric (mean or median) is estimated for the samples belonging to the same class, rather than the whole dataset.

Example 3.4 (Dealing with Missing Data)
Problem: Consider the example dataset of population statistics, it is missing multiple values in different columns, performs the data preprocessing to drop the rows of missing data, or fills the missing values with mean to provide an approximate value.
Solution: Almost all real-world data has missing values due to many reasons as errors in data collection. But for efficient modeling of a machine learning algorithm, the input data must be uniform and the missing values in a dataset must be handled before the training. When a dataset is loaded, first the missing values in each column are identified and then deal with it by filling missing values with mean values, as shown in Table 3.3, which can be applied for variable type columns. The code snippet is shown below.

```
import pandas as pd
from numpy import nan
# load the dataset
dataset = pd.read_csv('path to population dataset', header=None)
# making missing values as missing or NaN
dataset[[columns]] = dataset[[columns]].replace(0, nan)
# fill missing values with mean column values
dataset.fillna(dataset.mean(), inplace=True)

```

```
#Note: Filling missing values with mean values can be applied for variable
    type columns only and can not be applied on categorical columns.
```

3.2.3 *Dealing with Imbalanced Datasets*

Another pivotal challenge in the domain of machine learning deployed in realistic applications including security, autonomous systems, and other classification applications is lack of sufficient data samples for some classes, referred to as *imbalanced data*. For instance, in the case of anomalous detection problems, the number of anomalous samples does not need to be always the same as the normal samples or vice versa. One way to deal with such a challenge is to deploy Ensemble techniques (discussed in Sect. 3.6).

Oversampling and undersampling are standard techniques that are employed to address the imbalanced data challenge.

Furthermore, instead of relying on the accuracy as the performance metric, the literature describes utilizing other metrics such as precision, recall, moving the threshold for the ROC curves are seen as the evaluation metrics to overcome the data imbalance challenge.

Post addressing the data preprocessing challenges, the data can be fed to machine learning algorithms for the purpose of classification or prediction or other purposes. In the following sections, we describe some of the popular and widely used supervised learning strategies for prediction and classification applications.

3.3 Regression

Regression analysis indicates the significant relationship between input and output, as well as the strength and impact of the independent variables on the (output) dependent variable. There are various types of regression methods to make predictions such as linear regression, polynomial regression, and logistic regression.

Regression is one of the most vital and traditional supervised learning techniques utilized for predicting continuous values in various applications such as time-series analysis. For instance, in a time-varying series, time is an independent variable and the value of the series is a dependent variable. The data prediction for future time instances can be performed through regression and building the model for the time-series or the data samples under experimentation. Regression exploits the relationship between dependent and independent variables of the signal, based on which the analysis for new input data samples is carried out and validated whether the relationship holds or not. Regression is mainly employed to forecast and predict the signal by exploiting the dependency between dependent and independent variables. Multiple regression techniques exist, which are employed depending

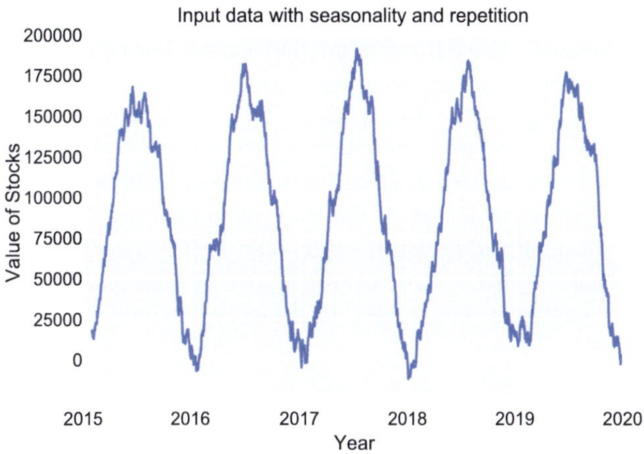

Fig. 3.5 Representing the seasonality in the stock market price of a firm in the past 5 years. Based on the seasonal pattern of the data future stock market price can be predicted.

on the application. The prominent and widely utilized supervised learning-based regression techniques for signal analysis (prediction) are described below.

Example 3.5 (Regression)
Problem: Consider a stock market prediction problem. Based on the input of the previous data, predict the future stock market value.
Solution: Regression is popularly used for predicting time series. Time series regression is a statistical method used for predicting the future based on the past. Here, past data serves as input to the regression model predicting the future. The input data may follow certain seasonality in repeating the pattern during different quarters of a year, as shown in Fig. 3.5. This is the important information observed by the regression model for predicting the future.

3.3.1 Linear Regression

Linear regression is the basic analysis of the change of one variable with respect to another variable. Linear regression is widely used to model the relationship between input and output variables in a plethora of linear problems. Linear regression can be further categorized as:

- *Simple linear regression*: When there is only one independent variable and one dependent variable, simple linear regression is used. For instance, the prediction of weight value based on height attribute.
- *Multiple linear regression*: In this case, there will be multiple independent variables and one dependent variable. For instance, predicting weight based on age and height of a person is one simple example of multiple linear regression.
- *Multivariate linear regression*: If there are more than one dependent variables to predict, the multivariate linear regression is effective. This type of linear regression is also called *multi-target prediction*. For instance, predicting weight and obesity based on age, height is a simple example of the multivariate regression.

Consider a series of data samples with independent variable X and dependent variable Y (X, Y can be matrices or vectors). The linear regression analysis [11–13] can be mathematically defined as

$$Y = f(X, \beta),$$

where β indicates the set of unknown parameters, often termed as weights. In general, β is calculated by minimizing the error in the least-square sense. For a particular case of linear regression, β is derived based on the input X and the output Y, and can be given as

$$\beta = (X^T X)^{-1} X^T Y. \tag{3.2}$$

For a multi-variable dependent regression, i.e., multiple linear regression, the regression is given as

$$Y = f(X_1, X_2, \ldots, X_n, \beta_1, \beta_2, \ldots \beta_n).$$

The function $f(\cdot)$ could be linear or a polynomial, depending on the class of regression considered. This regression analysis is widely adopted for interpolation, extrapolation, and prediction of the time-series sequences.

Example 3.6 (Linear Regression)
Problem: Consider the dataset from Kaggle.[a] Given two variables `X` and `Y`, find the relation between them and model as a regression problem.
Solution: With the linear regression model, one can predict the change in depending variable with respect to the independent variable, as shown in the code snippet below. In this example, `X_train` represents the independent features of a property, and `Y_train` is the depending variable changing based on the input features. Here, the `x_train` and `y_train` represent the

(continued)

Example 3.6 (continued)
training samples of independent and dependent variables, respectively. The future predictions `y_pred` is obtained for the data `X_test`.

[a]https://www.kaggle.com/andonians/random-linear-regression.

The python code snippet looks as follows: (Fig. 3.6)

```
#importing the libraries
import pandas as pd
import numpy as np
import matplotlib.pyplot as plt
from sklearn.linear_model import LinearRegression

#loading train and test datasets
train = pd.read_csv('...../input/train.csv')
test = pd.read_csv('....../input/test.csv')

#Looking into the dataset info
train.info()
test.info()

#droping empty values from training set
train.dropna(inplace= True)

#Reshaping the data to split into x_train, y_train, x_test, y_test
x_train =train.iloc[:,0].values.reshape(-1,1)
y_train =train.iloc[:,1].values.reshape(-1,1)

x_test =test.iloc[:,0].values.reshape(-1,1)
y_test =test.iloc[:,1].values.reshape(-1,1)

#defining the model and fitting it for the input data
model = LinearRegression();
model.fit(x_train,y_train)

#Predicting on the test set
y_pred = linear.predict(x_test)

plt.title('Ploting the prediction of test data')


plt.scatter(x_train,y_train,label='real data',color='blue')
plt.scatter(x_test,y_pred,label='predicted data',color='red')
plt.xlabel('x')
plt.ylabel('y')
plt.legend()
plt.show()

#Evaluating the model
accuracy_score = linear.score(y_test,y_pred)
print(accuracy_score)

> 0.9883886222259362
```

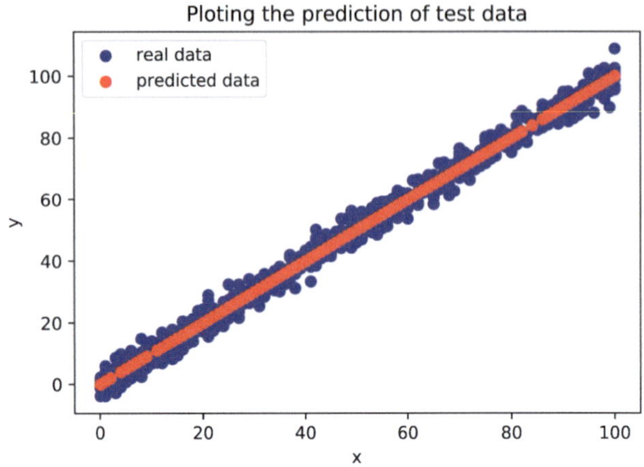

Fig. 3.6 Prediction of the test data for the Example 3.6

3.3.2 Multi-Variable Linear Regression

The most commonly used linear regression is multi-variable linear regression. Multi-variable linear regression [14–16], also known as multivariate regression, exploits the relation between dependent and independent variables, with more than 2 independent variables:

$$y = \beta_0 + \beta_1 x_1 + \beta_2 x_2 + \ldots + \beta_N x_N, \tag{3.3}$$

where y is the output, i.e., the dependent variable with dependence on N input variables x_i, $i = 1, 2, \ldots, N$. The weights are denoted by β.

Example 3.7 (Multi-Variable Linear Regression)
Problem: Construct a multi-variable linear regression model that can predict the charges to be paid for insurance based on a person's age, `bmi` of a person.
Solution: The code snippet for the multivariate linear regression is similar to the previously depicted code snippet, except that there will be a vector of variables in the `X_train` rather than a single column. In the case of the multivariate linear regression both `X_train` and `Y_train` are matrices. In this example, `X_train` represents the independent features of a property and `Y_train` is the depending variable changing based on the input features. With the linear regression model, we can predict the change in depending variable with respect to the independent variable, as shown in the code snippet below.

(continued)

Example 3.7 (continued)

The performance of the prediction is shown below.[a]

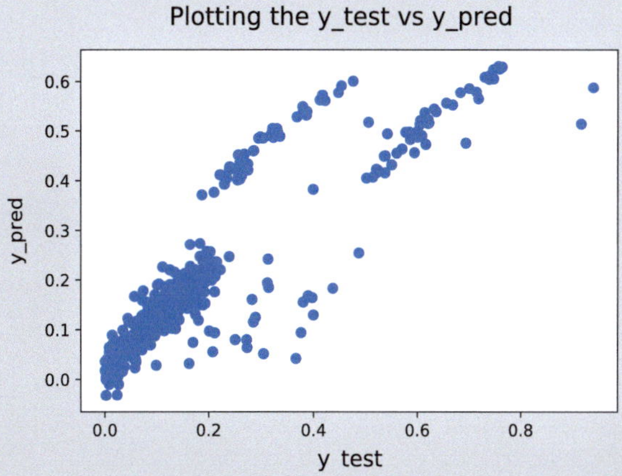

[a]https://www.kaggle.com/mirichoi0218/insurance.

```python

import warnings
warnings.filterwarnings('ignore')

#Import all important libraries
import numpy as np
import pandas as pd
import matplotlib.pyplot as plt
%matplotlib inline
import seaborn as sns
from sklearn.metrics import r2_score

#Load the dataset and check initial entries of the dataset
df=pd.read_csv('...../input/insurance.csv')
df.head()

df.sex=df.sex.apply(lambda x: 1 if x=='male' else 0)
df.smoker=df.smoker.apply(lambda x: 1 if x=='yes' else 0)
df.head()

# Lets convert region as dummy variables
region = pd.get_dummies(df['region'], drop_first = True,prefix='region')
df = pd.concat([df, region], axis = 1)

#Dropping season variable
df.drop('region',axis=1,inplace=True)
df.head()

from sklearn.model_selection import train_test_split
df_train, df_test = train_test_split(df, train_size = 0.7, test_size = 0.3,
    random_state = 100)

from sklearn.preprocessing import MinMaxScaler
scaler = MinMaxScaler()

```

```
# Scaling columns of numerical type
col = ['charges', 'age', 'bmi']
df_train[col] = scaler.fit_transform(df_train[col])
df_train.head()

y_train = df_train.pop('charges')
X_train = df_train

from sklearn.linear_model import LinearRegression

# Fitting LinearRegression onto the train data
model = LinearRegression()
model.fit(X_train, y_train)

col = ['charges', 'age', 'bmi']
df_test[col] = scaler.transform(df_test[col])

y_test = df_test.pop('charges')
X_test = df_test

y_pred = model.predict(X_test)

# Plotting y_test and y_pred to understand the spread

fig = plt.figure()
plt.scatter(y_test, y_pred)
# Plot heading
fig.suptitle('y_test vs y_pred', fontsize = 20)
plt.xlabel('y_test', fontsize = 18)
plt.ylabel('y_pred', fontsize = 16)
plt.show()

score = r2_score(y_test,y_pred)
print(score)

> 0.7772310511733103
```

3.3.3 *Multi-Variable Adaptive Regression Splines (MARS)*

Multi-variable Adaptive Regression Splines (MARS) [17, 18] is an extension of the traditional linear regression, with added capabilities to effectively capture nonlinearities and interactions between dependent and independent variables (Fig. 3.7).

As a simple example, consider a univariate linear regression with $x, x \in R$ being the independent variable and $y, y \in R$ being the dependent variable. A traditional linear regression (as discussed in the previous section) will model the behavior as

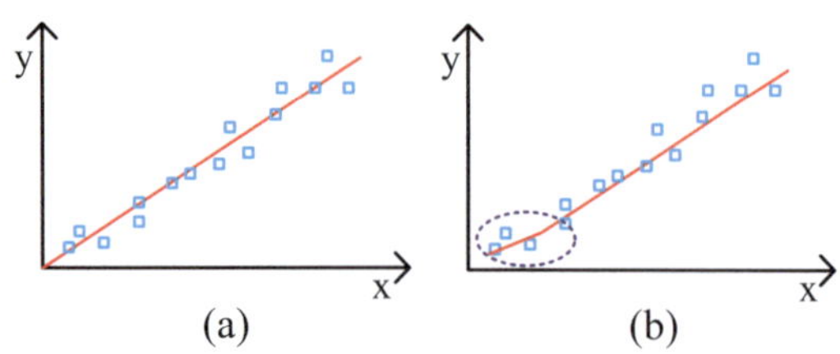

Fig. 3.7 Regression with: (**a**) Traditional linear regression; (**b**) MARS

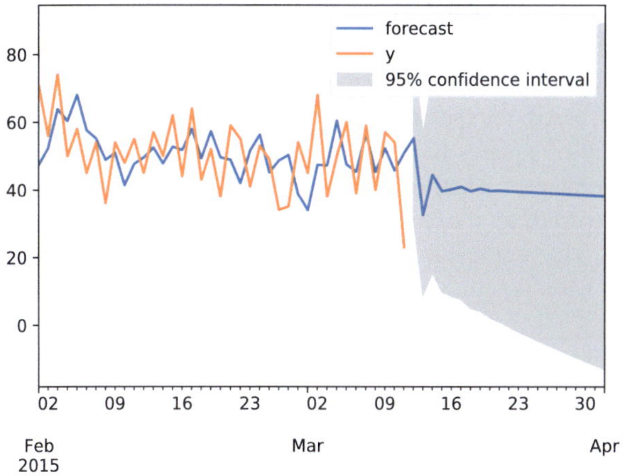

Fig. 3.8 Showing the prediction of Boston housing costs and adding an error interval for the predictions

$$\widehat{y} = ax + b,$$

where a and b are constants and $\widehat{y}$ is the estimated value of y. Figure 3.8a shows the solution of linear regression as a line passing through most of the data points (indicated by boxes). For the sake of explanation, simple linear regression is considered, but, multi-variable and multivariate regression techniques also follow the same methodology. In contrast to linear regression, MARS builds the model considering the nonlinearities in the data. As such, a model built with MARS looks like

$$\widehat{y} = a \ \max(0, x - c) + b,$$

where a, c, and b are constants. Note, that this formula is a simple example of MARS. However, the hinge function $\max(0, x - c)$ could be in a different form or could be the sum of multiple hinge functions, as discussed below. Thus, any deviation from perfect linearities is captured effectively, as shown in Fig. 3.8b. In general, the model of a MARS is given as follows:

$$\widehat{y} = \sum_{i=1}^{k} a_i B_i(x), \tag{3.4}$$

where a_i is a constant and $B_i(x)$ is a basis function, dependent on x. The basis function can be a constant or a hinge function, or it can be a product of two or more hinge functions. A *hinge function* is a function defined by a variable and a knot, and takes the following possible forms: $max(0, x - c)$ or $max(0, c - x)$, where x is the variable and c is a constant called *knot*.

This MARS model is built in two steps: forward pass step and backward pass step. In forward pass, the modeling starts with an initial intercept term, i.e., a constant (which is the mean of the response values). Further, it keeps adding new basis functions based on the input. New basis functions that give a maximum reduction in sum-of-squares of residual error are often added. This is carried out in a brute force manner. This forward pass is followed by backward pass step. In order to avoid overfitting, the least effective basis functions are found and removed from the model. The user can limit the number of terms in the forward pass and the total number of basis functions, i.e., k in (3.4).

Example 3.8 (Multi-Variable Adaptive Regression Splines)
Problem: Consider a problem where we predict the number of houses being sold based on their distance to the Walmart. Perform the model fitting.
Solution: In the real world, most of the problems cannot be solved by the linear regression model. The real-world data is often nonlinear, even for problems that are considered linear. For example, the problem of fitting the number of houses being sold based on their distance to Walmart can be seen as a simple linear regression problem, but to achieve better predictions MARS is employed for model fitting. In the example code snippet, we can observe the implementation of MARS model using the py-earth package. A code snippet to perform MARS regression is shown below:

```
#importing libraries
from numpy import mean
from numpy import std
from sklearn.datasets import make_regression
from sklearn.model_selection import cross_val_score
from sklearn.model_selection import RepeatedKFold
from pyearth import Earth

# defining the dataset
X, y = make_regression(n_samples=10000, n_features=20, n_informative=15,
     noise=0.5, random_state=7)

# define the model
model = Earth()

# defining the evaluation proceduce
cv_func = RepeatedKFold(n_splits=10, n_repeats=3, random_state=1)

# evaluating the model and collecting results
scores = cross_val_score(model, X, y, scoring='neg_mean_absolute_error', cv=
     cv_func, n_jobs=-1)
print('MAE: %.3f (%.3f)' % (mean(scores), std(scores)))
```

3.3.4 AutoRegressive Moving Average

Autoregressive Moving Average (ARMA) [19, 20] regression models the temporal data to analyze the data and/or predict the future series. ARMA is mostly applied especially when the data is non-stationary (a series whose probability distribution changes when shifted in time).

For better understanding, AutoRegressive (AR) and Moving Average (MA) models are individually presented first, followed by the ARMA model.

AutoRegressive: AutoRegressive (AR) models are utilized for predicting univariate time-series or ordered data, purely based on its previous values. An autoregression is expressed as

$$\begin{aligned} y_t &= f(y_{t-1}, y_{t-2}, \ldots, y_{t-p}, \epsilon_t) \\ &= c + \sum_{i=1}^{p} \phi_i y_{t-i} + \epsilon_t, \end{aligned} \tag{3.5}$$

where y_t represents value of variable y at time instant t; ϵ_t is the prediction error at time instant t; p represents the order of AR model and ϕ_i represents the AR parameters, i.e., weights for the model; c is a constant and $f(\cdot)$ represents the model that fits the time series or the ordered variable set.

Moving average: Moving Average (MA) modeling is another regression technique efficient for univariate time-series or signal prediction. The model for moving average (MA(q)) regression, with order q is given by

$$\begin{aligned} y_t &= g(\epsilon_t, \epsilon_{t-1}, \ldots, \epsilon_{t-q}) \\ &= \gamma + \epsilon_t + \sum_{i=1}^{q} \theta_i \epsilon_{t-i}, \end{aligned} \tag{3.6}$$

where $\theta_1, \theta_1, \ldots, \theta_q$ are the parameters for the moving average model, γ is a constant, q denotes the order of the moving average model, and ϵ_t denotes the error, as in (3.5). The order of the MA model indicates the number of error terms considered from the previous predictions. Thus, moving average can be seen as a linear regression of present and previous error terms.

Autoregressive Moving average: Based on the above-presented autoregression and moving average models, ARMA can be given as

$$y_t = \gamma + \epsilon_t + \sum_{i=1}^{p} \phi_i y_{t-i} + \sum_{i=1}^{q} \theta_i \epsilon_{t-i}. \tag{3.7}$$

ARMA is generally represented as ARMA(p, q) with p and q representing the orders of AR and MA, respectively. The main advantage of ARMA is its capability to capture the impacts of noise from previous predictions effectively, resulting in an improved prediction accuracy. This method is useful, especially for lower-order polynomials.

To build ARMA regression, the data is initially modeled to estimate the AR and MA models based on the previous values and errors, and then an ARMA model is formed. Further, based on the differences in predicted signal and the actual signal, the signal can be evaluated. The major drawback of ARMA is the amount of computation increases proportional to the selected order, though the prediction accuracy might not increase at the same scale.

Example 3.9 (AutoRegression)
Problem: Consider the data in [19],[a] perform an ARMA on time-series analysis in fractional order systems.
Solution: In the presence of a diffusion or dispersion component, the fractional behavior forces the estimated poles and zeros of the ARMA model to form an alternating chain. This chain can be detected and consecutively be compressed to finally result in a fractional order version of the ARMA model. The plot for the regression is shown in Fig. 3.8.

[a]https://github.com/ispmarin/arima_example/blob/master/data/f_series.csv.

```
%matplotlib inline
%reload_ext rpy2.ipython

import pandas as pd
import matplotlib.pyplot as plt
import statsmodels.api as sm
import numpy as np
import warnings
warnings.simplefilter(action='ignore', category=FutureWarning)
warnings.simplefilter(action='ignore', category=UserWarning)


df = pd.read_csv('...../data/f_series.csv')
date_index = pd.date_range('2015-01-01', periods=len(df))
df = df.set_index(date_index)
df.y = df.y.astype(float)

model = sm.tsa.ARIMA(df, (2,1,0)).fit()

fig = model.plot_predict(start='2015-02-01', end='2015-04-01')
model.predict(start=70, end=80)
```

The above code snippet shows the technique for ARIMA regression. The fitting model uses a ARIMA with a lag of 5 for prediction, MA model of 0, and makes the series stationary with a difference order of 1.

3.3.5 Bayesian Linear Regression

Bayesian linear regression (in short referred to as Bayesian regression) [21–23] is a multivariate linear regression, where the output presents the probability of the regression value for a particular point based on the given data. Other similar approaches such as Maximum Likelihood Estimation (MLE) suffer from overfitting, especially when the amount of trained data has a large variance and biasing when the variance in the dataset is too less [24]. The basic principle of Bayesian regression is to provide an inference of how good or bad (in terms of probability) the predicted values are.

Consider a simple linear regression case, given by

$$y_i = x_i^T \beta + \epsilon_i \tag{3.8}$$

here x_i and β are vectors of size $k \times 1$, and output is y_i. β represents the model fitting parameter vector. ϵ_i values are independent and normally distributed random variables, i.e., it follows a Gaussian distribution with 0 mean ($\epsilon_i \sim N(0, \sigma^2)$). The likelihood function for the output at an input with a given variance (variance is set by user or already known) is given by

$$p(\mathbf{y}|\mathbf{X}, \beta, \sigma^2) \propto (\sigma^2)\mathbf{exp}(-\frac{1}{2\sigma^2}(\mathbf{y} - \mathbf{X}\beta)^T(\mathbf{y} - \mathbf{X}\beta)), \tag{3.9}$$

where $\mathbf{X}$ is a $n \times k$ matrix and each row is a vector x_i^T. $\mathbf{y}$ is the column n-vector ($[y_1, y_2, \ldots, y_n]^T$) with variance σ. β, denoting the weights, can be derived similar to that in (3.2). For more in-depth analysis of how these are derived, please refer to [22, 23, 25].

In practice, the objective of Bayesian linear regression is to maximize β (3.9) for the given data. Hence, it is often desired to have higher probability for the estimated output for a given input. A low probability indicates that the formed model does not fit the data, hence needs to be further optimized. In case of time-series or signal analysis, consider that a model is formed as given in (3.8) with output y_i. However, it is not known how good the derived model is and how reliable the output is. In such cases, the Bayesian regression analysis is carried out to estimate the quality of the derived model. A low probability to have the y_i based on the given data indicates that the model is not accurate and needs to be improved.

Example 3.10 (Bayesian Linear Regression)
Problem: Use Bayesian Linear Regression to predict the housing cost of the houses from Boston dataset.
Solution: First, we need to write a custom function definition for Bayesian Linear Regression model. Further, one can load the Boston dataset from sklearn datasets. Model and fit the data using predefined Bayesian Linear Regression models and can perform the model evaluation.

```
#importing libraries
import numpy as np
from scipy import stats

#Defining Bayesian Linear Regression function
class BLR:

    def __init__(self, n_features, alpha, beta):
        self.n_features = n_features
        self.alpha = alpha
        self.beta = beta
        self.mean = np.zeros(n_features)
        self.cov_inv = np.identity(n_features) / alpha

    def learn(self, x, y):

        # Update the inverse covariance matrix
        cov_inv = self.cov_inv + self.beta * np.outer(x, x)

        # Update the mean vector
        cov = np.linalg.inv(cov_inv)
        mean = cov @ (self.cov_inv @ self.mean + self.beta * y * x)

        self.cov_inv = cov_inv
        self.mean = mean

        return self

    def predict(self, x):

        # Obtain the predictive mean
        y_pred_mean = x @ self.mean

        # Obtain the predictive variance
        w_cov = np.linalg.inv(self.cov_inv)
        y_pred_var = 1 / self.beta + x @ w_cov @ x.T

        return stats.norm(loc=y_pred_mean, scale=y_pred_var ** .5)

    @property
    def weights_dist(self):
        cov = np.linalg.inv(self.cov_inv)
        return stats.multivariate_normal(mean=self.mean, cov=cov)


#Importing libraries
from sklearn import datasets
from sklearn import metrics

#Importing dataset
```

```
X, y = datasets.load_boston(return_X_y=True)

#Defining the model
model = BLR(n_features=X.shape[1], alpha=.3, beta=1)

# Predicting and evaluating the model
y_pred = np.empty(len(y))

for i, (x_i, y_i) in enumerate(zip(X, y)):
    y_pred[i] = model.predict(x_i).mean()
    model.learn(x_i, y_i)

print(metrics.mean_absolute_error(y, y_pred))

> 3.784125061857545
```

3.3.6 Logistic Regression

In contrast to other discussed regression techniques, logistic regression [26–29] is used for the purpose of classification only. The underlying basic principle is similar to that of a regression. A dependency among dependent and independent variables is initially exploited, following the application of the logistic function (hence the name logistic regression).

For a given data $D = \{(X_1, y_1), (X_2, y_2), \ldots, (X_n, y_n)\}$, with $X_i \in R^N$, i.e., N-dimensional vector and y_i represents the class to which it belongs, i.e., $y_i = \{0, 1\}$ for the binary case. The logistic function

$$\sigma(a) = \frac{1}{1 + e^{-a}} \tag{3.10}$$

has an output between 0 and 1, depending on a representing the hypothesis function of the matrix form of $\beta^T X$, such that β^T is the transpose of the weight matrix and X is the input matrix. This means that logistic regression returns a probabilistic estimate, which may be compared against a threshold to generate a binary output (i.e., 0 or 1):

$$Y_i = \begin{cases} 0, & \text{if } \sigma(a) < threshold \\ 1, & \text{if } \sigma(a) \geq threshold. \end{cases} \tag{3.11}$$

This is very similar to *Perceptron Learning Algorithm*, which takes an input, calculates the weighted sum, and returns 1 if the weighted sum is greater than a threshold, or 0 otherwise.

Logistic regression produces a measure of uncertainty in occurrence of a binary output. The output depends on the hypothesis function (a linear model). Therefore, the logistic regression works for the given data as follows:

$$Y_i \sim P_r(\sigma(\beta^T \cdot X_i)), \tag{3.12}$$

where P_r denotes the Bernoulli distribution (i.e., $P(X = 1) = 1 - P(X = 0)$, where P indicates the probability) and β indicates the weight vector, similar to that described in (3.2). Vectors β and X are of the same dimension. Based on the probability value, the class to which the input belongs to can be determined. The above-mentioned example is for a simple binary classification, but it can be extended for multiclass classification.

Example 3.11 (Logistic Regression)
Problem: Consider the Iris dataset and try to perform the classification of the data using a minimal set of features with logistic regression.
Solution: To perform the classification of the Iris dataset, we first need to load the dataset and choose the first two features from the dataset in this example. Further, the data is fit into the logistic regression model using the `model.fit()` command, similar to the previous cases. However, we consider the first two features in this example, but, one can perform selection through the aforementioned feature selection techniques. The code snippet is shown below.

```
# Import the dependencies
import matplotlib.pyplot as plt
import seaborn as sns
from sklearn.linear_model import LogisticRegression
from sklearn.metrics import classification_report
from sklearn.metrics import accuracy_score
from sklearn.model_selection import train_test_split

# import some data to play with
iris = datasets.load_iris()
X = iris.data[:, :2]  # we only take the first two features.
Y = iris.target

#Split the data into 80% training and 20% testing
x_train, x_test, y_train, y_test = train_test_split(X, Y, test_size=0.2,
    random_state=42)

#Train the model
model = LogisticRegression()
model.fit(x_train, y_train) #Training the model

#Test the model
predictions = model.predict(x_test)
print(accuracy_score(y_test, predictions))

> 0.9
```

The above code snippet provides information on using the logistic regression for classifying the data x into multiple classes by splitting the data into train and test datasets. It needs to be noted that the provided code snippets are mere examples that outline the main usage of the techniques, and other parts such as importing data or preprocessing are not included for the sake of conciseness.

3.4 Artificial Neural Networks

Many real-world problems cannot be categorized under linear problems or aforementioned nonlinear regression techniques. To address complex classification challenges, Artificial Neural Networks (ANN) are introduced. ANNs are basic representations of a human brain that replicate nonlinear learning through a network of neurons. In Neural Networks, these neurons are called *artificial neurons*. The basic building block of artificial neural networks (ANNs) is the neuron. First, we discuss the modeling of a neuron following which we describe the integration and working of the artificial neural networks.

3.4.1 Modeling of Neuron

In the year 1943, McCulloch and Pitts first proposed the oversimplified model of the biological neuron, popularly known as M-P model, which is widely adopted and further enhanced in recent times. The M-P model of the neuron can be split into two parts. Firstly, the neuron sums all the incoming signals coming in and aggregates them, as shown in right-side zoom-out box of Fig. 3.9, mathematically described as follows:

$$g(x) = \sum_{i=1}^{n} x_i. \tag{3.13}$$

Here, $g(x)$ represents the aggregated sum of the inputs x_i. Further, the aggregated sum is passed through an activation function $f(x)$, thus the model of a neuron is

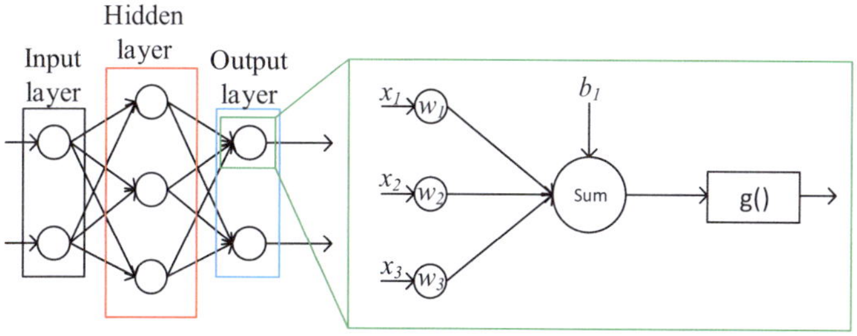

Fig. 3.9 A single hidden layer neural network with 2 inputs and 2 outputs

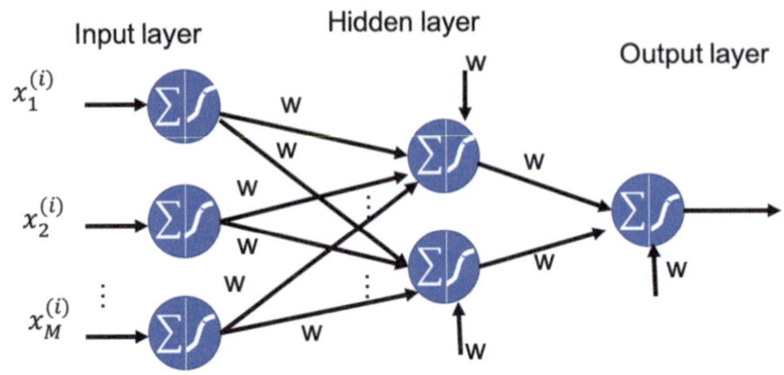

Fig. 3.10 Artificial neural network

provided as $f(g(x))$. Over time, the models have improved, such advancements include including the bias, weighted sum of the inputs, and working with non-binary inputs (x_i).

An artificial neuron aggregates incoming input data using a combination function, and the output of the combination function is fed to an activation function (usually a nonlinear function such as *sigmoid*, *binary step function*, or *softmax*) and produces an output that is channeled to the neurons in other layers of the network downstream. Figure 3.10 shows a neural network in which each neuron aggregates the incoming inputs and scales the output of the combination function (Σ) using an activation function. The types of activation functions are discussed later in this book.

One of the requirements of ANN is to standardize the attributes of the dataset (categorical or continuous) to take a value between 0 and 1. In Python, sklearn object *MinMaxScalar* can be used to scale continuous variables. For categorical variables, flag variables or one-hot encoding can be applied if there are not too many categorical features. An example of using a flag variable to encode a categorical attribute is representing a *male* gender as 0 and a *female* gender as 1.

In a neural network, there are neurons in the input layer, hidden layer, and output layer. The input layer is the layer that receives the attributes of the dataset, and the output later is the last layer in the network that produces the outcome of the classifier. The layers between the input and output layers are called *hidden layer* (see Fig. 3.10). The number of neurons in these layers does not have to be always same. This number can change case by case and depending on the problem. For instance, we may have one output node in a neural network, or multiple output nodes depending on the classification problem. For binary classification problems, one neuron is used in the output layer. Also, if the output classes are ordered such as *first-winner, second-winner, so forth*, still one neuron is sufficient to perform multiclass classification. However, if the output labels are not ordered, more than one neuron should be used in the output layer. This is called *1-of-n output encoding*. The neural network in Fig. 3.10 is a fully connected network since there is a connection from every neuron to all other neurons in another layer downstream. The connecting edges are associated with weights. Learning artificial neural network is about tuning these weights that serve as the model parameters, which will be described in the later sections.

3.4.2 Implementing Logical Gates with ANN

Neural Networks or Artificial Neural Networks can be used to emulate the functionality of logic gates. Figure 3.12 shows the truth table for different types of logical gates. Let us understand how a logical gate functionality is implemented using an ANN. We show two scenarios of emulating *AND* gate and a *XOR* gate with a 2-layered (Perceptron) and 3-layered neural network (Multi-layer Perceptron), illustrated in Fig. 3.11a and b, respectively.

In part (a), there is a single neuron, emulating the functionality of logical *AND* gate with a threshold or hard limiting activation function such as *Sigmoid* function, which scales the output to a desirable range between 0 and 1. Such scaled output of the *Sigmoid* function is 1 if the output is greater than or equal to a threshold, and it is 1 if the output is less than the specified threshold. In other words, the output of the *Sigmoid* function represents the probability of the outcome, being either 0 or 1. In this example, the threshold is 0.5.

The variable z is the input of the activation function $g(z)$. This makes z represent the output of the combination function.

$$y = \frac{1}{1 + e^{-z}} = \frac{1}{1 + e^{-g(w^T x)}}. \tag{3.14}$$

The input x_i is obtained from the dataset and combined through the combination function Σ. Then, the output z is fed to the activation function to produce a binary outcome. To implement the functionality of logical gate that generates a binary output, the *Sigmoid* function uses a threshold. The value of the threshold is set a priori to separate the outcome of classification. The threshold can be fine-tuned heuristically.

The truth table for part (a) in Fig. 3.11 shows how the output y is computed based on the weights w_0, w_1, w_2 and the inputs. In the input layer, the attribute values from the dataset are fed to the single neuron N in the output layer; x_0 is the bias and is equal to 1. A similar procedure is applied in part (b). What makes part (b) different from the Perceptron implementation of *AND* gate is the presence of a hidden layer in the Artificial Neural Network. In this 3-layered ANN, there are two neurons in the hidden layer (i.e., N_1 and N_2). The data in the dataset (i.e., the attribute values for an observation) from the input layer enter the neurons at the hidden layer, then combined using the Σ combination function, then go to an activation function (here, *Sigmoid* function). It is also possible to choose another activation function such as *Binary Step Function*. Experiments show that replacing the neuron N_1 and N_2 with *NAND* and *OR* gate functionality, respectively, facilitates the functionality of a *XOR* gate. In fact, with the help of an additional layer (i.e., hidden layer) we made a more

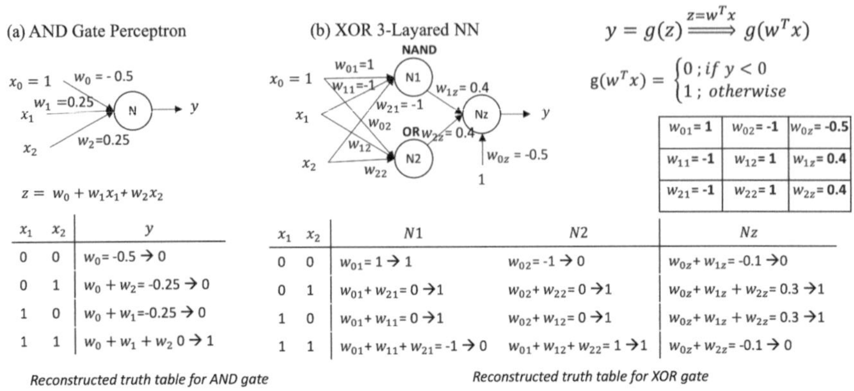

Fig. 3.11 Implementing logical gates with artificial neural networks

GATE	AND			NAND			OR			NOR			XOR			XNOR			NOT	
	AB			$\overline{AB}$			$A+B$			$\overline{A+B}$			$A\oplus B$			$\overline{A\oplus B}$			$\overline{A}$	
Truth Table	A	B	AB	A	B	$\overline{AB}$	A	B	$A+B$	A	B	$\overline{A+B}$	A	B	$A\oplus B$	A	B	$\overline{A\oplus B}$	A	$\overline{A}$
	0	0	0	0	0	1	0	0	0	0	0	1	0	0	0	0	0	1	0	1
	0	1	0	0	1	1	0	1	1	0	1	0	0	1	1	0	1	0	1	0
	1	0	0	1	0	1	1	0	1	1	0	0	1	0	1	1	0	0		
	1	1	1	1	1	0	1	1	1	1	1	0	1	1	0	1	1	1		

Fig. 3.12 Truth table for logical gates

complex model to emulate the functionality of logical *XOR* gate. The outcome of N_1 and N_2 neurons from the hidden layer is fed to the single neuron output layer to finally generate the outcome of *XOR* gate truth table according to Fig. 3.12.

3.4.3 Multi-Layer Perceptron

As aforementioned, the multi-layer Perceptron (MLP) or ANN comprises at least three layers: an input layer, one or more hidden layers, and an output layer. In a fully connected neural network, all the neurons (or nodes) in a succeeding layer are connected to all the nodes in the preceding layer. Each node has inputs and outputs and performs an operation based on its activation function. Different layers could have different activation functions. The learning process in neural networks refers to the adaptation of weights between layers and the update of activation functions. A traditional neural network with one hidden layer is illustrated in Fig. 3.9.

As seen earlier, the output of a neuron in a hidden or output layer of an ANN or MLP is given as

$$y_j = g(\sum_{i=1}^{n} w_i x_i + b_j), \tag{3.15}$$

where y_j indicates the output of the j-th neuron with n inputs, given by x_i, and b_j denotes the bias. The activation function of the node is given by $g(\cdot)$.

To improve the performance of the MLPs or ANNs, a common way is to tune different parameters of the neural networks. The main parameters that are tunable in neural networks are:

- Number of hidden layers (if the number of hidden layers is more than 1, it is usually classified under a deep neural network; otherwise, it is classified as a shallow neural network).
- Number of nodes in input, and hidden layers.
- Activation functions in the hidden and output layers.
- Update procedure for the weights between different layers (learning methodology).

The number of nodes in the output layer denotes the number of output classes the user expects from the data analysis. In the case of binary classification, the number of neurons is 1, where the output is either zero or one. To classify a dataset into one of the five categories, the neural network architecture will have five nodes in the output layer and the number of input neurons depends on the input data dimensions. The used activation functions and the methodology to update the weights in neural network are discussed here.

Activation Functions

In neural networks, the activation function can be defined as an abstraction representing the rate of action potential firing in the cell, i.e., the activation function indicates the effectiveness with which the presented data will be helpful in determining final output. Different activation functions are used in the literature. The mathematical representation of some of the most widely used activation functions in neural networks are presented in Table 3.4. The scikit learn library supports a majority of the activation functions. The most widely used activation functions are $tanh(x)$ and ReLU functions. In Table 3.4, x_0 is the x-value of sigmoid's midpoint, L is the curve's maximum value, and k represents the steepness of the curve.

The graphs for most of the activation functions presented in Table 3.4 are shown in Fig. 3.13. The shapes of logistic and sigmoid functions look similar, except the change in the bounds of the function. Similarly, from the perspective of shapes, the hyperbolic tangent, arctan, and softsign look similar with differences in the slope of the curves and the bounds of the functions. It needs to be noted that the graphs shown are a simpler representation of their corresponding shapes, not drawn to the scale.

Table 3.4 Activation functions in a neural network

Name	Formula
Identity	$f(x) = x$
Binary step	$f(x) = \begin{cases} 0, & \text{if } x < 0 \\ 1, & \text{otherwise} \end{cases}$
Logistic	$f(x) = \frac{L}{1+e^{-k(x-x_0)}}$
Sigmoid	$f(x) = \frac{1}{1+e^{-x}}$
Hyperbolic tangent	$f(x) = \tanh(x) = \frac{2}{1+e^{-2x}} - 1$
ArcTan	$f(x) = \tan^{-1}(x)$
Softsign	$f(x) = \frac{x}{1+\|x\|}$
Exponential linear unit	$f(x) = \begin{cases} \alpha(e^x - 1), & \text{if } x < 0 \\ x, & \text{otherwise} \end{cases}$
SoftPlus	$f(x) = \ln(1 + e^x)$
Sinusoid	$f(x) = \sin(x)$
Gaussian	$f(x) = e^{-x^2}$
ReLU	$f(x) = \begin{cases} 0, & \text{if } x < 0 \\ x, & \text{otherwise} \end{cases}$

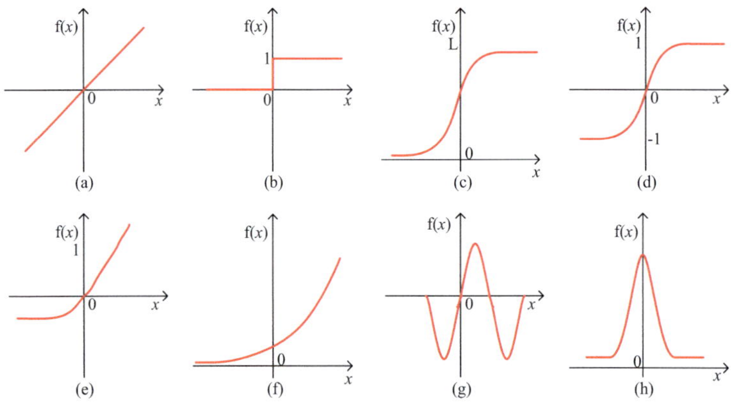

Fig. 3.13 Activation functions: (**a**) identity; (**b**) binary step; (**c**) sigmoid; (**d**) hyperbolic tangent; (**e**) exponential linear unit; (**f**) soft plus; (**g**) sinusoid; (**h**) Gaussian

Activation functions are often the most computationally expensive part in neural networks. Reducing the implementation complexity of these functions often reduces the overall computational complexity. These activation functions could be implemented in different manners, such as using Look-Up-Tables (LUTs), BRAMs, utilizing CORDIC algorithms to compute the hyperbolic and trigonometric functions [30, 31], and approximate computations [32, 33] such as Piecewise-Linear-Approximations (PLAs).

Example 3.12 (Multi-Layer Perceptron)
Problem: Perform a simple XOR function using a multi-layer Perceptron.
Solution: The code snippet below shows the use of 2 hidden layers ANN with X and Y as input and outputs, respectively. The parameter `alpha` represents the learning rate of the ANN. The `hidden_layer_sizes` shows the sizes of the two hidden layers used. In order to predict for the new incoming sample or for test data, `clf.predict()` function is used, where the inputs are passed to the function. The parameter solver indicates the solver used for the ANN. In this case, the LBFGS solver is used. Different solvers can be used as discussed below. An example of XOR problem being solved with the help of MLP can be observed in the code snippet below.

```
from sklearn.neural_network import MLPClassifier

X = [[1, 1], [1, 0]]
y = [0, 1]
clf = MLPClassifier(solver='lbfgs', alpha=2e-5, hidden_layer_sizes=(4, 2),
    random_state=1)
clf.fit(X, y)
print(clf.predict([[0, 0]]))

> [1]
```

3.4.4 Training of MLPs

Weights in a neural network can be updated in multiple ways. One of the most widely used rules is the Delta rule [34]:

$$\Delta w_{ij} = \alpha(t_j - y_j)x_i, \tag{3.16}$$

where i indicates the index of input x_i and j represents the index of the j-th neuron with t_j and y_j as target and actual outputs. α denotes the learning rate. Different techniques and optimizers exist for training the MLP, as discussed below.

Stochastic Gradient Descent (SGD)

Backpropagation means updating the weights in the network architecture. In backpropagation neural networks (BPNN) [13, 35–38], the input data is fed from the input layer to the hidden layer and then to the output layer. However, the weights are updated in a reverse manner. One popular method to update the weights of individual nodes is the Delta rule (Eq. (3.16)). The weights between the hidden layer and the output layer are updated first based on the target outputs of the system, followed by updating the weights between the hidden layer and the input layer. The overall architecture is also termed popularly as a feed-forward neural network.

The stochastic gradient descent (SGD) can be performed using Scikit learn libraries. The following code snippet shows the configuration and deployment of SGD in ANNs:

```
from sklearn.linear_model import SGDClassifier
X = [[1, 1], [1, -1]]
y = [0, 1]
clf = SGDClassifier(loss="hinge", penalty="l2", max_iter=15)
clf.fit(X, y)
print(clf.predict([[0.8, -1.6666]]))

> [1]
```

Here, a simple classification is performed using SGD classifier with 5 iterations. A similar strategy can be adopted in ANNs.

Example 3.13 (Stochastic Gradient Descent)
Problem: Show how to perform classification using SGD classifier.
Solution: The above code snippet has `X` and `Y` as input and outputs, respectively. The parameter `loss` represents the loss function of the classifier. In order to predict for the new incoming sample or for test data, `clf.predict()` function is used, where the inputs are passed to the function.

The code below shows the updation of the gradient parameter for each sample in the training data by evaluating the gradient at each training sample.

```
import numpy as np

for i in range(epochs):
    np.random.shuffle(data)
    for each_sample in data:
        gradient_parameter = evaluate_gradient(loss_function, each_sample,
         parameter)
        parameter = parameter - learning_rate * grad_parameter
```

Mini-Batch Gradient Descent

SGD updates parameters after considering each single data entry. The frequent updating makes convergence faster, yet each update is noisier and the parameters maintain a high variance. Contrarily, Batch gradient descent, which is also known as vanilla gradient descent, updates parameters only after computing the gradient of the cost function w.r.t. the entire training set. It makes smooth updates on the parameters, but takes a long time for making a single update.

Mini-batch gradient descent makes a compromise between batch gradient descent and SGD, balancing between the speedy convergence and the smoothness of updates. It updates the parameters after the computation of the cost function of each batch of the dataset. The speed of convergence and the variance of parameters can be tuned by adjusting the way to divide the dataset into various batches. By now, mini-batch gradient descent is considered as the best gradient descent algorithm due to its flexibility and robustness.

Unfortunately, mini-batch gradient descent has not been implemented in classic data science libraries such as Scikit Learn. Its pseudocode is presented below.

```
for i in nb_epochs:
    shuffle(train_data)
    for mini in divide_batches(train_data, batch_size):
        params_grad = evaluate(cost_function, mini, params)
        params = params - learning_rate * params_grad)
```

Masters and Luschi [39] suggest that using small batch sizes between 2 and 32 achieves the best performance on both training stability and generalization in most scenarios. But it is recommended to review the trade-offs between the model validation error and the training time under different batch-size settings.

Adagrad

One of the challenges faced by gradient descent algorithms is that a predefined learning rate is applied to all parameter updates. It does not adapt neither to a dataset's characteristics nor to different features within the dataset. When we have a sparse dataset and the features' frequencies vary heavily, we would like to have relatively larger updates for rarely occurring features than for frequent features.

Adagrad is a gradient-based optimization that adapts the learning rates to the parameters. For parameters associated with features that occur frequently, it performs small updates, while for parameters associated with rarely occurring features, it makes larger updates. Comparing to gradient descent algorithms, it suits better for sparse data, which is very common in the real world. A well-known neural network learned to recognize cats in videos trained at Google used Adagrad and found that it greatly improved the robustness of SGD [40].

One of the benefits brought by Adagrad is that the need to manually tune the learning rate is removed. The default value of 0.01 can be simply adopted and the algorithm will update the learning rate automatically. However, Adagrad has its shortcomings. The greatest weakness is that the learning rate will keep shrink and eventually decay, at which point the algorithm stops learning.

Adagrad can be implemented using TensorFlow with Keras library. Please refer to the API of TensorFlow section tf.keras to find out the details.

Adam

To solve the decaying learning rate problem of Adagrad, multiple methods have been developed. AdaDelta and RMSprop are two of the most famous and widely-used ones. They are independently developed but both take the route of an exponentially decaying average of past squared gradients. Similar to Adagrad, with these two optimizations we do not need to set a default learning rate since it is not included in the update rule. The difference from Adagrad is that the learning rate does not decay.

Another optimization that computes adaptive learning rates for each parameter is Adaptive Moment Estimation (Adam) [41]. It stores an exponentially decaying average of past squared gradients like AdaDelta and RMSprop do, and also keeps track of an exponentially decaying average of past gradients. This method converges very rapidly while performs well in practice.

Adam can be implemented with TensorFlow and keras too. The default values for β_1 and β_2 are 0.9 and 0.999 respectively, and default ϵ is 10^{-8}.

Which Optimizer Should We Use?

So, which optimizer should we really pick out from the multiple algorithms that are listed above, as well as tens of those not even mentioned here? Long story short, Adam is the best optimizer at this point of time. As we stated in Section 2-5-2-3, most of the practical datasets are sparse, and for sparse data, we should use the optimizers with dynamic learning rates since they are capable of capturing the variance in the rarely occurring features better than the optimizers without adaptive learning rates.

Among the optimizers with the adaptive learning rates, Adam not only deals with the problem of diminishing learning rate faced by Adamgrad, but also outperforms AdaDelta and RMSprop toward the end of optimization due to its consideration toward bias-correction and momentum. Despite that optimizers without the adaptive learning rates such as SGD can be implemented easily with Scikit learn libraries, we should use Adam if we care about fast convergence and the neural network we are training is deep or complex.

If under some certain circumstances in which gradient descent algorithms are more favorable, such as when computational capacity is lacking, mini-batch gradient descent is the best optimizer due to its higher flexibility and robustness comparing to vanilla gradient descent and SGD.

3.4.5 Inference

During the inference phase, the obtained hyperparameters will be utilized to perform the predictions. In most of the classification examples, the softmax is the widely considered and used output layer's activation function.

3.4.6 Issues with Multi-Layer Perceptron

Despite MLPs being widely deployed and effective for a wide range of applications, careful consideration of the challenges is non-trivial to ensure the application and MLP architectures are well matched. Four major challenges in the design of MLP designs are the overfitting, underfitting, scaling of the inputs, and existence of multiple optimum points.

Overfitting

Overfitting is a common issue with many machine learning including the previously discussed regression, ANN, and other techniques. Given the learning capability of the neural networks, they are often trained with complex data or highly nonlinear and complex to extract the relationship between dependent and independent variables. In such scenarios, when a model tries to fit the data that is highly noisy it overfits. The overfitting of a model indicates lack of generalization of the data and does not reflect the underlying trends and relationship between the variables in the data. In other words, the model performs well on the seen data and performs poorly in the case of unseen data. There exist multiple solutions that aid in preventing a neural network (or machine learning models, in general) from overfitting. We discuss a few of them below.

Lowering the Model Complexity

As a practice, the programmers assume that providing a large model can capture complex data. However, this is not often the case and can lead to overfitting concerns. Thus, a simple solution to overcome the overfitting challenge is lower the complexity of the network by reducing the number of hidden layers or reducing the number of neurons in the hidden layer. However, there is no standard rule on modifying the number of layers or the number of neurons. Thus, by experimenting and verifying the performance and loss plots, one needs to fine-tune the model to alleviate overfitting challenge.

Regularization

The training of the model happens with respect to the loss of the model, i.e., the loss value is fed back during the backpropagation based on which the model hyperparameters will be tuned to fit the data. In the case of regularization, a penalty term is added to the loss function to reduce the complexity of the model, leading to better generalization. Two types of regularization terms are widely used in practice, namely L_1 and L_2 regularization, as shown in Equations below.

$$L(x, y) = \sum_{i=1}^{n} (y_i - h_\theta(x_i))^2 + \lambda \sum_{i=1}^{n} |\theta_i| \tag{3.17}$$

$$L(x, y) = \sum_{i=1}^{n} (y_i - h_\theta(x_i))^2 + \lambda \sum_{i=1}^{n} |\theta_i|^2. \tag{3.18}$$

The first term indicates the loss and the last term indicates the regularization term. As seen, the L_1 penalty aims to minimize the absolute value of the weights, whereas L_2 aims to minimize the square value of the weights.

Limited Training

Termination of training the model within a limited number of iterations is considered as a form of regularization. By limiting the number of training iterations one can ensure that the model learns to fit the data well and capture the relationship between the variables. Once the required performance is achieved, further trying to improve the fitting accuracy often leads to minimizing the generalization capabilities. Thus, limiting the training iterations will help to avoid overfitting scenarios.

Data Augmentation

Given that the data available during the training could be limited or biased to one class often leads to overfitting. To avoid such a scenario, data augmentation could be utilized to increase the training data and avoid the model to be fit to a specific set of data. Some of the popular data augmentation techniques in computer vision domain include flipping, translation, rotation, scaling, changing brightness, adding noise to the existing data. A sample image with data augmentation is shown below.

```
from keras.preprocessing.image import ImageDataGenerator

datagen = ImageDataGenerator(
    rotation_range = 45,
    shear_range = 0.5,
    zoom_range = 0.5,
    horizontal_flip = True,
    brightness_range = (0.5, 1.5)
)
```

Dropout

Dropout is another technique that is seen as an anti-dote to overfitting challenge. In the case of dropout, the connections between the neurons in the hidden layers are removed, i.e., no propagation of values. The aforementioned techniques such as regularization perform the modification of the cost function to minimize the overfitting. However, the dropout technique randomly drops the neurons in the ANN. Different neural networks with different dropout values will overfit in different manners, thus, the net impact of overfitting is mitigated.

In addition to addressing the overfitting challenge, it also helps to prevent underfitting. Performing dropout in every iteration will improve the generalization capability by ensuring that the information leading to underfitting is randomly removed or not learned. Reducing the dropout also aids in improving the complexity of the model and helps to better learn the training data.

Underfitting

In addition to overfitting challenges, underfitting is another hurdle in fitting the machine learning classifiers, including neural networks to the input data. Underfitting occurs when the neural network is unable to perform well on the training data, i.e., not able to capture the relationship between dependent and independent variables during the training phase. The underfitting challenge signifies that the deployed neural network is unable to learn and exploit the relationship in the underlying data, making it inefficient to use. Poor training performance and high variations in the inference are seen as indicators of the underfitting. To overcome the challenge of underfitting multiple solutions are suggested as discussed below.

Adding More Layers or Neurons

Given that the underfitting leads to capturing the complexity of the fitting, adding hidden layers and/or neurons will aid in improving the learning of the underlying data and limiting the underfitting challenge. Additionally, increasing the number of input parameters can also help in better understanding the complex data and enables better fitting. For instance, instead of reducing the features through the techniques such as principal component analysis (PCA), feeding the raw data or more preprocessed features will help in avoiding the underfitting challenge.

Decreasing Regularization Parameter

As seen in Eqs. (3.17) and (3.18), adding the penalty term to the loss function aids in tuning the complexity of the model based on the penalty term λ. Thus, by tuning the penalty term, ideally, reducing the penalty term helps in improving the fitting and reduce the bias.

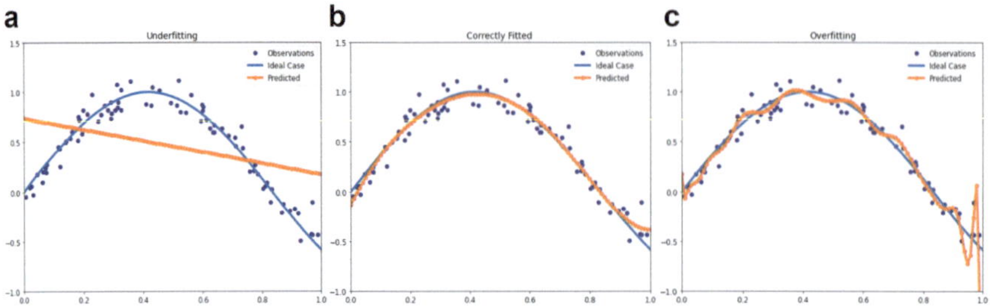

Fig. 3.14 (**a**) Underfitting. (**b**) Correctly fitted. (**c**) Overfitted

Increasing the Training Data

Another solution to avoid underfitting of the model is to increase the training samples. As underfitting indicates the inability to learn the training data and exploit the underlying (complex) relationship, increasing the training data that better represents the variance of parameters in the dataset at large will avoid underfitting of the model.

Overfitting vs Underfitting

Overfitting and underfitting are two common problems in the training of neural networks and machine learning algorithms. Underfitting refers to the model's inability to learn the training data, whereas overfitting refers to the inability of the model to generalize well. As shown in Fig. 3.14, underfitting leads to improper classification, whereas overfitting makes the model highly specialized to the data and unable to generalize well. Thus, a network that is not overly fitted or underly fitted, i.e., a learning technique that learns the training data well and showcases capability to generalize well is the machine learning model that is anticipated to perform better.

Scaling of Inputs

Scaling of inputs is not a challenge unlike overfitting or underfitting but can degrade the performance of the neural networks. Given that the weights of the neural network are initialized with small random values at the beginning of the training and updated via training algorithms based on loss value, the scale of inputs and outputs plays a critical role. Unscaled input data can lead to slower or unstable learning due to inability to adapt to the gradients. Similarly, unscaled output data can result in larger gradients leading to learning failures. Thus, it is non-trivial to scale the data for efficient learning and convergence. Multiple scaling techniques such as normalization and standardization to scale the input and output variables are widely supported by the machine learning libraries. Most of the scaling techniques scale the data to be in the range of [0,1].

Data normalization technique converts the data to be in the range of 0 to 1. In order to perform this, the following way is adopted.

$$y = (x - min)/(max - min). \tag{3.19}$$

Here, the unscaled value is represented by x, which has a minimum of min and maximum value of max. The scaled value is represented by y. Libraries such as scikit extend support to perform the data normalization.

```
from sklearn.preprocessing import MinMaxScaler
data = iris.dat
scaler = MinMaxScaler()
scaler.fit(data)
normal_value = scaler.transform(data)
```

Here, the `MinMaxScaler` is used to estimate the minimum and maximum values from the whole training dataset. The transform function normalizes the data into the desired range. The range of scaling can be specified in `MinMaxScaler` as follows:

```
scaler = MinMaxScaler(feature_range=(-1,1))
```

Here, the range of the data will be set between -1 and 1.

Standardization is another data scaling technique whose goal is to rescale the distribution of the values such that the mean of the data distribution is 0 with a standard deviation of 1. The underlying assumption is that the data distribution closely fits the Gaussian distribution. The standardization is performed as follows:

$$y = (x - mean)/(standard\_deviation). \tag{3.20}$$

Here $mean$ and $standard\_deviation$ represent the mean and standard deviation of the data x, respectively. Scikit library supports the data standardization through `StandardScaler` function as shown in the following code snippet:

```
from sklearn.preprocessing import MinMaxScaler
data = iris.data
scaler = StandardScaler()
scaler.fit(data)
normal_value = scaler.transform(data)
```

Batch normalization is another form of normalization technique, which not only scales the inputs and outputs but also scales the values and weights in the intermediate layers, leading to better learning efficiency and convergence.

```
from keras.models import Sequential
from keras.layers import Activation, BatchNormalization, Dense
model = Sequential
model.add(Dense(32))
model.add(BatchNormalization())
model.add(Activation('relu'))
...
```

Multiple Optimum Points

Given the fact that the ANN is primarily focused on minimizing the loss function during the training phase, existence of multiple optimum points can lead to inefficient training. Furthermore, searching solely for the global optima can lead to non-convergence or large training times. As a solution, one can utilize opting for the local minima as the local and global minima can be close and furthermore, the neural networks are resilient to errors.

3.4.7 *Instances of Deep Neural Networks*

Previously presented MLP technique assumes there is a single hidden layer, often termed as shallow neural networks. Such technique is sufficient as long as the dataset is small and has low levels of noise and irregularities. But, the presence of noise and large datasets adversely affect the performance of the traditional MLP techniques. Deep learning has been introduced to address these issues.

Deep learning [42] can be seen as an extension traditional MLP with abstractions and processing at multiple levels. Deep neural networks are an extended version of neural networks with an increased number of processing layers (hidden layers). Different deep learning architectures [42, 43] including convolutional neural networks [44, 45], recursive neural networks [46, 47], deep stacking networks [48, 49], and so on are introduced depending on the application, with architecture, i.e., the connectivity between and within layers, and the kernels as the main differentiating factor. For instance, deep stacking networks are a straightforward extension of traditional 3-layer networks. They contain a stack of hidden layers with the outputs of one layer forming the inputs to the succeeding. As such, the weights for the individual hidden layer networks are calculated. This hierarchical processing has the advantages of abstracting the input data at different levels.

Convolutional Neural Networks

As the processing of images of large feature vectors that comprise few thousand of input features will increase the computational complexity of the deep neural networks or MLPs, convolutional neural networks are introduced. A Convolutional Neural Network (CNN) is a variant of feed-forward neural networks with deep learning [50–53]. The input provided for CNN will be tiled into several smaller segments, with two successive segments having an overlap. The amount of overlap can be modified depending on the application and the input data size. Each input neuron/node will be fed with this smaller segment of input data and convoluted with a mask. A mask is a filter with which the convolution is performed. Consider an object recognition task, the mask could be a small matrix that represents a

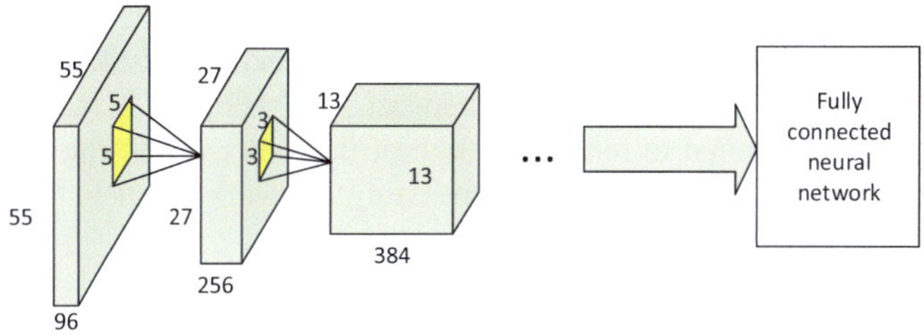

Fig. 3.15 Convolutional neural network architecture

subset of the image. The main necessity of masks also termed as filters is for extracting different levels of features in image processing applications. A similar set of parallel layers are arranged with each layer having specific masks, i.e., different filters. Further, the output from each layer is transferred to the next layer. Thus, the whole CNN comprises a set of parallel layers arranged in a serial manner. The output of each node is smaller in dimension compared to the input. This dimension reduction continues. Finally, this reduced dimensional output is provided to a fully connected network, i.e., a traditional neural network for final processing. CNN finds applications for multi-dimensional data classification and object recognition.

A convolution-based neural network implementation is shown in Fig. 3.15. The input is convoluted with different masks at the beginning states, and the convolution formed with different masks are concatenated. The yellow color rectangle represents the mask or the activation function used in the layer. This procedure is repeated and the output is finally obtained by providing the convoluted results to a fully connected neural network. It needs to be noted that the input and outputs of the CNN layers are often termed as input feature map (*ifmap*) and output feature map (*ofmap*). The filters are termed as activation maps. The process of convolution is mathematically defined as follows:

$$O[m,n] = \sum_j \sum_k h[j,k] f[m-j, n-k]. \tag{3.21}$$

Here, the $o(m, n)$ denotes the output feature map of the convolution of the input feature map f and kernel h. In the above scenario, the convolution is a two-dimensional convolution, i.e., the input feature map has only single channel. In many applications, the input can comprise multiple channels, thus a 3D convolution is performed, defined as follows:

$$o[m,n,c] = \sum_i \sum_j \sum_k h[i,j,k] f[m-i, n-j, c-k]. \tag{3.22}$$

Here, the c represents the channel index. For instance, for RGB images, there are three channels at the input.

As the convolution operations can lead to large dimensional data, it is advised to reduce the dimensionality for minimizing the complexity. In order to reduce the dimension, pooling stages are used in between. For example, a moving average filter, mean filter, applied to reduce the size of the data can be seen as a Pooling stage. CNNs are widely used in image processing applications where the amount of data to be processed is often large.

Example 3.14 (Convolutional Neural Network)
Problem: Classify the `Fashion_MNIST dataset` into different classes using a CNN.
Solution: We have 10 different classes in `Fashion_MNIST` dataset each class representing a piece of clothing. With the help of CNN, we can classify this image data into their respective classes. For which we need to load the training and test set from the dataset, reshape the size of each image so that we can train them through a defined CNN model.

A simple code snippet for CNN is shown below.

```
#Importing Libraries
import tensorflow as tf
import numpy as np
import matplotlib.pyplot as plt

#Loading Data of Fashion MNIST dataset
fashion_mnist = tf.keras.datasets.fashion_mnist
(train_images, train_labels), (test_images, test_labels) = fashion_mnist.
     load_data()

#Defining the class names
class_names = ['T-shirt/top', 'Trouser', 'Pullover', 'Dress', 'Coat',
               'Sandal', 'Shirt', 'Sneaker', 'Bag', 'Ankle boot']

#Reshaping the training and test set
train_images = train_images / 255.0
test_images = test_images / 255.0

#Defining the CNN model
model = tf.keras.Sequential([
    tf.keras.layers.Flatten(input_shape=(28, 28)),
    tf.keras.layers.Dense(128, activation='relu'),
    tf.keras.layers.Dense(10)
])

#Compiling and fitting the model
model.compile(optimizer='adam',
              loss=tf.keras.losses.SparseCategoricalCrossentropy(from_logits=
     True),
              metrics=['accuracy'])

model.fit(train_images, train_labels, epochs=10)

#Evaluating the model
test_loss, test_acc = model.evaluate(test_images,  test_labels, verbose=2)
test_acc

> 0.86949
```

The above code is a snippet for Alexnet implementation using the Keras framework. As shown, one can add convolution layers using the `add` command. Furthermore, post convolution layer, batch normalization and max-pooling are performed to reduce the dimensionality. Some of the frameworks such as Keras and Tensorflow allow users to use the state-of-the-art directly without requiring to design the network from scratch.

Radial Basis Function (RBF) Neural Networks

Radial basis functions (RBF) have proven their efficiency in many applications such as predictions, classifications, approximations, system controls, and many others [12, 13, 54–56]. [57] proposes a neural network with RBF as activation functions. Similar to traditional neural networks, the architecture consists of three kinds of layers. However, the activation functions in the hidden layers use RBF, and the output layer simply can be seen as the linear combination of outputs from the last hidden layer, i.e., RBFs and the neuron parameters such as bias. Different applications of this hybrid technique can be seen as fault detection [58, 59], control optimization [60], and so on.

```
class RBFNet(object):
    """Implementation of a Radial Basis Function Network"""
    def __init__(self, k=2, lr=0.01, epochs=100, rbf=rbf, inferStds=True):
        self.k = k
        self.lr = lr
        self.epochs = epochs
        self.rbf = rbf
        self.inferStds = inferStds

        self.w = np.random.randn(k)
        self.b = np.random.randn(1)
```

Recurrent Neural Networks

So far we have discussed different deep neural networks; however, most of the aforementioned techniques lack the temporal information. To consider the temporal information and understand the temporal variations, different classes of DNNs are introduced. Recurrent neural networks (RNNs) [61–63] are deep neural networks with the notion of memory; hence, they remember the previous state of the network and can be seen as a class of Elman networks. The state of a node is defined by the input weights of the neuron. An RNN is modeled as

$$h_t = f(h_{t-1}, x_t), \tag{3.23}$$

where h_t represents the state of the neural network at time t, x_t is the input at time t, and $f(\cdot)$ represents the next-state function. The state of a neural network is defined as the weights in the network.

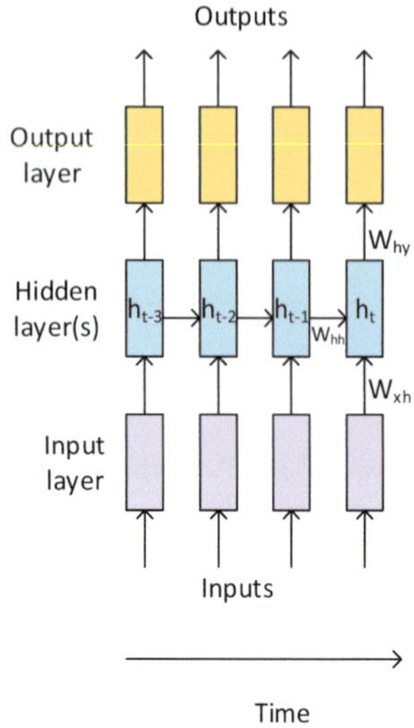

Fig. 3.16 Recurrent neural network architecture

Figure 3.16 illustrates the RNN. The X-axis represents the notion of time and Y-axis represent the network. Each block can be seen as a layer (or several layers) in the network, i.e., a set of nodes. The current state and the previous state of the hidden layers determine the output at a time instant. In the following equation, the hyperbolic tangent is considered as the activation function of the hidden layer node, and it takes the product of the previous state (h_{t-1}) and a weight vector (W_{hh}) along with the input x_t and the weight vector W_{xh}. Based on the current state and the output weight vector W_{hy}, the output y_t is determined.

$$\begin{aligned} h_t &= \tanh(W_{hh}h_{t-1} + W_{xh}x_t) \\ y_t &= W_{hy}h_t. \end{aligned} \tag{3.24}$$

Note, that the number of states that are considered to determine h_t is variable. In this example, only the previous state (h_{t-1}) is considered, but more history states $h_{t-1} \cdots h_{t-k}$ could be used. If a large number of previous states is considered, the system may require a significant amount of memory.

In general, the current state h_t with $tanh()$ activation functions can be given as

$$h_t^l = \tanh\left(W^l \begin{pmatrix} h_t^{l-i} \\ h_{t-1}^l \end{pmatrix}\right), \tag{3.25}$$

where l represents the hidden layer number. W^l is the weight vector of the current layer (W^l can be seen as $[W_{xh} W_{hh}]$, for this case, the h_t^{l-1} is the input x_t, as only one hidden layer is used). This indicates that the current state is derived based on the previous state of the current layer and the current state of the one layer below the hidden layer in the hierarchy. It is not necessary to consider the state of lower-level hidden layers of all the times, which, however, depends on the application.

RNNs are employed widely in applications like image captioning, predicting strings, predicting words or characters, language translation, with training set comprised of few words or characters [61, 64], and so on.

Example 3.15 (Recurrent Neural Networks)
Problem: Construct a model that can do image classification of malignant and benign tumor images.
Solution: To build a model that can do the image classification of malignant and tumor images, we not only need a CNN classifier but also RNN to add the memory element, for making the connection between various images. To do this we take a pretrained model ResNet50 on Imagenet and stack RNN on top of it.

```
#Example showing the combination of CNN and RNN for image classification

from keras.applications.resnet50 import ResNet50
from keras.models import Model, Sequential
from keras.layers import GlobalAveragePooling2D, Dense, RNN

#Initializing CNN pretrained model for transfer learning
pretrained_model = ResNet50(weights='imagenet', include_top=False, pooling=
    None)

x = pretrained_model.output
x = GlobalAveragePooling2D()(x)
pretrainedmodel.outputs = Dense(1024, activation='relu')(x)

#Stacking RNN on the pretrained CNN model
X = Model(inputs=base_model.input, outputs=pretrainedmodel.outputs)
model = Sequential()
model.add((X))
model.add(RNN(64, return_sequences=True, stateful=True))
model.add(Dense(N, activation='softmax'))
```

The above code snippet explains the model architecture that uses a pretrained CNN model and stacking it on an RNN model for image classification. CNN adds the features extracted in the images and RNN combines them with memory elements between the images. Based on this information, CNN and RNN are useful in applications related to medical image processing.

Vanishing Gradient Problem

As we have seen in Stochastic Gradient Descent, information travels from front input neuron to output neuron, the error is calculated and each neuron's weight is updated by using backpropagation. Similar to this, in RNN we have a gradient updation process, but here the information travels in terms of time and we can calculate the error at each time point. The problem in RNN is that the weight update process does not occur one neuron at a time, but all the neurons have participated in the calculation of the error function. So weights of all the neurons far back in time are updated at a time. In backpropagation with the increase in the number of epochs, the number of updations increases and the gradient gets smaller. At some point, the gradient gets much smaller, and at some point, the gradient is not useful for updating weights of neurons that are way back in time. Due to which lower half neurons are not updated, and for the prediction, the model depends on all the neurons. Thus, the performance of the model degrades because of the untrained neurons from the vanishing gradient problem.

Long-Short-Term Memory Neural Networks

Long short term-memory (LSTM) neural network [62, 65, 66] is an extension of RNNs. The main issue in RNNs is the need for large storage to remember a large number of states and computations performed on each of the previous states. As a remedy, LSTM is activated with forget and remember gates. During the training the network evaluates the impact of the previous and current states. The current input and the states that have impact on the output are kept while the rest is discarded. It can be mathematically given as

$$\begin{pmatrix} i \\ f \\ o \\ g \end{pmatrix} = \begin{pmatrix} sigm \\ sigm \\ sigm \\ tanh \end{pmatrix} W^l \begin{pmatrix} h_t^{l-i} \\ h_{t-1}^l \end{pmatrix}$$

$$c_t^l = f \odot c_{t-1}^l + i \odot g$$

$$h_t^l = o \odot tanh(c_t^l). \tag{3.26}$$

The notations in the aforementioned equation are the same as that in Eq. (3.24). Here, $\odot$ represents the point-wise computations. LSTMs are very popular in language processing. Similar to RNNs, LSTMs combined with CNNs are employed for image captioning [67] purposes.

Example 3.16 (Long-Short-Term Memory Neural Networks)
Problem: Predict the next character that may occur in a sentence from the IMDB dataset.

(continued)

Example 3.16 (continued)
Solution: To solve this problem, we need to start by importing the IMDB dataset, then building the model using LSTM, which can observe the pattern and predict the future characters. Then, we fit the input padded `X_train` data onto the defined model.

```
# Importing the libraries
import numpy as np
from tensorflow import keras
from tensorflow.keras import layers

max_features = 50000
max_len = 200

# Input for variable-length sequences of integers
inputs = keras.Input(shape=(None,), dtype="int32")
# Embed each integer in a 128-dimensional vector
x = layers.Embedding(max_features, 128)(inputs)

#Adding the LSTM layers
x = layers.LSTM(64, return_sequences=True)(x)
x = layers.LSTM(64)(x)

# Adding a classifier for the model
outputs = layers.Dense(1, activation="sigmoid")(x)
model = keras.Model(inputs, outputs)
model.summary()

#Importing the data from IMDB dataset
(x_train, y_train), (x_val, y_val) = keras.datasets.imdb.load_data(num_words=
    max_features)

#Padding the sentences in training and validation set to maximum length
x_train = keras.preprocessing.sequence.pad_sequences(x_train, maxlen=max_len)
x_val = keras.preprocessing.sequence.pad_sequences(x_val, maxlen=max_len)

model.compile("adam", "binary_crossentropy", metrics=["accuracy"])
model.fit(x_train, y_train, batch_size=32, epochs=2, validation_data=(x_val,
    y_val))
```

3.5 Support Vector Machines

Support Vector Machines (SVMs) [13, 68, 69] are another class of ML, which is used for classification and regression analysis. An SVM builds a model based on the input data with labels such that it could be classified as clear as possible (as provided in labels). Here, the label indicates the class or a group to which the input data belongs to. For example, consider an ECG signal, here the label could be the name of the component at a particular time instant. Every new input is mapped to the corresponding category. SVMs operate on vectors rather than individual points, making them robust. For estimating the distances, SVMs make use of dot products.

In SVMs, the objective is to find a hyper-plane that could separate the positive and negative samples (considering a simple binary classification) with a clear boundary. Let us assume a sample space with vector w being normal to the hyper-plane that separates the dataset and a constant b. Then the following equation has to be satisfied:

$$t_i(w^T \cdot x_i + b) \geq 1,$$

where x_i is a sample, and t_i is the target output, i.e., the class to which it should belong. $t_i = +1$ for the sample belonging to positive class and $t_i = -1$ for the sample belonging to the negative class. The width of the hyper-plane is given by

$$max(\frac{1}{\|w\|}),$$

where w represents the normal vector to the hyper-plane. Considering all these assumptions, it boils down to optimizing the following equations that could be performed using Karush–Kuhn–Tucker (KKT) conditions [70].

$$\begin{aligned} \lambda_i(1 - t_i(w^T \cdot x_i + b)) &= 0 \\ 1 - t_i(w^T \cdot x_i + b) &\leq 0 \\ \lambda_i &\geq 0, \end{aligned} \tag{3.27}$$

where λ_i is a Lagrange multiplier. The main objective here is to find w and b, such that for any sample, the classification can be carried out.

A simple example, explaining this SVM method is shown in Fig. 3.17. The goal is to perform the classification of new input based on the existing data. In this figure, the positive samples are indicated by circles and negative with squares. We first find the normal vector w to the median of the hyper-plane and the width of the hyper-plane. Further, the hyper-plane is formed and the classification can be easily performed by calculating the distance of the sample to the plane, as given by constraint 1 in Eq. (3.27).

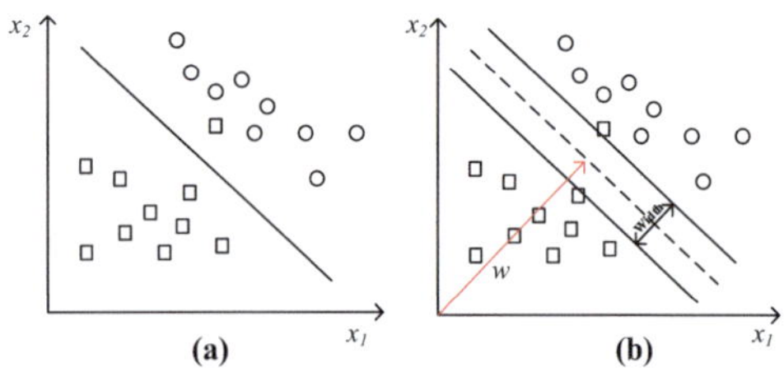

Fig. 3.17 Binary classification with (**a**) other classifiers; (**b**) SVM

Support vector machines (SVMs) [71–73] are employed in classification applications. The main objective of SVM is to compute the hyper-plane based on KKT conditions as given in Eq. (3.27).

The basic process of SVM starts with training of labeled input data (labels represent the class to which the data belongs). Further, based on the training data, the normal vector w is calculated. Once the normal vector is obtained, the position of new input data (test data) relative to the obtained hyper-plane has to be computed, and based on the relative position, the class of the new input is determined.

The traditional SVM classifiers are capable of performing binary classification, i.e., only two possible labels are possibles for a given input. Consider a simple case of binary classification, i.e., the input data has two classes. A hyper-plane will be formed in between those two classes. Class A is on the left side of hyper-plane and class B to the right side, similar to that in Fig. 3.17. Any new input data whose distance is negative, i.e., left side of the hyper-plane belongs to class A and the input data whose distance to the hyper-plane is positive belongs to class B.

In the above code snippet, the SVM classifier is used from the Scikit Learn library. The `X` and `y` represent the input data and corresponding labels. The SVM with the linear kernel is used in the above snippet for fitting the data and further use for inference.

As a simple illustrative example, consider the problem of images that contain either apple or pear picture and labeled accordingly. Given a new image the task of SVM is to classify the image into one of these classes. Once all the input images are converted into data matrices, the SVM takes the previously seen images and corresponding labels to fit the model, as described by the `fit(X,y)`. For predicting the new (unseen) image, the image is converted into a matrix and fed to the trained model. The SVM uses a kernel function (Linear) in the above code snippet to obtain the manifold that best separates both the classes. Different kernels can be used, which are discussed below.

Example 3.17 (Support Vector Machines)
Problem: Classify the malignant and benign data from cancer dataset using SVM classifier.
Solution: To do this, import the required libraries as shown in code below. Then load the cancer dataset from datasets in sklearn. Then we need to split the dataset to train and classify the data using the SVM classifier. Next we predict on the test train and find the accuracy of the model.

```
#Importing required libraries
from sklearn import datasets
from sklearn.model_selection import train_test_split
from sklearn import svm
from sklearn import metrics

#Loading the cancer dataset
```

```
cancer_data = datasets.load_breast_cancer()

# Spliting dataset into training set and test set
X_train, X_test, y_train, y_test = train_test_split(cancer_data.data,
     cancer_data.target, test_size=0.3,random_state=109)

#Creating a svm Classifier model
clf = svm.SVC(kernel='linear') # Linear Kernel

#Train the model using the training sets
clf.fit(X_train, y_train)

#Predict the response for test dataset
y_pred = clf.predict(X_test)

# Model Accuracy: how often is the classifier correct?
print("Accuracy:",metrics.accuracy_score(y_test, y_pred))

> 0.96491
```

3.5.1 SVM Kernels

SVM algorithms utilize different sets of Kernels to build the classification model. The performance of the SVM classification highly depends on the type of Kernel used. Kernel is a mathematical function that takes data as input and transforms it into the required format for classification. Kernel functions output the dot product between the input point and a point in the feature space. Different SVM kernels include linear, nonlinear, and polynomial kernels. Widely used SVM kernel is the radial basis function due to its localized and finite response. Some of the popularly used SVM kernels include:

1. Polynomial Kernel: Used for image processing applications. It is mathematically given by

$$k(x, y) = (x \cdot y + 1)^d. \tag{3.28}$$

   Here, the x, y represent two data points and d represents the polynomial degree.
2. Gaussian kernel: Widely used for general data for which distribution is assumed to be Gaussian. Also used when no prior data assumptions are available. It is mathematically given by

$$k(x, y) = exp(-\frac{||x - y||^2}{2\sigma^2}). \tag{3.29}$$

   Here, σ denotes the standard variance.
3. Gaussian radial basis function (RBF): This is similar to the Gaussian kernel and is used when no prior data information is available. It is mathematically given as

$$k(x, y) = exp(-\gamma||x - y||^2).for\gamma > 0 \tag{3.30}$$

4. Laplace RBF kernel: Similar to the above two kernels, Laplacian RBF kernel is also deployed when no prior information regarding the data is available. The kernel is mathematically defined as

$$k(x, y) = exp(-\frac{||x - y||}{\sigma}. \tag{3.31}$$

5. Hyperbolic Tangent kernel: This is used in neural networks, given by

$$k(x, y) = tanh(a \cdot x \cdot y + c), \tag{3.32}$$

where $a > 0$ and $c < 0$.

6. Sigmoid Kernel: The kernel takes the shape of a sigmoid function and is used as a proxy for the neural networks. This is mathematically defined as

$$k(x, y) = tanh(\alpha x^T y + c). \tag{3.33}$$

7. ANOVA RBF Kernel: To meet the requirements of the regression problems, ANOVA kernels are used. It is mathematically given by

$$k(x, y) = \sum_{k=1}^{n} exp(-\sigma(x^a - y^a)^2)^d. \tag{3.34}$$

Here, a denotes the degree.

8. Linear Spikes Kernel: To perform the classification or regression of sparse data vectors, this kernel is deployed. Text classifications are one of the applications that use this kernel. This is mathematically given by

$$k(x, y) = 1 + xy + xy \min(x, y) - \frac{x + y}{2} \min(x, y)^2 + \frac{1}{3} \min(x, y)^3. \tag{3.35}$$

3.5.2 Multiclass Classification

SVMs are primarily introduced for binary classification. However, the SVMs can be extended to multiclass classification as well. Unlike neural networks where training the data from multiple classes and modifying the architecture of the neural network (specifically the output later) aids in performing multiclass classification, SVMs need a different approach.

In the case of SVMs, the multiclass classification happens through one vs all or one vs rest approach. In this approach, the initial classification is modified to be a binary classification, where the classifier first considers the classification problem into two classes where one is related to a specific class and the rest of the classes are grouped into one class. This process happens iteratively to enable multiclass classification.

SVMs are primarily supervised learning algorithms that are trained to classify unseen or unknown data. Similar to neural networks, SVMs are deployed in a wide range of applications. Some of the common applications of SVM include face detection, text classification, object detection and classification and bioinformatics such as cancer detection or pathology cell detection. Other applications include handwriting recognition and predictive control applications. The efficiency of the SVMs depends on the application and kernel used.

Example 3.18 (Multiclass SVM Classification)
Problem: Perform a multiclass classification on the Iris dataset using SVM classifier.
Solution: To perform the multiclass classification on Iris data, we first need to import the libraries. Then, we load the dataset from datasets available through sklearn. We split the dataset in training and test sets, then define the model that is fitted onto the training set and do predictions on the test set. We then observe the accuracy and confusion matrix; confusion matrix gives the information of number of correct and wrong classified values per class.

```
# importing necessary libraries
from sklearn import datasets
from sklearn.metrics import confusion_matrix
from sklearn.model_selection import train_test_split
from sklearn.svm import SVC

# loading the iris dataset
iris = datasets.load_iris()

X = iris.data
y = iris.target

# dividing X, y into train and test data
X_train, X_test, y_train, y_test = train_test_split(X, y, random_state = 0)

# training a linear SVM classifier
model = SVC(kernel = 'linear', C = 1).fit(X_train, y_train)
y_pred = model.predict(X_test)

# model accuracy for X_test
accuracy = model.score(X_test, y_test)
print(accuracy)

# creating a confusion matrix
conf_m = confusion_matrix(y_test, y_pred)
print(conf_m)

> 0.9736842105263158

> [[13  0  0]
   [ 0 15  1]
   [ 0  0  9]]
```

3.6 Ensemble Learning

Despite individual classifiers or predictors above discussed are efficient for a wide range of applications, the presence of noisy data or smaller amounts of data or features can lead to inefficiency of the individual classifiers. To address such scenarios, ensemble learning is introduced. Ensemble learning refers to using multiple classifiers trained on different instances of data for the purpose of classification. It fully exploits complementary information of different classifiers to improve the decision accuracy and performance. In the field of machine learning, the ensemble learning and joint decision procedure are widely used to devise learning methods to achieve more accurate predictions and stronger generalization performance. Multiple ensemble learning techniques exist, which are discussed below.

The idea of ensemble learning is to use the weak learners to build blocks to design more complex models. This method tries to reduce the bias and/or variance of these weak learners by combining them together to form a stronger learner. Ensemble learning imitates our nature to seek several opinions before making a decision. It involves designing two or more classifiers (or regressors) and combining them to obtain a classifier (regressor) that outperforms each one individually.

There are three primary ways of combining weak learners to design an ensemble meta-algorithm:

- Bagging: A homogeneous set of weak learners that are trained independently and in parallel are combined using some form of averaging process called aggregation.
- Boosting: A homogeneous set of weak learners that are trained sequentially in an adaptive way (i.e., a base model depends on the previous ones) are combined using a deterministic strategy.
- Stacking: A heterogeneous set of weak learners that are learned in parallel are combined by training a meta-model to output a prediction based on the different weak model predictions.

3.6.1 Bagging

Bagging, also known as Bootstrap Aggregation is an ensemble learning model that is used for classification and regression problems. It is a statistical prediction technique where a statistical value like a mean is estimated from multiple random samples of training data that are drawn with replacement and used to train different ML models. Each model is then exploited to make a prediction, and the results are averaged to give a more robust and generalized prediction. Figure 3.18 illustrates the overview of bagging model. Bagging is a technique that is best used with models with low bias and high variance, in which the predictions of base learners are highly dependent on the data from which they were trained. The most used algorithm for bagging that fits the requirement of high variance is decision trees.

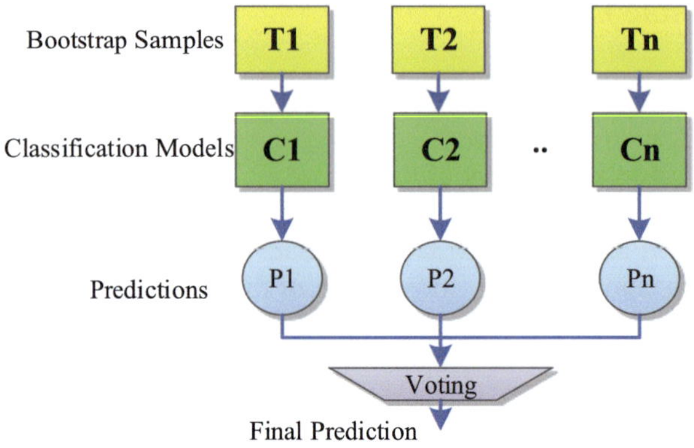

Fig. 3.18 Procedure of bagging

Bagging algorithm is performed as follows: a) Create multiple datasets out of existing dataset; b) build multiple classifiers—based on the created multiple subsets; and c) combine the predictions—the predictions from individual classifiers are combined using statistical techniques such as mean, median, or deviation depending on the nature of problem. Using a large number of classifiers can lead to better performance.

The Scikit implementation of the Bagging technique is presented below. Here, the number of samples is set as 1000 with each having 8 features. The number of estimators is set to 20.

Example 3.19 (Bagging)
Problem: Consider the dataset `load_digits` and perform bagging for 100 estimators classification. Also, display the classification accuracy of the model.
Solution: The first step to solve the problem is to load `load_digits()` dataset from sklearn dataset library using `load_digits()` command, followed by splitting the dataset into train and test sets in ratio of 80:20. Bagging is performed using `BaggingClassifier` and accuracy is obtained by `score` function. `n_estimators`=100 is used to specify the number of estimators. The code that performs above specified task is shown below.

```
from sklearn.ensemble import BaggingClassifier
from sklearn.model_selection import train_test_split
from sklearn.datasets import load_digits

digits = load_digits(n_class=10)
X = digits.data
y = digits.target
```

```
X_train, X_test, y_train, y_test = train_test_split(X, y, test_size=0.2,
    random_state=50)
cls = BaggingClassifier(n_estimators=100).fit(X_train, y_train)
cls.score(X_test, y_test)

> 0.9222
```

3.6.2 AdaBoost

Adaptive Boosting (AdaBoost) is one of the most commonly used ensemble learning methods for enhancing the performance of ML algorithms. In AdaBoost, each base classifier is trained on a weighted form of the training set in which the weights depend on the performance of the previous base classifier. Each instance in the training data is weighed with initial weight set to $weight(x_i) = 1/n$, where x_i denotes the i-th training sample and n denotes the number of training instances. Once all the base classifiers are trained, they are combined to produce the final classifier. Each training instance in the dataset is weighted and the weights are updated based on the overall accuracy of the model and whether an instance was classified correctly or not. Subsequent models are trained and added until a minimum accuracy is achieved, or no further improvement is possible. In simple words, a set of weak classifiers or models are added together to form a strong model. The process continues until desired performance is achieved. Figure 3.19 depicts the AdaBoost technique, where the individual weak models are added sequentially to form a strong and efficient model.

In AdaBoost, for an incoming data sample that needs to be classified, each weak learner predicts the value as 1 or -1. Further, as aforementioned, each of the predictions from the individual classifiers is weighted, and the prediction for the ensemble model is the weighted sum of the individual predictions. Including a large number of noisy samples or outliers can enforce the AdaBoost to non-convergence and inefficiency. Thus, preprocessing of the data is required to address these concerns in AdaBoost. A majority of today's ensemble learning techniques are built on top of the AdaBoost algorithm, most notably stochastic gradient boosting machines.

Example 3.20 (AdaBoosting Classification)
Problem: Consider the dataset `load_digits` and perform Gradient boosting to train weaker models with 100 number of estimators. Also, finding the classification accuracy of the model and estimate the expected error using cross-validation.

(continued)

Fig. 3.19 Block diagram of Adaboost ensemble learning

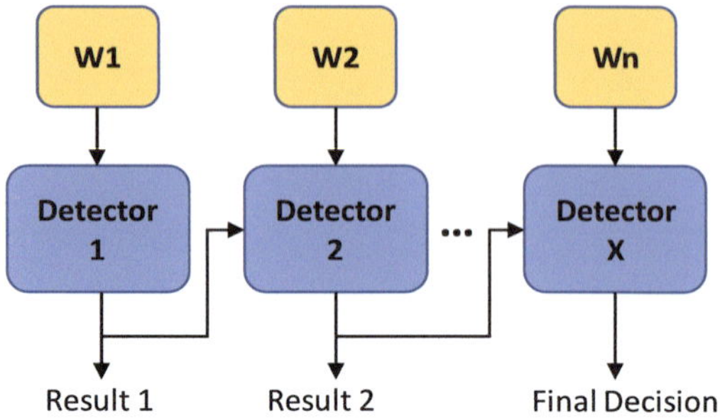

Example 3.20 (continued)
Solution: The dataset is imported from sklearn dataset library, training is performed with train and test dataset split in a ratio of 80:20. Boosting is performed using `AdaBoostClassifier` and accuracy is obtained by `score` function.

```
from sklearn.ensemble import AdaBoostClassifier
from sklearn.datasets import load_digits

digits = load_digits(n_class=10)
X = digits.data
y = digits.target

X_train, X_test, y_train, y_test = train_test_split(X, y, test_size=0.2,
     random_state=50)

clf = AdaBoostClassifier(n_estimators=100, random_state=0)
clf.fit(X_train, y_train)

from sklearn.model_selection import cross_val_score
all_accuracies = cross_val_score(estimator=clf.fit(X_train, y_train), X=
     X_train, y=y_train, cv=5)
#Print accuracies
print(all_accuracies)
print(all_accuracies.mean())

> [0.35763889 0.26041667 0.27526132 0.26132404 0.24390244]
> 0.2797
```

The above code snippet shows the AdaBoost-based classification using Scikit libraries. The number of estimators is set to 200 with each data sample having 10 features.

3.6.3 Bootstrap

Bootstrap is another ensemble learning technique that performs data sampling to estimate the statistical properties through data replacement, unlike cross-validation. This is primarily used for data splitting that can be used to train and test different classifiers and use them for ensembling. The statistical properties include mean or standard deviation. Bootstrap method is a technique to estimate the dataset characteristics despite having few data samples. This technique involves resampling of the data iteratively to estimate the data statistics. Such iterative process with sample replacement policy facilitates that each data sample be involved in estimating the data statistics more than once.

The two main variables that impact the performance of the bootstrap are the size of the samples and the number of repetitions of the data samples. Size of the samples is similar to the traditional ML classifiers that signify the size of the original dataset. Similarly, the number of repetitions signifies the number of iterations used for estimating the data statistics such as mean, variance, and standard deviation or errors. As a general rule, 20 to 30 repetitions are set; however, it could even be few hundreds or thousands depending on the size of dataset. Also, the number of iterations could be few hundreds to thousands depending on the time complexity.

Once the distribution of the data is determined, it could be used for training or testing the classifiers. The bootstrap process is performed as follows: (a) Determine the number of bootstrap samples; (b) extract the samples randomly and add it to the observation, if the number of samples is less than the threshold; (c) the bootstrap can estimate the properties of the data. These steps are performed iteratively.

Example 3.21 (Bootstrap)
Problem: Consider the scikit library function `Bootstrap` and perform resample for the bootstrap technique.
Solution: The bootstrap splits the data with 9 values into three sets where each set has 5 training samples and 4 test samples. For this purpose, one can utilize the `bootstrap` command from the Sklearn library.

```
from sklearn import cross_validation
b = cross_validation.Bootstrap(9, random_state=0)
Bootstrap(9, num_bootstraps=3, num_train=5, num_test=4, random_state=0)
```

3.6.4 Gradient Boosting

Gradient Boosting is another popular boosting algorithm just like AdaBoost; it works by sequentially adding predictors to an ensemble, each one correcting its predecessor. However, instead of tweaking the instance weights at every iteration, gradient boosting tries to fit the new predictor to the residual errors made by the previous predictor. Gradient Boosting is a procedure that can be used for both regression and classification problems in a variety of areas including Web search ranking and ecology.

At each stage m, $1 \leq m \leq M$, assume an imperfect model $F_m(x)$. Gradient boosting improves on $F_m(x)$ by constructing a new model that adds an estimate h(x) to produce a better model,

$$F_m + 1(x) = F_m(x) + h(x) = y. \tag{3.36}$$

Here, Gradient boosting fits h(x) to the residual $y - F_m(x)$. And $F_m(x)$ will attempt to correct the errors of the predecessor $F_m - 1(x)$.

The scikit library provides a library function with the methods to perform both the classification and regression via gradient boosted decision trees.

Example 3.22 (Gradient Boosting Classification)
Problem: Consider the dataset `load_digits` and perform Gradient boosting to train weaker models with learning rate 0.1 and 100 estimators. Also, find the classification accuracy.
Solution: The dataset is imported from sklearn dataset library, training is performed with train and test dataset split ratio of 60:40. Boosting is performed using `GradientBoostingClassifier` and accuracy is obtained by `score` function.

```
from sklearn.datasets import load_digits
from sklearn.ensemble import GradientBoostingClassifier
from sklearn.model_selection import train_test_split

digits = load_digits(n_class=10)
X = digits.data
y = digits.target

X_train, X_test, y_train, y_test = train_test_split(X, y, test_size=0.4)
clf = GradientBoostingClassifier(n_estimators=100, learning_rate=1.0,
    max_depth=1, random_state=0).fit(X_train, y_train)
clf.score(X_test, y_test)

> 0.840055
```

Example 3.23 (Gradient Boosting Regression)
Problem: Consider the dataset `load_digits` and perform Gradient boosting to train weaker models with a learning rate 0.1 and 100 number of estimators. Also, find the classification accuracy.
Solution: The dataset is imported from sklearn dataset library, training is performed with a train and test dataset split ratio of 60:40. Boosting is performed using `GradientBoostingRegressor` and accuracy is obtained by `score` function.

```
from sklearn.ensemble import GradientBoostingRegressor
from sklearn.datasets import load_digits
from sklearn.model_selection import train_test_split

digits = load_digits(n_class=10)
X = digits.data
y = digits.target

X_train, X_test, y_train, y_test = train_test_split(X, y, test_size=0.4,
     random_state=50)

clf = GradientBoostingRegressor(n_estimators=100, learning_rate=1.0,
     max_depth=1, random_state=0).fit(X_train, y_train)
clf.score(X_test, y_test)

> 0.60865
```

3.6.5 Stacking

Instead of using trivial functions (such as hard voting) to aggregate the predictions of the predictors in an ensemble, a model is trained to perform this aggregation. Stacking is a method for combining estimators to reduce their biases. More precisely, the predictions of each individual estimator are stacked together and used as input to a final estimator to compute the prediction. This final estimator is trained through cross-validation.

Suppose we have an ensemble of three predictors performing a regression task on a new instance. Each predictor predicts a different value, and the final predictor called the blender takes these three predictions as an input and makes the final prediction. In order to train the blender shown in Fig. 3.20, first, the training set is split into two subsets. The first subset is used to train the predictors in the first layer. These first layer predictors are used to make predictions on the second training set. This ensures that the predictions are clean. A new training set is created using the three predicted values as input features and keeping the target values. And the blender is trained on this new training set. It learns to predict the target value given the first layer's predictions.

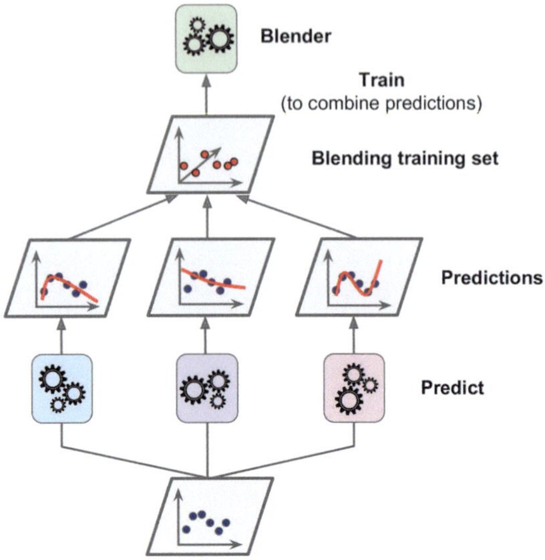

Fig. 3.20 Block diagram of Stacking ensemble learning

Example 3.24 (Stacking)
Problem: Consider the Iris dataset and perform Stacking to train with 500 estimators. Also, find the classification accuracy.
Solution: The dataset `load_iris` is imported from sklearn dataset library, training is performed on train, and test datasets split in the ratio of 60:40. Stacking is performed using `StackingClassifier`, which allows to stack the output of individual estimators like `RandomForestClassifier` `LinearSVC` with the final classifier and accuracy is obtained by `score` function.

```
from sklearn.ensemble import StackingClassifier
from sklearn.datasets import load_iris
from sklearn.model_selection import train_test_split
from sklearn.ensemble import RandomForestClassifier
from sklearn.svm import LinearSVC
from sklearn.linear_model import LogisticRegression
from sklearn.preprocessing import StandardScaler
from sklearn.pipeline import make_pipeline

dataset = load_iris()
X_train, X_test, y_train, y_test = train_test_split(dataset.data, dataset.
    target, test_size=0.2)

estimators = [('rf', RandomForestClassifier(n_estimators=10, random_state=42)
    ),('svr', make_pipeline(StandardScaler(),LinearSVC(random_state=42)))]
clf = StackingClassifier(estimators=estimators, final_estimator=
    LogisticRegression())
clf.fit(X_train, y_train).score(X_test, y_test)

> 0.96666
```

3.7 Other Machine Learning Techniques

In addition to above discussed techniques, there exist a plethora of additional classification and prediction techniques. This section provides a glimpse of few more popular supervised learning techniques that are widely used in different applications.

3.7.1 *Bayesian Model Combination*

Bayesian model combination (BMC) is another ensemble learning technique that utilizes Bayesian properties. It is seen as an extension of the Bayesian model averaging (BMA) technique. Bayesian model averaging (BMA) makes predictions using an average over several models with weights given by the posterior probability of each model given the data [15]. BMA is known to generally give better answers than a single model, obtained, e.g., via stepwise regression, especially where very different models have nearly identical performance in the training set but may otherwise perform quite differently.

In the BMC, the samples from the ensemble space are combined together instead of sampling each model individually. The samples are extracted with model weightings drawn from a Dirichlet distribution. BMC provides an advantage of better convergence compared to the BMA technique, where it is possible that all the weight could be assigned to a single model. The weights for individual models are computed using Bayes' law. The variation of the data distribution for each of the models determines the weights assigned to each of the models. In other words, equal weights are assigned to all the models if the data distribution for all the models is same. The final ensemble result can be approximated through cross-validation.

3.7.2 *Random Forest*

Random forest technique can be seen as a tree-based ensemble learning technique, where it builds an ensemble of decision trees. This is usually trained with the bagging technique, which is presumed to improve the performance of the ensemble classifier. In simple words, multiple decision trees are integrated together for better performance and to get accurate predictions. The random forest can be employed for both classification and prediction problems alike. Random forest instead of searching for pivotal features among all the trees while splitting the nodes, it searches for pivotal features among a random subset of features. This reduces the computational complexity and convergence challenges. Furthermore, this increases the diversity in the chosen features and leads to a better learning model. The diversity can be further enhanced by employing random thresholds for each feature rather than setting optimal thresholds.

Scikit learn provides support to showcase the importance of each of the features by observing the utilization of a particular feature by the trees. Furthermore, it computes the score for each feature post-training and eventually scales the results to make the sum of all the feature importance to one. Though random forest is seen as a collection of decision trees, there exist significant differences between both techniques. For a given dataset with annotations (labels), the decision tree will formulate a set of rules for the purpose of predictions or classifications. However, the random forest randomly selects the features and observations to build the decision trees and then average results. For instance, in the case of product recommendations, the decision trees look at the items viewed by the shoppers and formulate the rules for the recommendation.

In addition, deep decision trees suffer from overfitting, whereas random forests avoid overfitting by creating random subsets of features and forming smaller subsets of trees. This can be seen as analogous to pruning technique discussed earlier. However, building random forest with a large set of decision trees leads to computational complexity, thus, a trade-off has to be observed.

The following code snippet shows the implementation of the random forest algorithm using scikit library.

Example 3.25 (Random Forest)

Problem: Consider the `make_moons` dataset and use Random forest classification to build a strong classifier with a maximum depth of 5, 500 estimators and find the classification accuracy. Also, investigate the effect of the number of trees and the number of features used in the design of each tree on the performance of the classifier.

Solution: The dataset `load_moons` is imported from sklearn dataset library, training is performed on train, and test datasets split in the ratio of 80:20. Classification is performed using `RandomForestClassifier` out-of-bag estimate of ensemble classification error and accuracy is obtained by `score`, `oob_score_` function.

```
from sklearn.ensemble import RandomForestClassifier
from sklearn.datasets import make_moons

X, y = make_moons(noise=0.3, random_state=0)
X_train, X_test, y_train, y_test = train_test_split(X, y, test_size=0.2)
cls = RandomForestClassifier(oob_score=True, max_depth=5, n_estimators=500)
cls.fit(X_train, y_train)
print("accuracy", cls.score(X_test,y_test))
print("oob score", cls.oob_score_)

> accuracy 0.8
> oob score 0.9125
```

In the above code snippet, the random forest is built with 4 levels of depth. The number of samples is 2000 with each having 8 features. The model fits `x` and `y` using the fit command and uses the random forest classifier to do the fitting.

3.7.3 Tree-Based Methods

Tree-based algorithms is a popular class of supervised classification techniques used. The decision trees are a popular tree-based classification algorithm that works for both categorical and continuous input and output data variables. In the case of decision trees, the dataset is split into two or more homogeneous sets based on the differentiability of the input variables.

Consider a simple example of determining the probability that a student plays tennis in a class of 30 students. Each student has three variables, namely gender (Male or Female), class, and height (4ft to 5ft). 20 students play tennis in their leisure time. Thus, in this problem, the decision tree segregates students who play tennis based on the significant variable among the three aforementioned features. The decision tree will help in determining the pivotal variable among the tree that can be used to determine the probability of a student playing tennis.

The advantages of decision trees include ease to understand, derive the relationships, and extract the rules for categorization. Furthermore, it does not require in-depth knowledge of statistics to understand and the graphical representation makes them easy to understand. Compared to other techniques, tree-based techniques do not impose constraints on the data type. Determining the variable importance is obtained based on Gini and Chi-Square features.

Example 3.26 (Decision Trees)
Problem: Consider the Iris dataset and use Decision tree classification to build a classifier and report the classification accuracy. Also plot the trees.
Solution: The dataset `load_iris` is imported from sklearn dataset library, training is performed on the train, and test datasets split in the ratio of 80:20. Classification is performed using `DecisionTreeClassifier`, and accuracy is obtained by `score` and plotting the trees with the help of `plot_tree` function from sklearn.

```
from sklearn.datasets import load_iris
from sklearn import tree

X, y = load_iris(return_X_y=True)
X_train, X_test, y_train, y_test = train_test_split(X, y, test_size=0.2)
cls = tree.DecisionTreeClassifier()
cls = cls.fit(X_train, y_train)
print("accuracy",cls.score(X_test, y_test))
```

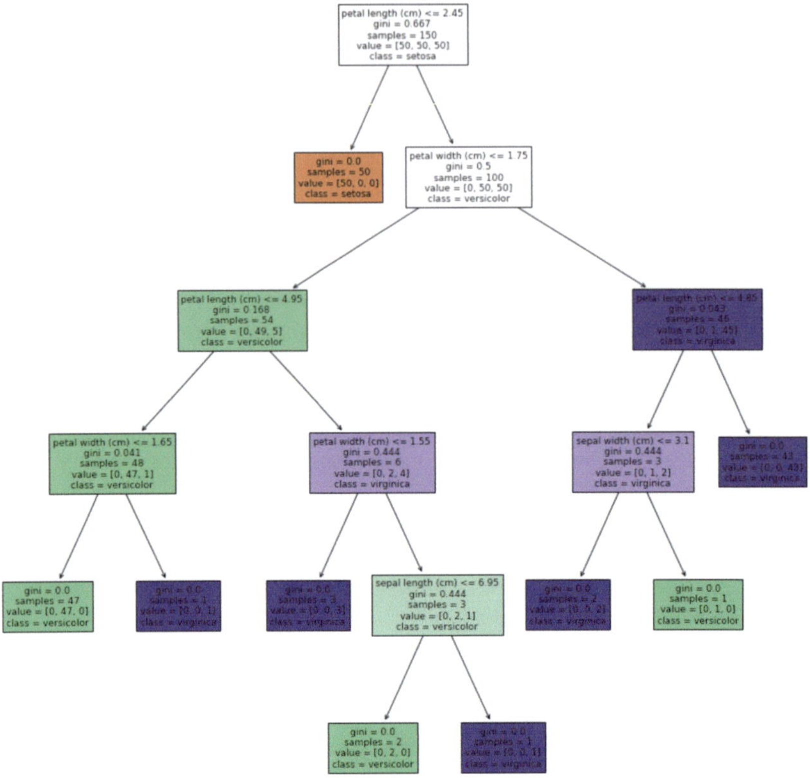

Fig. 3.21 Decision tree for classifying Iris data

```
tree.plot_tree(cls)

> accuracy 0.9
```

In the above example, the Iris dataset is used and the decision tree determines the features needed to classify the input images into different classes. The splitting of features based on the decision tree is shown in Fig. 3.21.

Example 3.27 (Extra Trees)
Problem: Consider the Iris dataset and use Decision tree classification to build a strong ensemble classifier and find the classification accuracy.
Solution: The dataset `load_iris` is imported from sklearn dataset library, training is performed on train, and test datasets split in the ratio of 80:20. Classification is performed using `ExtraTreesClassifier`, and accuracy is obtained by `score` and plotting the trees using `plot_tree` function from sklearn.

```
from sklearn.datasets import load_iris
from sklearn.ensemble import ExtraTreesClassifier

X, y = load_iris(return_X_y=True)
X_train, X_test, y_train, y_test = train_test_split(X, y, test_size=0.2)

cls = ExtraTreesClassifier(n_estimators=50, max_depth=3, min_samples_split=2,
      oob_score=True, bootstrap=True)
cls = cls.fit(X_train, y_train)
print("accuracy",cls.score(X_test, y_test))
print("oob score", cls.oob_score_)

> accuracy 0.9666666666666667
> oob score 0.95
```

3.7.4 AutoEncoder

Autoencoders are an unsupervised learning technique that utilizes neural networks for representation learning. Autoencoders encode the input data based on the correlation between the features. In other words, if the provided input data features were highly independent of each other, the compression is tedious. However, in the scenario of correlated features, the structural dependencies can be exploited for better compression.

The autoencoders are primarily used as preprocessing elements or for compressing the input data to remove the redundancies among the features and utilize the reduced features for further processing, i.e., transform high-dimensional inputs to lower dimensionality. The converse of the autoencoders is the auto-decoders that can reconstruct the data that was compressed by the autoencoders.

In contrast to the existing dimensionality reduction techniques such as principal component analysis (PCA) and other techniques, autoencoders are sensitive to the input data and its distribution, leading to accurate reconstruction. It is insensitive to the inputs, so that the model does not merely memorize the encoding nor overfit the trained data. In order to overfit the autoencoders, the regularization term is added to alleviate any memorizing or overfitting concerns. Compared to traditional PCAs, autoencoders are capable of learning the nonlinear manifolds. A manifold is defined as a continuous non-intersecting surface that separates two or more classes.

An autoencoder is composed of two parts: an encoder that converts the inputs to an internal representation, followed by a decoder that converts internal representation to the outputs . A sample architecture of autoencoder representing the chess memory experiment is shown in Fig. 3.22. One of the simplest techniques for constructing the autoencoders is to limit the number of nodes in the hidden layers of the network. Given that autoencoders encode based on the penalty, adding a regularization term will aid in minimizing the reconstruction errors.

There are two different types of autoencoders structures.

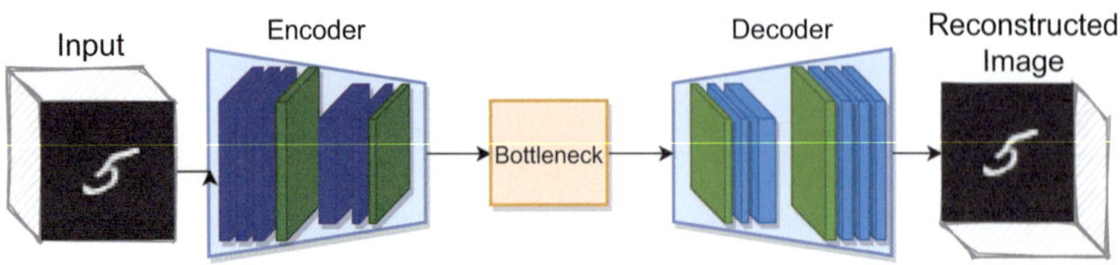

Fig. 3.22 Simple autoencoder

- Overcomplete Autoencoder: The structure has a larger hidden layer than the input layer. Each hidden unit could copy a different input component and no compression in hidden layer.
- Undercomplete Autoencoder: The structure has a smaller hidden layer than the input layer. It compresses well for the input training set.

There exist multiple types of autoencoders as defined below.

Example 3.28 (Simple Autoencoder)
Problem: Consider the MNIST dataset and design a simple autoencoder using fully connected layers and display the reconstructed images. Also, compare the reconstructed images with the original images.
Solution: First step is to load the MNIST dataset using the command `load_data`, which is imported from keras library. Design the encoder and decoder models using fully connected layers; the models can be built using `keras.Model` and to build these model based on a fully connected neural network layer `keras.layer.Dense` is used. After building the model, in order to train it, first, the model needs to compile and configure to use `Adam` optimizer, `binary_crossentropy`loss function. Training can be done by setting the hyperparameter values such as `epoch` to 10 and `batch_size` to 512. To visualize and compare the input image with the reconstructed image, `plt.imshow` command from `matplotlib` library is used (Figs. 3.23 and 3.24).

```
#loading the data set
from keras.datasets import mnist
import numpy as np
(x_train, _), (x_test, _) = mnist.load_data()
x_train = x_train.astype('float32') / 255.
x_test = x_test.astype('float32') / 255.
x_train = x_train.reshape((len(x_train), np.prod(x_train.shape[1:])))
x_test = x_test.reshape((len(x_test), np.prod(x_test.shape[1:])))
print(x_train.shape)
print(x_test.shape)

#Simple fully connected autoencoder
import keras
```

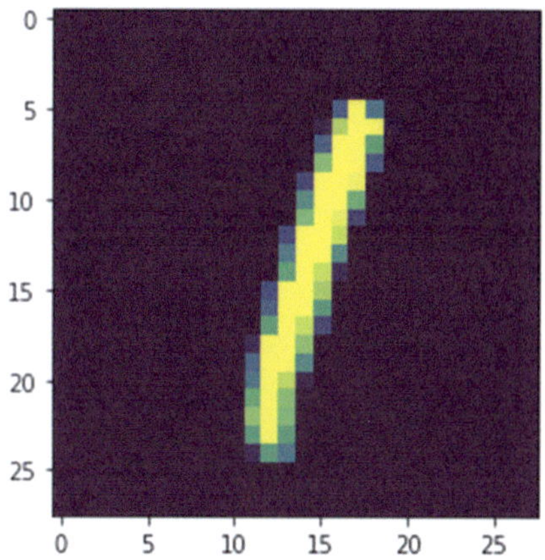

Fig. 3.23 Input image to an autoencoder

```
from keras import layers
encoding_dim = 32
input_img = keras.Input(shape=(784,))
encoded = layers.Dense(encoding_dim, activation='relu')(input_img)
decoded = layers.Dense(784, activation='sigmoid')(encoded)
autoencoder = keras.Model(input_img, decoded)

# seperate encoder and decoder models
encoder = keras.Model(input_img, encoded)
encoded_input = keras.Input(shape=(encoding_dim,))
decoder_layer = autoencoder.layers[-1]
decoder = keras.Model(encoded_input, decoder_layer(encoded_input))

#training the model
autoencoder.compile(optimizer='adam', loss='binary_crossentropy')
autoencoder.fit(x_train, x_train,
                epochs=10,
                batch_size=512,
                shuffle=True,
                validation_data=(x_test, x_test))

# encode and decode images
encoded_imgs = encoder.predict(x_test)
decoded_imgs = decoder.predict(encoded_imgs)

#plot the images
import matplotlib.pyplot as plt
plt.imshow(x_test[5].reshape(28, 28))
plt.imshow(decoded_imgs[5].reshape(28, 28))
```

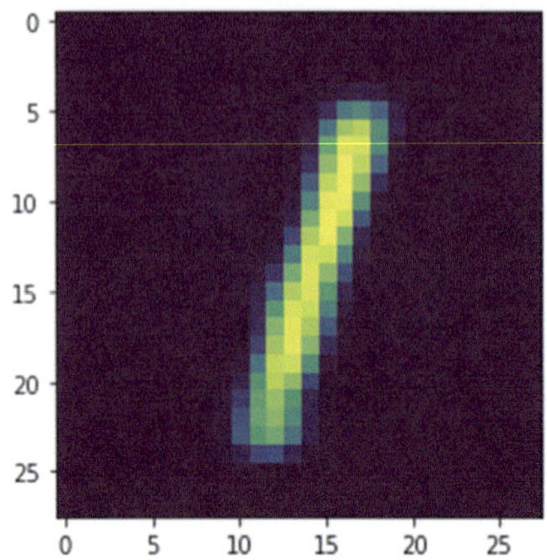

Fig. 3.24 Reconstructed image by the autoencoder discussed earlier

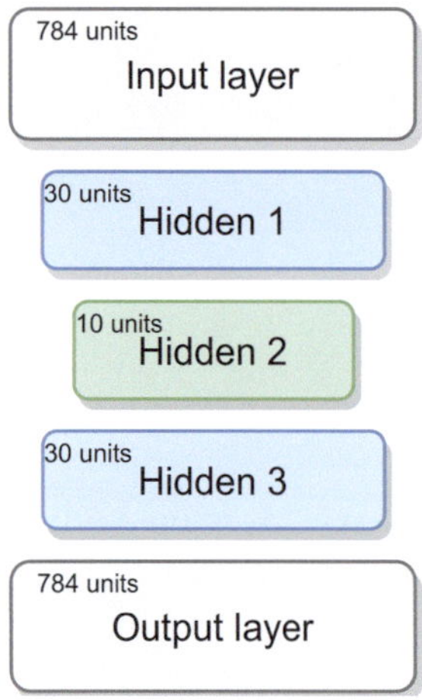

Fig. 3.25 Stacked autoencoder

Stacked Autoencoders

Autoencoders can have multiple hidden layers. In this case they are called stacked autoencoders or deep autoencoders. Adding more layers helps the autoencoder learn more complex coding. It is often faster to train one shallow autoencoder at a time, then stack all of them into a single stacked autoencoder. First the autoencoder learns to reconstruct the inputs. Then it learns to reconstruct the output for the first autoencoder layer. Finally, stack all the layers of the autoencoders as shown in Fig. 3.25.

The architecture of a stacked autoencoder is typically symmetrical with regard to the central hidden layer (the coding layer). To put it simply, it looks like a sandwich as shown in Fig. 3.25.

Example 3.29 (Stacked Autoencoder)
Problem: Consider the MNIST dataset and design a simple autoencoder encompassing fully connected layers and display 10 reconstructed images. Compare the reconstructed images with the original images.
Solution: First step is to load the MNIST dataset using the command `load_data`, which is imported from keras library. Design the encoder and decoder models using fully connected layers, the models can be built using `keras.Model` and to build these models based on a couple of fully connected neural network layers, `keras.layer.Dense` layer from keras library is used. After building the model, in order to train it first, the model needs to compile and configure to use `Adam` optimizer, `binary_crossentropy`loss function. Training can be done by setting the hyperparameter values such as `epoch` to 100 and `batch_size` to 256. To visualize and compare the input image with the reconstructed image, `plt.imshow` command from `matplotlib` library is used (Fig. 3.26).

```
#loading the dataset
from keras.datasets import mnist
import numpy as np
(x_train, _), (x_test, _) = mnist.load_data()
x_train = x_train.astype('float32') / 255.
x_test = x_test.astype('float32') / 255.
x_train = x_train.reshape((len(x_train), np.prod(x_train.shape[1:])))
x_test = x_test.reshape((len(x_test), np.prod(x_test.shape[1:])))
print(x_train.shape)
print(x_test.shape)

#Autoencoder code
import keras
from keras import layers
encoding_dim = 32
input_img = keras.Input(shape=(784,))
encoded = layers.Dense(128, activation='relu')(input_img)
encoded = layers.Dense(64, activation='relu')(encoded)
encoded = layers.Dense(32, activation='relu')(encoded)

decoded = layers.Dense(64, activation='relu')(encoded)
decoded = layers.Dense(128, activation='relu')(decoded)
decoded = layers.Dense(784, activation='sigmoid')(decoded)
autoencoder = keras.Model(input_img, decoded)

#training the model
autoencoder.compile(optimizer='adam', loss='binary_crossentropy')
autoencoder.fit(x_train, x_train,
                epochs=100,
                batch_size=256,
                shuffle=True,
                validation_data=(x_test, x_test))

decoded_imgs = autoencoder.predict(x_test)
import matplotlib.pyplot as plt
n = 10
plt.figure(figsize=(20, 4))
for i in range(n):
```

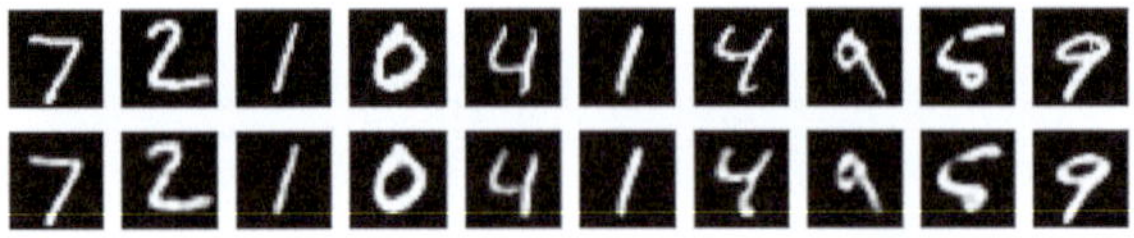

Fig. 3.26 Input and reconstructed output images of stacked autoencoder

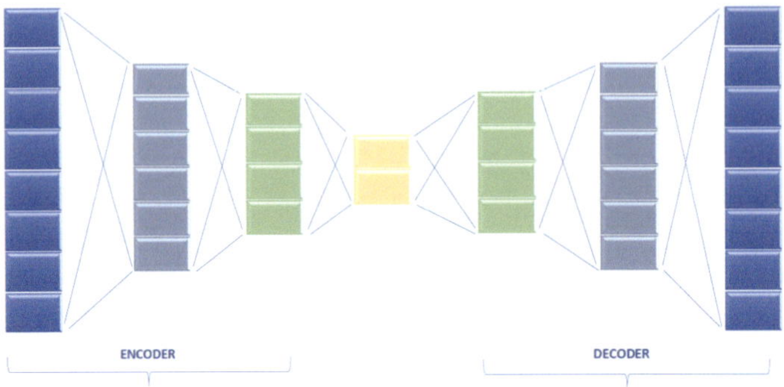

Fig. 3.27 Convolutional autoencoder

```
    # Display original
    ax = plt.subplot(2, n, i + 1)
    plt.imshow(x_test[i].reshape(28, 28))
    plt.gray()
    ax.get_xaxis().set_visible(False)
    ax.get_yaxis().set_visible(False)

    # Display reconstruction
    ax = plt.subplot(2, n, i + 1 + n)
    plt.imshow(decoded_imgs[i].reshape(28, 28))
    plt.gray()
    ax.get_xaxis().set_visible(False)
    ax.get_yaxis().set_visible(False)
plt.show()
```

Convolutional Autoencoders

When dealing with images the simple autoencoders are not effective. So far, it is observed that the Convolutional neural networks (CNN) are most suitable for image processing applications, so in order to build an autoencoder for image applications a convolutional autoencoder must be built. The architecture of the convolutional autoencoder contains the encoder that is regular CNN composed of convolutional layers and pooling layers that does the dimensionality reduction of the input dataset and the decoder must do the reverse, i.e., upscale the images. An example for Convolutional Autoencoders is as shown in Fig. 3.27.

Example 3.30 (Convolutional Autoencoder)
Problem: Consider the MNIST dataset and design a simple autoencoder with Convolutional layers, fully connected layers for reconstruction of the images. Also, plot the original and reconstructed images and visualize loss per epoch.
Solution: First step is to load the MNIST dataset using the command `load_data`, which is imported from Keras library. The models can be built using `keras.Model` and build the model based on Convolutional, Max-pooling, fully connected neural network layers, `keras.layer.Conv2D`, `keras.layer.Maxpooling2D`, `keras.layer.Dense` layers from keras library are used. After building the model, we train using the `Adam` optimizer and `binary_crossentropy`loss function. Training is performed by setting the hyperparameter values such as `epoch` to 100 and `batch_size` to 256. To visualize and compare the input image with the reconstructed image, `plt.imshow` command from `matplotlib` library is used and to visualize the loss per epoch `plt.plot()`, `plt.show()` commands are used.

```
#loading the dataset
from keras.datasets import mnist
import numpy as np
(x_train, _), (x_test, _) = mnist.load_data()
x_train = x_train.astype('float32') / 255.
x_test = x_test.astype('float32') / 255.
x_train = x_train.reshape((len(x_train), np.prod(x_train.shape[1:])))
x_test = x_test.reshape((len(x_test), np.prod(x_test.shape[1:])))
print(x_train.shape)
print(x_test.shape)

#Autoencoder code
import keras
from keras import layers
input_img = keras.Input(shape=(28, 28, 1))

x = layers.Conv2D(16, (3, 3), activation='relu', padding='same')(input_img)
x = layers.MaxPooling2D((2, 2), padding='same')(x)
x = layers.Conv2D(8, (3, 3), activation='relu', padding='same')(x)
x = layers.MaxPooling2D((2, 2), padding='same')(x)
x = layers.Conv2D(8, (3, 3), activation='relu', padding='same')(x)
encoded = layers.MaxPooling2D((2, 2), padding='same')(x)

# at this point the representation is (4, 4, 8) i.e. 128-dimensional

x = layers.Conv2D(8, (3, 3), activation='relu', padding='same')(encoded)
x = layers.UpSampling2D((2, 2))(x)
x = layers.Conv2D(8, (3, 3), activation='relu', padding='same')(x)
x = layers.UpSampling2D((2, 2))(x)
x = layers.Conv2D(16, (3, 3), activation='relu')(x)
x = layers.UpSampling2D((2, 2))(x)
decoded = layers.Conv2D(1, (3, 3), activation='sigmoid', padding='same')(x)

autoencoder = keras.Model(input_img, decoded)

#training the model
autoencoder.compile(optimizer='adam', loss='binary_crossentropy')
```

```
history = autoencoder.fit(x_train, x_train, epochs=100, batch_size=256,
     shuffle=True,validation_data=(x_test, x_test))

decoded_imgs = autoencoder.predict(x_test)
import matplotlib.pyplot as plt
n = 10
plt.figure(figsize=(20, 4))
for i in range(n):
    # Display original
    ax = plt.subplot(2, n, i + 1)
    plt.imshow(x_test[i].reshape(28, 28))
    plt.gray()
    ax.get_xaxis().set_visible(False)
    ax.get_yaxis().set_visible(False)

    # Display reconstruction
    ax = plt.subplot(2, n, i + 1 + n)
    plt.imshow(decoded_imgs[i].reshape(28, 28))
    plt.gray()
    ax.get_xaxis().set_visible(False)
    ax.get_yaxis().set_visible(False)
plt.show()

plt.plot(history.history['loss'])
plt.plot(history.history['val_loss'])
plt.title('model loss')
plt.ylabel('loss')
plt.xlabel('epoch')
plt.legend(['train', 'test'], loc='upper left')
plt.show()
```

Sparse Autoencoders

In contrast to merely reducing the number of nodes in the hidden layers, sparse autoencoders provide an alternative for encoding the data. In the sparse encoders, the loss function is constructed in such a way that the penalty is happening in the activations within the layer. This is a different approach toward regularization where the weights are normalized rather than the activation functions.

The opacity of the node denotes the level of activation. It needs to be noted that different inputs lead to the activation of different nodes through the network. This will result in limiting the memorization of the input data without confining the network to extract the features from the data. This facilitates the regularization of the network and the latent space representation of the data independent of each other. Two different ways by which one can impose the sparsity constraint in the hidden layers are by regularization. The regularization terms include L_1 regularization and KL-divergence term. The L_1 regularization is discussed in the previous sections.

The KL-divergence term refers to measure of difference between two different probability distributions. Thus, one can define a sparsity parameter that denotes average activation of a neuron over a collection of samples. Thus, by constraining the activation functions of neurons to a subset of neurons, the network forces to only fire for a particular subset of neurons rather than all the features. Compared to the autoencoders with no regularization term, the sparse autoencoders train the model according to the reconstruction loss ensuring the networks fits the data in a good manner.

Example 3.31 (Sparse Autoencoders)
Problem: Consider the MNIST dataset and design a simple sparse autoencoder for reconstruction. Also, plot the original and reconstructed images and visualize loss per epoch.
Solution: First step is to load the MNIST dataset using the command `load_data`, which is imported from keras library. Add a dense layer with `L1 activity regularizer`, which acts as the sparsity for the model. Define the autoencoder model, compile, and fit on the dataset. To visualize and compare the input image with the reconstructed image, `plt.imshow` command from `matplotlib` library is used and to visualize the loss per epoch `plt.plot(),plt.show()` commands are used.

```

# Loading the libraries
from keras.layers import Input, Dense
from keras.models import Model
from keras import regularizers
from keras.datasets import mnist
import numpy as np
import matplotlib.pyplot as plt

#Loading the data
(x_train, _), (x_test, _) = mnist.load_data()

#Reshaping the input images to feed the model
x_train = x_train.astype('float32') / 255.
x_test = x_test.astype('float32') / 255.
x_train = x_train.reshape((len(x_train), np.prod(x_train.shape[1:])))
x_test = x_test.reshape((len(x_test), np.prod(x_test.shape[1:])))

#Model definition
encoding_dim = 32
input_img = Input(shape=(784,))
# add a Dense layer with a L1 activity regularizer
encoded = Dense(encoding_dim, activation='relu',
                activity_regularizer=regularizers.l1(10e-9))(input_img)
decoded = Dense(784, activation='sigmoid')(encoded)

autoencoder = Model(input_img, decoded)

encoder = Model(input_img, encoded)
# create a placeholder for an encoded (32-dimensional) input
encoded_input = Input(shape=(encoding_dim,))
# retrieve the last layer of the autoencoder model
decoder_layer = autoencoder.layers[-1]
# create the decoder model
decoder = Model(encoded_input, decoder_layer(encoded_input))

#Compiling the model
autoencoder.compile(optimizer='adam', loss='binary_crossentropy')
autoencoder.fit(x_train, x_train,
                epochs=50,
```

```
                batch_size=256,
                shuffle=True,
                validation_data=(x_test, x_test))

#Doing predictions on the test set
encoded_imgs = encoder.predict(x_test)
decoded_imgs = decoder.predict(encoded_imgs)

# Plotting the predictions of 10 images
n = 10
plt.figure(figsize=(20, 4))
for i in range(n):
    # display original
    ax = plt.subplot(2, n, i + 1)
    plt.imshow(x_test[i].reshape(28, 28))
    plt.gray()
    ax.get_xaxis().set_visible(False)
    ax.get_yaxis().set_visible(False)

    # display reconstruction
    ax = plt.subplot(2, n, i + 1 + n)
    plt.imshow(decoded_imgs[i].reshape(28, 28))
    plt.gray()
    ax.get_xaxis().set_visible(False)
    ax.get_yaxis().set_visible(False)
plt.show()

encoded = Dense(encoding_dim, activation='relu',
                activity_regularizer=regularizers.l1(10e-5))(input_img)
```

Recurrent Autoencoders

For applications like time-series sequence, text sequence, dimensionality reduction traditional network, or a dense network are not well suited. Therefore, for processing such sequences recurrent network is well suitable.

Building a recurrent autoencoder is not complex, it is a straightforward task; the encoder is typically a sequence-to-vector recurrent neural network that compresses the input sequence down to a single vector. The decoder is a vector-to-sequence recurrent neural network that does the reverse.

Example 3.32 (Recurrent Autoencoders)
Problem: Construct a simple recurrent autoencoder to predict the next elements in a sequence.
Solution: To construct a recurrent autoencoder, we import the relevant libraries initially. Create a sequence with some relation between elements, like a mathematical progression and give it as input to the model. Then define the model that we fit on the training and do predictions on the test set. We preprocess and train our model in such a way that it will predict all the next 8 elements in the series with given one element as input.

```
# Import necessary libraries
from numpy import array
from keras.models import Sequential
from keras.layers import SimpleRNN
from keras.layers import Dense
from keras.layers import RepeatVector
from keras.layers import TimeDistributed
from keras.utils import plot_model

# define input sequence
seq_in = array([0.2, 0.4, 0.6, 0.8, 1.0, 1.2, 1.4, 1.6, 1.8])

# data preprocessing
n_in = len(seq_in)
seq_in = seq_in.reshape((1, n_in, 1))
# prepare output sequence
seq_out = seq_in[:, 1:, :]
n_out = n_in - 1


# define model
model = Sequential()
model.add(SimpleRNN(100, activation='relu', input_shape=(n_in,1)))
model.add(RepeatVector(n_out))
model.add(SimpleRNN(100, activation='relu', return_sequences=True))
model.add(TimeDistributed(Dense(1)))
model.compile(optimizer='adam', loss='mse')

# fit model
model.fit(seq_in, seq_out, epochs=300, verbose=0)

# demonstrate prediction
y_pred = model.predict(seq_in, verbose=0)
print(y_pred[0,:,0])

> [0.40000004 0.6000001  0.79999995 1.         1.2000002  1.4
 1.6        1.8000002 ]
```

Denoising Autoencoders

The sparse or traditional autoencoders aim to compress the inputs and produce the outputs after decoding as close to the input as possible. However, it is also needed that the autoencoders need to be sometimes independent of the training data, i.e., insensitive to recreate original data and be generalizable enough to recreate the data despite being fed with noisy or slightly corrupt input data. To facilitate this, the denoising autoencoders learn a vector field to map the input data to a lower dimension manifold and then recreate the data from the lower dimension manifold. It needs to be noted that projecting to a lower dimension leads to cancellation of the noise leading to an output that is noise-free and follows a clean distribution of the original data.

This method prevents the autoencoder from trivially copying the inputs to the outputs, and it ends up finding the patterns in the data. Adding noise to the inputs and training to recover the original noise-free inputs is a useful method to learn the features in the data. The architecture of this autoencoder is as shown in Fig. 3.28.

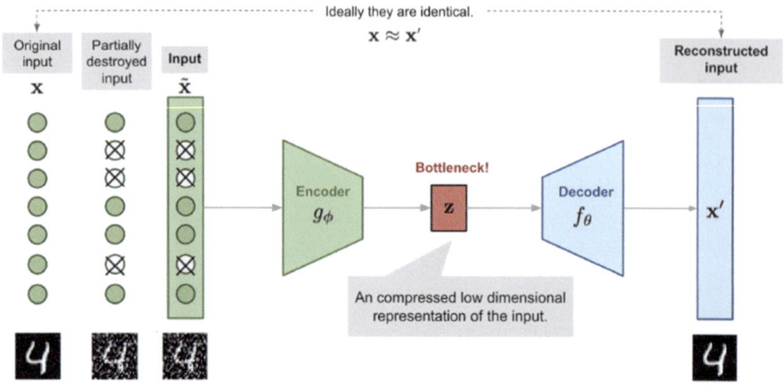

Fig. 3.28 Denoising autoencoder

Example 3.33 (Image Denoising Autoencoder)
Problem: Consider the MNIST dataset and design an autoencoder using Convolutional layers, fully connected layers to work on image denoising problem. Also, plot the original, noisy, and reconstructed images and visualize loss per epoch by the network.
Solution: First step is to load the MNIST dataset using the command `load_data`, which is imported from Keras library. Design the encoder and decoder models using `keras.Model` and build a model based on convolutional autoencoder `keras.layer.Conv2D`, `keras.layer.Dense`, `keras.layer.Maxpooling2D` layers from keras library. After building the model, in order to train it first, the model needs to compile and configure to use `Adam` optimizer, `binary_crossentropy`loss function. Training can be done by setting the hyperparameter values such as `epoch` to 100 and `batch_size` to 256. To visualize and compare the input image with the reconstructed image, `plt.imshow` command from `matplotlib` library is used and to visualize the loss per epoch `plt.plot()`, `plt.show()` commands are used.

```
#loading the dataset
from keras.datasets import mnist
import numpy as np
(x_train, _), (x_test, _) = mnist.load_data()
x_train = x_train.astype('float32') / 255.
x_test = x_test.astype('float32') / 255.
x_train = x_train.reshape((len(x_train), np.prod(x_train.shape[1:])))
x_test = x_test.reshape((len(x_test), np.prod(x_test.shape[1:])))
print(x_train.shape)
print(x_test.shape)

#adding noise to the data
```

```
noise_factor = 0.5
x_train_noisy = x_train + noise_factor * np.random.normal(loc=0.0, scale=1.0,
        size=x_train.shape)
x_test_noisy = x_test + noise_factor * np.random.normal(loc=0.0, scale=1.0,
      size=x_test.shape)

x_train_noisy = np.clip(x_train_noisy, 0., 1.)
x_test_noisy = np.clip(x_test_noisy, 0., 1.)

#Autoencoder code
import keras
from keras import layers
input_img = keras.Input(shape=(28, 28, 1))

x = layers.Conv2D(16, (3, 3), activation='relu', padding='same')(input_img)
x = layers.MaxPooling2D((2, 2), padding='same')(x)
x = layers.Conv2D(8, (3, 3), activation='relu', padding='same')(x)
x = layers.MaxPooling2D((2, 2), padding='same')(x)
x = layers.Conv2D(8, (3, 3), activation='relu', padding='same')(x)
encoded = layers.MaxPooling2D((2, 2), padding='same')(x)

x = layers.Conv2D(8, (3, 3), activation='relu', padding='same')(encoded)
x = layers.UpSampling2D((2, 2))(x)
x = layers.Conv2D(8, (3, 3), activation='relu', padding='same')(x)
x = layers.UpSampling2D((2, 2))(x)
x = layers.Conv2D(16, (3, 3), activation='relu')(x)
x = layers.UpSampling2D((2, 2))(x)
decoded = layers.Conv2D(1, (3, 3), activation='sigmoid', padding='same')(x)

autoencoder = keras.Model(input_img, decoded)

#training the model
autoencoder.compile(optimizer='adam', loss='binary_crossentropy')
history = autoencoder.fit(x_train_noisy, x_train, epochs=100, batch_size=256,
      shuffle=True,validation_data=(x_test, x_test))

decoded_imgs = autoencoder.predict(x_test_noisy)
import matplotlib.pyplot as plt
n = 10
plt.figure(figsize=(20, 4))
for i in range(n):
    # Display original
    ax = plt.subplot(2, n, i + 1)
    plt.imshow(x_test[i].reshape(28, 28))
    plt.gray()
    ax.get_xaxis().set_visible(False)
    ax.get_yaxis().set_visible(False)

    # Display noisy data
    ax = plt.subplot(1, n, i)
    plt.imshow(x_test_noisy[i].reshape(28, 28))
    plt.gray()
    ax.get_xaxis().set_visible(False)
    ax.get_yaxis().set_visible(False)

    # Display reconstruction
    ax = plt.subplot(2, n, i + 1 + n)
    plt.imshow(decoded_imgs[i].reshape(28, 28))
    plt.gray()
    ax.get_xaxis().set_visible(False)
    ax.get_yaxis().set_visible(False)
plt.show()

plt.plot(history.history['loss'])
plt.plot(history.history['val_loss'])
plt.title('model loss')
plt.ylabel('loss')
```

```
plt.xlabel('epoch')
plt.legend(['train', 'test'], loc='upper left')
plt.show()
```

Variational Autoencoders (VAE)

The task for the autoencoders is to train to encode and decode with as little information loss as possible. In the case of lower-dimensional latent space the encoded data can be decoded by the autoencoder with zero loss; this leads to overfitting problem. To address this problem, a special autoencoder called, Variational Autoencoder (VAE) was introduced in 2014 by Diederik Kingma and Max Welling. VAE is quite different from traditional autoencoders; it is a deep generative model-based autoencoder that learns from the input dataset by using a latent variable model is introduced.

Because VAE is a generative model, there is a slight modification in the encoder–decoder process of the VAE. The encoder of the VAE behaves like a generative autoencoder, meaning it can generate new instances with the available input sample dataset from a point in latent space. This encoding of the input is then distributed over the latent space. Then during the decoding process, the VAE decoder decodes the outputs based on a probabilistic approach and computes the reconstruction errors (as opposed to denoising autoencoders, which use randomness only during training).

During the training, the model parameters are trained *via* two loss functions:

- The usual reconstruction loss that pushes the autoencoders to produce outputs that match with the initial inputs.
- The latent loss that

After training a VAE can easily generate a new instance by using just the samples from the random Gaussian distribution and decode it.

Example 3.34 (Variational Autoencoder)
Problem: Consider the MNIST dataset and design a variational autoencoder with Convolutional layers, fully connected layers to work on the image denoising problems. Also, plot the original, noisy, and reconstructed images and visualize loss per epoch by the network.
Solution: First step is to load the MNIST dataset using the command `load_data`, which is imported from keras library. Design the encoder and decoder models using `keras.Model` and build a model based on convolutional autoencoder `keras.layer.Conv2D`, `keras.layer.Dense`, `keras.layer.Maxpooling2D` layers from keras library are used. After building the model, in order to train it, first, the model needs to compile and configure to use `Adam` optimizer, `binary_crossentropy` loss

(continued)

Example 3.34 (continued)
function. Training can be done by setting the hyperparameter values such as `epoch` to 100 and `batch_size` to 256. To visualize and compare the input image with the reconstructed image, `plt.imshow` command from `matplotlib` library is used and to visualize the loss per epoch `plt.plot(),plt.show()` commands are used.

```
#loading the dataset
from keras.datasets import mnist
import numpy as np
(x_train, _), (x_test, _) = mnist.load_data()
x_train = x_train.astype('float32') / 255.
x_test = x_test.astype('float32') / 255.
x_train = x_train.reshape((len(x_train), np.prod(x_train.shape[1:])))
x_test = x_test.reshape((len(x_test), np.prod(x_test.shape[1:])))
print(x_train.shape)
print(x_test.shape)

#adding noise to the data
noise_factor = 0.5
x_train_noisy = x_train + noise_factor * np.random.normal(loc=0.0, scale=1.0,
      size=x_train.shape)
x_test_noisy = x_test + noise_factor * np.random.normal(loc=0.0, scale=1.0,
     size=x_test.shape)

x_train_noisy = np.clip(x_train_noisy, 0., 1.)
x_test_noisy = np.clip(x_test_noisy, 0., 1.)

#Autoencoder code
import keras
from keras import layers
input_img = keras.Input(shape=(28, 28, 1))

x = layers.Conv2D(16, (3, 3), activation='relu', padding='same')(input_img)
x = layers.MaxPooling2D((2, 2), padding='same')(x)
x = layers.Conv2D(8, (3, 3), activation='relu', padding='same')(x)
x = layers.MaxPooling2D((2, 2), padding='same')(x)
x = layers.Conv2D(8, (3, 3), activation='relu', padding='same')(x)
encoded = layers.MaxPooling2D((2, 2), padding='same')(x)

x = layers.Conv2D(8, (3, 3), activation='relu', padding='same')(encoded)
x = layers.UpSampling2D((2, 2))(x)
x = layers.Conv2D(8, (3, 3), activation='relu', padding='same')(x)
x = layers.UpSampling2D((2, 2))(x)
x = layers.Conv2D(16, (3, 3), activation='relu')(x)
x = layers.UpSampling2D((2, 2))(x)
decoded = layers.Conv2D(1, (3, 3), activation='sigmoid', padding='same')(x)

autoencoder = keras.Model(input_img, decoded)

#training the model
autoencoder.compile(optimizer='adam', loss='binary_crossentropy')
history = autoencoder.fit(x_train_noisy, x_train, epochs=100, batch_size=256,
     shuffle=True,validation_data=(x_test, x_test))

decoded_imgs = autoencoder.predict(x_test_noisy)
import matplotlib.pyplot as plt
n = 10
plt.figure(figsize=(20, 4))
```

```
for i in range(n):
    # Display original
    ax = plt.subplot(2, n, i + 1)
    plt.imshow(x_test[i].reshape(28, 28))
    plt.gray()
    ax.get_xaxis().set_visible(False)
    ax.get_yaxis().set_visible(False)

    # Display noisy data
    ax = plt.subplot(1, n, i)
    plt.imshow(x_test_noisy[i].reshape(28, 28))
    plt.gray()
    ax.get_xaxis().set_visible(False)
    ax.get_yaxis().set_visible(False)

    # Display reconstruction
    ax = plt.subplot(2, n, i + 1 + n)
    plt.imshow(decoded_imgs[i].reshape(28, 28))
    plt.gray()
    ax.get_xaxis().set_visible(False)
    ax.get_yaxis().set_visible(False)
plt.show()

plt.plot(history.history['loss'])
plt.plot(history.history['val_loss'])
plt.title('model loss')
plt.ylabel('loss')
plt.xlabel('epoch')
plt.legend(['train', 'test'], loc='upper left')
plt.show()
```

3.8 Putting It All Together

We explored many supervised models in this chapter, spanning from regression through SVMs and autoencoders. Each of these strategies varies in complexity. Apart from the computational difficulty, the application and functioning of each of these strategies vary.

The primary application of supervised learning techniques such as regression and its derivatives is in prediction applications. The type of regression that is most appropriate for a given application is strongly dependent on the distribution of the underlying data. When the data distribution is uniform and the variables are highly correlated, low complex linear regression may perform better than the other complex regression techniques outlined. ANNs or MLPs can also be used for prediction tasks. In comparison to linear regression techniques, these strategies are superior at capturing nonlinear relationships between variables.

Another critical application in real-world applications is classification. Classification approaches such as ANNs, SVMs, and tree-based techniques are commonly employed. SVMs are typically used for binary classification tasks. However, using the aforementioned methodologies, they could be expanded to multiclass categorization. However, the complexity and training time for SVMs might rise as the number of classes increases. On the other hand, ANNs and DNNs are highly effective at extracting latent features and performing classification, even when dealing with large and complicated datasets. While ensemble learning approaches are computationally intensive, they can be used for classification when the input data is noisy or the sample size is small.

Random forest techniques are commonly employed in classification applications. These strategies begin by extracting latent features in order to establish rules and construct trees for classification purposes. The depth of the trees is proportional to the number of latent characteristics and the data's complexity. On the other hand, autoencoders are frequently employed for data denoising and classification applications. These strategies are commonly used in adversarial machine learning to improve robustness to adversarial samples. Denoising can be used to improve classification performance. However, the computational complexity involved may be greater than that of typical DNNs and ANNs.

3.9 Exercise Problems

Problem 3.1 Build a Support Vector Machine regressor (`sklearn.svm.SVR`) with various hyperparameters, such as different kernels and C values on the airline dataset discussed in Chap. 1. Note which combination produces the best output.

Problem 3.2 Consider the airline dataset discussed in Chap. 1 and build a regression model with fare as a dependent variable and class, guests, and age as independent factors. Evaluate and compare the prediction accuracy and loss with linear, logistic regression and SVM regression.

Problem 3.3 Build an AlexNet and perform the CIFAR-10, CIFAR-100, and MNIST digits data classification with dropout rate of (a) 0.1, (b) 0.2, and (c) 0.5. Compare the classification performance in terms of accuracy and the loss function. Also, plot the loss plots.

Problem 3.4 Build a ResNet-18 and perform CIFAR-10 and CIFAR-100 classification. Compare the results with the above provided problem. Show the differences in terms of loss plots and confusion matrices.

Problem 3.5 Perform the MNIST and Iris data classification with (i) a five-layer neural network and (ii) a 3 convolutional layer and 2 fully connected layer networks and compare the performance and loss function for each of them.

Problem 3.6 Consider the data in the problem of time series prediction from `Example3.9` AutoRegression and use the data to build a simple MLP. Discuss which one is better and why?

Problem 3.7 Consider the airline dataset discussed in Chap. 1 and build Bagging and Boosting based ensemble models and compare the performance with SVM and tree-based techniques.

Problem 3.8 Build different regression models using AdaBoost regressor, Bagging regressor, Gradient Boosting regressor, and XGBoost regressor on `make_regression` dataset from sklearn.datasets. Observe the Mean absolute error, Mean Squared error, and R2 score of all these regression models. Discuss which model has a better score and discuss which model is better.

Problem 3.9 Consider the `make_classification` dataset from sklearn.datasets and build a classification that contains model stacking with KNeighborsClassifier(), DecisionTreeClassifier(), SVC(), GaussianNB(), and RandomForestClassifier() as estimators and LogisticRegression() as `final_estimator`.Compare the accuracy of stacking model with individual classification models.

Problem 3.10 Build a similar stacking model similar to Problem 3.9, but, instead use regressor models in stacking with `make_regression` dataset from sklearn.datasets. Compare them in terms of MSE, MAE scores and discuss which is better.

Problem 3.11 Build a regressor using Random Forest on `make_regression` dataset from sklearn.datasets. Do predictions using different `max_depth` and `random_state` values and discuss which combination produces the best predictions and why?

Problem 3.12 Consider the code `Example 3.32` Recurrent Autoencoder, replace the SimpleRNN layers with LSTM and GRU layers, and observe which combination produces the best predicted sequence.

Problem 3.13 Complete a classification task using an autoencoder. Consider the dataset `make_classification` from sklearn.datasets library that has a binary classification task and build an autoencoder architecture to do this.

Problem 3.14 Build a basic autoencoder on `fashion_mnist` dataset. Compare the performance with the variational autoencoder.

Chapter 4
Unsupervised Learning

4.1 Introduction

Unlike supervised learning techniques, unsupervised techniques are provided with data that is not labeled, classified, and categorized. Hence, the "teacher" is absent for these techniques. Considering an example of a set of images of dogs and cats, the classifier is allowed to learn the patterns or the similarities and dissimilarities in the data points. For both the animals, the commonality is the number of legs, while the dissimilar feature could be their average weight/size, and so on and so forth. The classifier, based on a variety of techniques, learns the patters in the data to classify data into different categories. Again, the classifier does not know that the object class is either a "dog" or a "cat," yet it can segregate the data points into two main classes based on the patterns. Some of the techniques that help the classification in unsupervised scenarios are discussed next.

We will be discussing the clustering algorithms, their sub-categories, unsupervised neural networks, classification techniques, multi-dimensional scaling, assignment-based clustering, and Google's page ranking algorithm in this chapter.

4.2 Clustering

Clustering is an unsupervised learning technique that deals with finding a pattern or a structure in an unlabeled dataset. It clusters the objects based on "similar" and "dissimilar" features; in other words, classification is performed based on the similarity (and dissimilarity) of the objects. The decision of building clusters can be based on popular similarity measures such as the "Cosine Distance" and the "Euclidean Distance." Equation 4.1 shows the formula to calculate the cosine similarity between the two vectors A and B. The cosine distance is then calculated as $1 - Cosinesimilarity$. Another popular technique for clustering is based on the

S. Rafatirad et al., *Machine Learning for Computer Scientists and Data Analysts*,
https://doi.org/10.1007/978-3-030-96756-7_4

Euclidean distance as shown in Eq. 4.2. The Cosine distance is commonly used for the distance between two vectors, while the Euclidean distance is for the distance between two points in a space.

$$\cos(\theta) = \frac{\sum_{i=1}^{n} A_i B_i}{\sqrt{\sum_{i=1}^{n} A_i^2}\sqrt{\sum_{i=1}^{n} B_i^2}} \tag{4.1}$$

$$d(a, b) = \sqrt{\sum_{i=1}^{n} (b_i - a_i)^2}. \tag{4.2}$$

Figure 4.1 shows the clustering methodology. The algorithm is provided with a sample space that contains a mix of different shapes. Clustering tries to form groups or "clusters" of objects that are similar in characteristics and close—in distance—to each other. The result is shown for a cluster size of 4, but if we do not know the optimum size, we can use the Elbow method explained further, or experiment with some values until satisfactory results are achieved. For example, if the cluster size is set to 3, it might happen that clusters 1 and 4 combine to form one cluster, given some similarity in the object characteristics.

Let us consider a dataset containing different characteristics of humans, such as age, color of the eyes, hair, height, weight, bodily features, etc. An example of a problem that can be addressed by clustering is grouping people based on their different features. What features are considered depends on the outcome of clustering and the application. For instance, a cosmetic company's marketing department could target a cluster/group of people with a particular skin color for different shades the company produces. An apparel manufacturer could design garments for people with different heights, bodily features, their choice of fabric, price range, etc. That way, clustering algorithms are very crucial in today's day and age to help the user with clustering unlabeled data.

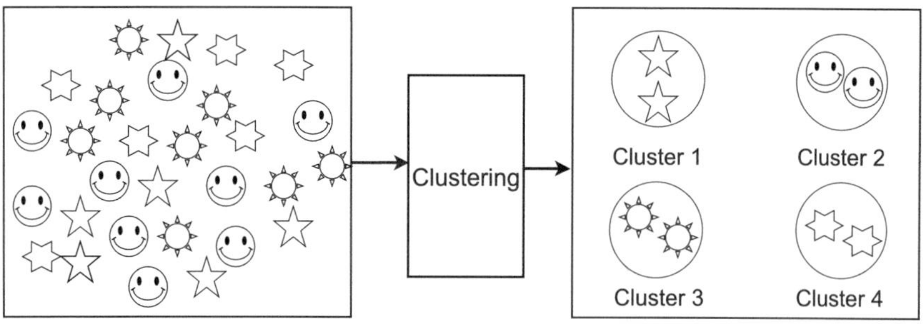

Fig. 4.1 Clustering at a high level

4.2.1 K-Means Clustering

The K-Means is a greedy algorithm for partitioning data points in a given space. K-Means is a popular yet simplest algorithm in the unsupervised learning category that helps partition the data points into a given number of clusters. The cluster size or the amount of clusters is fixed a prior, say "*c*" clusters. The centers of these clusters are placed as far as possible from each other to obtain better results. Initially, the position of the centroids in a given space is chosen randomly by the algorithm; the user selects the number of clusters, *c*. The data points are selected from the dataset and placed in one of the *c* clusters. The placement decision is based on the Euclidean distance between the data point and centroids; the data point is placed in a centroid that is the least distance away compared to the distances from other centroids. After all the points are placed, the process is repeated again by re-calculating new centroids. These new centroids are data points chosen from each cluster to become the *new centroids*. Again, the distances between data points and new centroids are calculated, based on which the points are re-located. The process repeats until there is no further change possible. The objective of the algorithm is to minimize the sum of the squared error function.

"How does one know what value to set initially for *c*?" is a common question when it comes to K-Means. It is worth noting that the results of K-Means vary with the number of clusters and the position of the initially chosen random centroids. One of the well-known methods is the "Elbow" method employed to derive the optimum number of clusters, *c*. The Elbow method plots a graph between a range of *c* values and the value of the sum of squared distances for each point to its assigned center. A sample Elbow plot is shown in Fig. 4.2 for *c* values from 1 to 10 on the X-axis, while the calculated sum of squared distances is plotted on the Y-axis. The point of inflection on the curve is the optimum value of *c*.

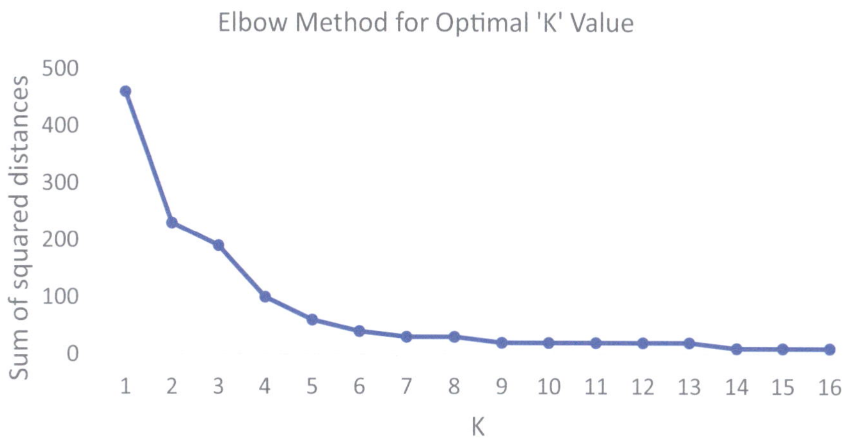

Fig. 4.2 Elbow method to find optimum value of *K*

The steps followed by the K-Means algorithm are outlined below:

1. A dataset contains a set of data points. Let X = $\{x_1, x_2, x_3, \ldots, x_n\}$, and $C = \{c_1, c_2, c_3, \ldots, c_c\}$ be the number of centers.
2. Select c clusters randomly.
3. Calculate the distance between data point and centroids.
4. Assign the data point to the centroid, the distance of which from the center is the minimum of all the other centroids.
5. Calculate the new centroid using Eq. 4.3, where C_c refers to the points in the c-th cluster.
6. If there are no further changes, then stop; else continue.

$$C_i = \frac{1}{c_i} \sum_{i=1}^{c} x_i. \tag{4.3}$$

K-Means being a simple yet powerful clustering technique finds itself in a variety of real-world applications, some of them include biological studies such as classifying plants, animals, and cells, business marketing applications to group customers, image clustering and document classification, crime location identification, and cyber-security applications.

Similar to other ML techniques, K-Means also has strengths and weaknesses as listed below:

- Low complexity in terms of understanding and implementation.
- Efficient in regard to time complexity. O(tkn), where n is the number of data points, k is the clusters, and the number of iterations is denoted by t. Hence, K-Means is also considered as a linear algorithm, when t and k are small.
- K-Means depends on the user to specify an optimum value of k.
- Sensitive to outliers.
- Sensitive to initial seed—initial points considered as clusters' centers.
- Generally, it is difficult to evaluate the results, meaning that the final result of the algorithm depends on various factors as explained above. There is no optimum method that finds the "best cluster." It really depends on the application of the algorithm and the expected result.

Example 4.1 (K-Means Clustering)

Problem: Generate a random dataset with 500 data points, perform clustering using the K-Means algorithm, and output a scatter plot.

Solution: To solve this problem, one can generate a random dataset using `make_blob` command with *n_samples*, *centers*, and *cluster_std* parameters. In the code snippet below, we first import the KMeans from sklearn package followed by matplotlib package required for plotting the graphs. Numpy is a

(continued)

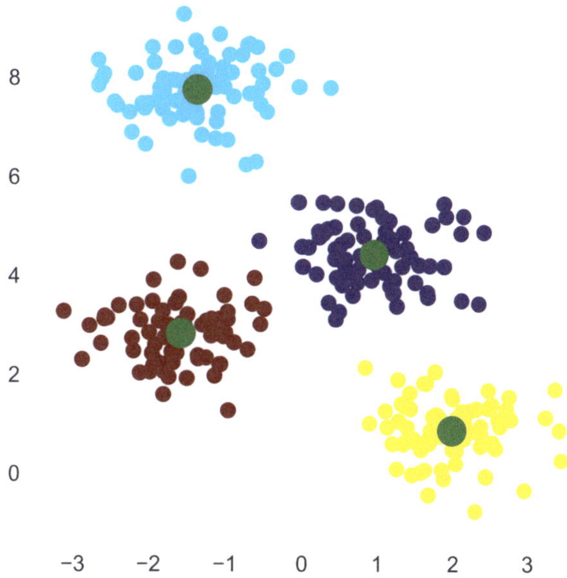

Fig. 4.3 Clusters with three centroids generated by a K-Means algorithm

Example 4.1 (continued)
commonly used package for data transformation, array processing, and much more. Hence, it is a common practice to import it. The make_blobs is imported to support data generation within the code. The X is assigned after generating data with the make_blobs function. The arguments of the function generate the data with the specified values for samples, centers, and the standard deviation of the data points. The generated data points are plotted with a scatter function to visualize the data. The next three lines fit the K-Means algorithm on the data and derive the predicted values in the y_kmeans. The data is plotted again to visualize the clusters, and the centers of the clusters are highlighted. The scatter plot is as shown in Fig. 4.3.

A sample code snippet for implementing K-Means clustering is shown below.

```
import numpy as np
from sklearn.cluster import KMeans
import matplotlib.pyplot as plt
from sklearn.datasets.samples_generator import make_blobs

X, y_label =  make_blobs(n_samples = 500, centers = 4,
                         cluster_std=0.70)
plt.scatter(X[:, 0], X[:, 1], s=10);
# K-Means technique is one of the clustering techniques, discussed in the
    following sections.
k_means = KMeans(n_clusters=4)
k_means.fit(X)
y_kmeans = k_means.predict(X)

```

```
plt.scatter(X[:, 0], X[:, 1], c=y_kmeans, s=50)

centers = k_means.cluster_centers_
plt.scatter(centers[:, 0],
            centers[:, 1], c='green', s=100, alpha=0.5);
```

Example 4.2 (K-Means Clustering)

Problem: Perform K-Means clustering on the Mall Customer dataset.[a] Load the libraries necessary in plotting the data, K-Means algorithm, numpy, etc. Load the dataset and generate an Elbow plot to decide on the optimum number of clusters for the `Age` and `Spending Score` features in the data. After deciding the number of clusters, proceed with fitting the data on the K-Means algorithm. Finally, generate a plot showing all the clusters with data points and their centroids.

Solution: A sample code snippet for implementing K-Means clustering is shown in the code below. The dataset consists of the gender, age, annual income, and spending scores for 200 customers. The problem statement is to find the optimum number of clusters for the K-Means algorithm. Another aspect of the problem is to use K-Means to cluster the customers according to their age and spending scores. An example to address the problem statement is presented below along with the plots after K-Means clustering on the dataset. The clustering is done on Age and Spending Score. The Elbow plot in Fig. 4.4 clearly shows that the optimum value of the clusters is 4. In Fig. 4.5, the Spending Score versus the Age of the customers is grouped in 4 clusters. The "*" markers in red color are the centroids for each cluster.

[a] https://www.kaggle.com/vjchoudhary7/customer-segmentation-tutorial-in-python.

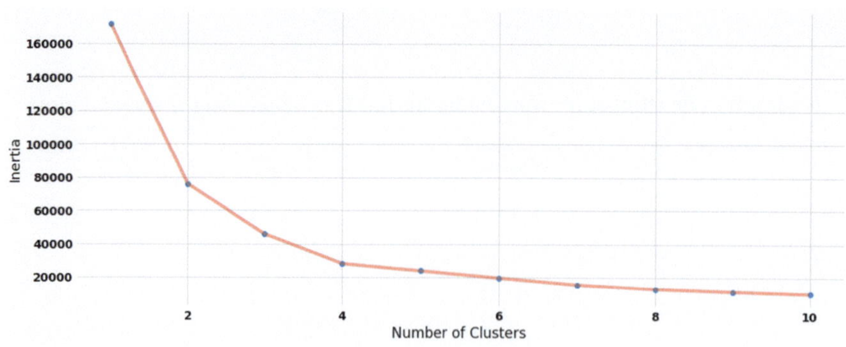

Fig. 4.4 Elbow method plot to find the optimum value of K

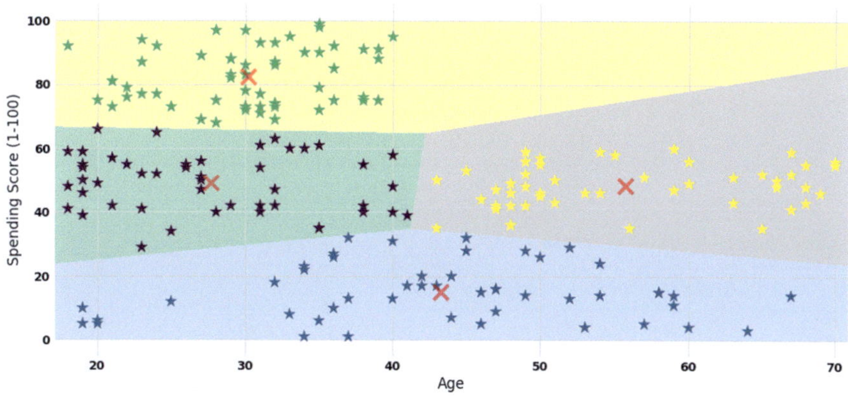

Fig. 4.5 K-Means clustering of age and spending score

```
## Importing the libraries
import numpy as np
import pandas as pd
import matplotlib.pyplot as plt
import seaborn as sns
import plotly as py
import plotly.graph_objs as go
from sklearn.cluster import KMeans
import warnings
import os

## Loading the Mall Customers dataset

df = pd.read_csv(r'../input/Mall_Customers.csv')
df.head()

## Clustering using K-Means on Age and Spending Score

## The lines of code below generate an Elbow plot

X1 = df[['Age' , 'Spending Score (1-100)']].iloc[: , :].values
inertia = []
for n in range(1 , 11):
    algorithm = (KMeans(n_clusters = n ,init='k-means++', n_init = 10 ,
     max_iter=300,
                        tol=0.0001,  random_state= 111  , algorithm='elkan')
     )
    algorithm.fit(X1)
    inertia.append(algorithm.inertia_)

plt.figure(1 , figsize = (15 ,6))
plt.plot(np.arange(1 , 11) , inertia , 'o')
plt.plot(np.arange(1 , 11) , inertia , '-' , alpha = 0.5)
plt.xlabel('Number of Clusters') , plt.ylabel('Inertia')
plt.show()

## The elbow plot shows the optimum number of clusters is '4'. Hence, proceed
      with the clustering using K-Means algorithm with four clusters

algorithm = (KMeans(n_clusters = 4 ,init='k-means++', n_init = 10 ,max_iter
     =300, tol=0.0001,  random_state= 111  , algorithm='elkan') )
algorithm.fit(X1)
```

```
labels1 = algorithm.labels_
centroids1 = algorithm.cluster_centers_

## Here we prepare to plot the Clusters with data points and their centroids

h = 0.02
x_min, x_max = X1[:, 0].min() - 1, X1[:, 0].max() + 1
y_min, y_max = X1[:, 1].min() - 1, X1[:, 1].max() + 1
xx, yy = np.meshgrid(np.arange(x_min, x_max, h), np.arange(y_min, y_max, h))
Z = algorithm.predict(np.c_[xx.ravel(), yy.ravel()])

plt.figure(1 , figsize = (15 , 7) )
plt.clf()
Z = Z.reshape(xx.shape)
plt.imshow(Z , interpolation='nearest',
           extent=(xx.min(), xx.max(), yy.min(), yy.max()),
           cmap = plt.cm.Pastel2, aspect = 'auto', origin='lower')

plt.scatter( x = 'Age' ,y = 'Spending Score (1-100)' , marker = '*', data =
     df , c = labels1 , s = 200 )
plt.scatter(x = centroids1[: , 0] , y =  centroids1[: , 1] , s = 300 , marker
      = 'x',c = 'red' , alpha = 0.5)
plt.ylabel('Spending Score (1-100)') , plt.xlabel('Age')
plt.show()
```

4.2.2 Hierarchical Clustering

Hierarchical clustering is similar to what we studied about clustering so far, but it is useful in scenarios where a hierarchy is more appropriate than "flat" clustering. Flat clustering here refers to the K-Means clustering. For hierarchical clustering, you may imagine it as a cluster-of-clusters. Hierarchical clustering (HC) delivers similarity between clusters (clusters built out of similar data points), and similarity between one cluster and other clusters, all of them belonging to a common root node. HC is usually accompanied by a dendrogram, which is a graphical representation of which samples are mostly similar, when were they clustered, and so on.

Figure 4.6 shows the hierarchical clustering on sample data points, while Fig. 4.6a–d shows a sample dendrogram with different features and samples. Only the top left of Fig. 4.6 would refer to a K-Means method, but it is seen that the hierarchical clustering takes it to another level by combining different clusters, in this case, clusters A through F. The clusters are combined in a hierarchy because the data samples fall under a hierarchy. Meaning, the data points show similar characteristics given the parent level characteristics; hence, the clustering method converges to a single large cluster. Figure 4.6a–d shows a dendrogram for a dataset. The link between the clusters, for example, between clusters A and B, and clusters D and E, shows that the clusters were grouped based on the distance between the data points. In the next step, referring to Fig. 4.6b, the clusters AB now combine with C, based on the characteristics found in the data. Likewise, all the clusters are combined to form one large cluster. The horizontal line depicts the combined clusters (the width of the link also depicts the similarity between the clusters), while

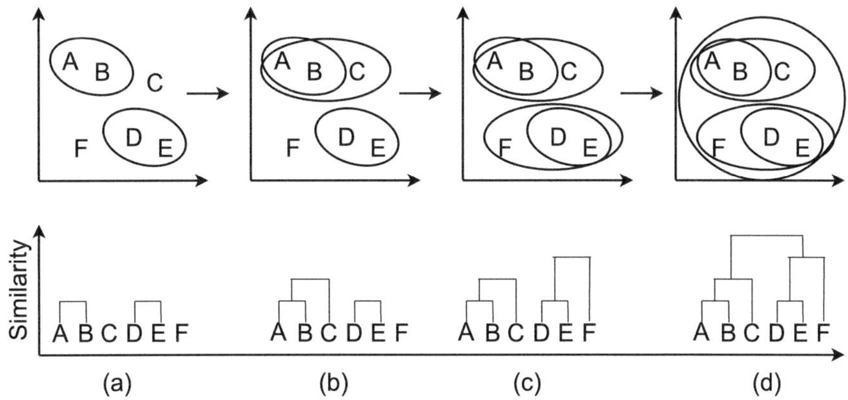

Fig. 4.6 Hierarchical clustering process and dendrogram representation of the clustering

the sequence of formation of the clusters is represented in the dendrogram by a tree-like structure. The algorithm for hierarchical clustering is presented below:

- Initiate the process by assigning all data points as being their own clusters.
- Find similar (closest data point) data point (also a cluster in case of hierarchical clustering) and merge it with another single cluster. The closest cluster can be found by Euclidean distance, Manhattan distance, and other methods.
- Find another nearby similar cluster and combine all clusters until only one cluster remains.

Hierarchical clustering is used for many applications. The applications include: Fake news identification: This technique works by considering the content of fake news (articles, prints, web pages, etc.) and then examining the words used to cluster the data. Based on the clustering, it then becomes possible to separate fake from true news. Spam filters: Spam filtering algorithms are used by email clients to group "spam" data from genuine emails. The clustering algorithm takes into account the header, content, originating email address, words, etc., to group similar emails together. Document analysis: Clustering is successfully employed in this domain to organize information in far less time compared to other methods. The algorithm considers the words in the document and then groups them into similar categories based on their characteristics. Some other applications of hierarchical clustering are in network traffic analysis, criminal activity trackers, and social media analysis.

The strengths and weaknesses of hierarchical clustering are listed below:

- No need to specify the number of clusters.
- The desired number can be obtained by pruning the resultant dendrogram at desired level.
- On the other hand, it leads to lower flexibility on the decision to combine two clusters.

- The clusters are sensitive to outliers, and irregular size of clusters is complex to handle.
- Not suitable for data that has no hierarchical structure.

Example 4.3 (Hierarchical Clustering)
Problem: Perform hierarchical clustering on Online Retail Dataset[a] to observe the resultant dendrograms. The dataset contains all the transactions occurring between a certain range for a UK-based and registered non-store online retail. The company mainly sells unique all-occasion gifts. Many customers of the company are wholesalers.
Solution: To solve the above-stated problem, follow the steps below:

1. Load the necessary libraries for generating plots, sklearn, numpy, etc.
2. Load the Online Retail Dataset from your local machine.
3. Clean the dataset by removing NaN (not a number) values. Prepare the data such that the Amount, Frequency, and Recency features are considered.
4. After individually preparing the data for Amount, Frequency, and Recency, merge all to generate one single dataframe.
5. Scale the dataset using StandardScaler.
6. Generate dendrogram plots for single, complete. and average linkages. Use Euclidean distance for generating all the plots.

Figures 4.7, 4.8, and 4.9 show the results of hierarchical clustering as explained in the example above. The figures correspond to dendrograms corresponding to single, complete, and average linkages.

[a]https://archive.ics.uci.edu/ml/datasets/online+retail.

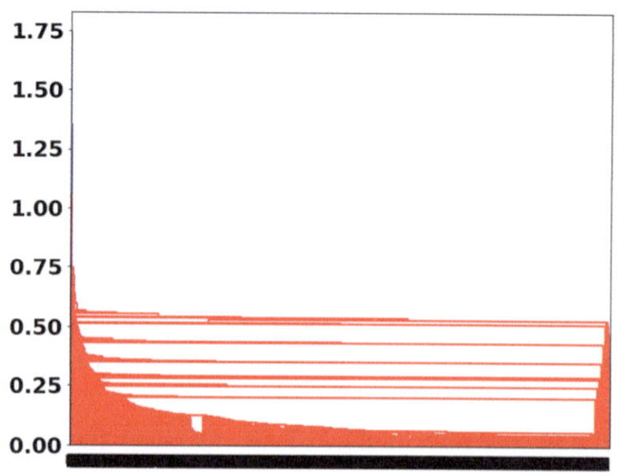

Fig. 4.7 Hierarchical clustering and dendrogram representation with single linkage

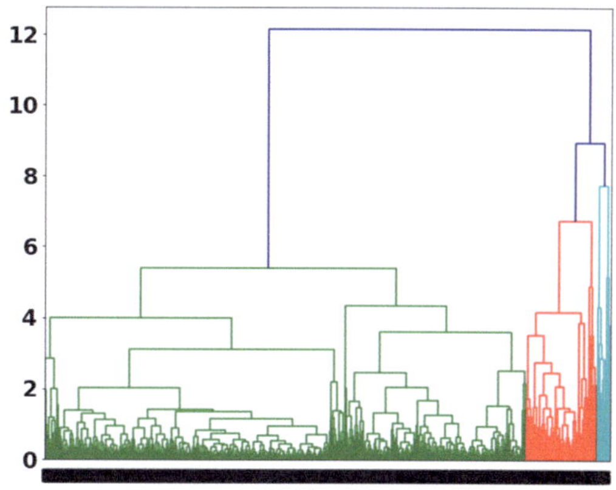

Fig. 4.8 Hierarchical clustering and dendrogram representation with complete linkage

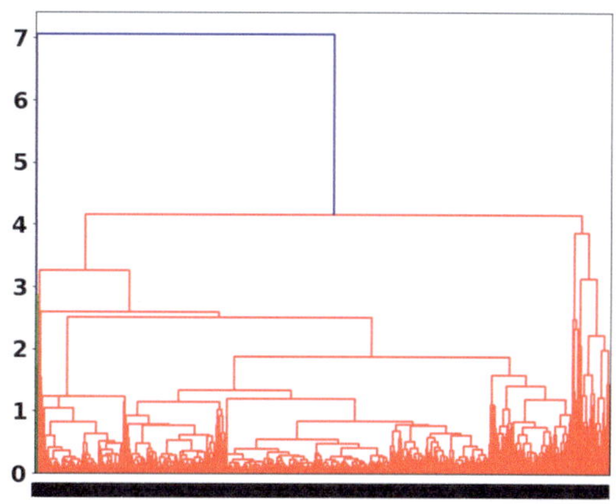

Fig. 4.9 Hierarchical clustering and dendrogram representation with average linkage

```
import numpy as np
import pandas as pd
import matplotlib.pyplot as plt
import seaborn as sns
import datetime as dt

# import required libraries for clustering
import sklearn
from sklearn.preprocessing import StandardScaler
from sklearn.cluster import KMeans
from sklearn.metrics import silhouette_score
from scipy.cluster.hierarchy import linkage
from scipy.cluster.hierarchy import dendrogram
from scipy.cluster.hierarchy import cut_tree

retail = pd.read_csv('../input/online-retail-customer-clustering/OnlineRetail
     .csv', sep=",", encoding="ISO-8859-1", header=0)
retail.head()

## Data cleaning

df_null = round(100*(retail.isnull().sum())/len(retail), 2)
df_null

```

```
retail = retail.dropna()
retail.shape
retail['CustomerID'] = retail['CustomerID'].astype(str)

## Data preparation

retail['Amount'] = retail['Quantity']*retail['UnitPrice']
rfm_m = retail.groupby('CustomerID')['Amount'].sum()
rfm_m = rfm_m.reset_index()
rfm_m.head()

# Frequency attribute

rfm_f = retail.groupby('CustomerID')['InvoiceNo'].count()
rfm_f = rfm_f.reset_index()
rfm_f.columns = ['CustomerID', 'Frequency']
rfm_f.head()

# Merge the dataframes

rfm = pd.merge(rfm_m, rfm_f, on='CustomerID', how='inner')
rfm.head()

# Convert to datetime to proper datatype

retail['InvoiceDate'] = pd.to_datetime(retail['InvoiceDate'],format='%d-%m-%Y
    %H:%M')

# Compute the maximum date to know the last transaction date

max_date = max(retail['InvoiceDate'])
max_date

# Compute the difference between max date and transaction date

retail['Diff'] = max_date - retail['InvoiceDate']
retail.head()

# Compute last transaction date to get the recency of customers

rfm_p = retail.groupby('CustomerID')['Diff'].min()
rfm_p = rfm_p.reset_index()
rfm_p.head()

# Extract number of days only

rfm_p['Diff'] = rfm_p['Diff'].dt.days
rfm_p.head()

# Merge tha dataframes to get the final RFM dataframe

rfm = pd.merge(rfm, rfm_p, on='CustomerID', how='inner')
rfm.columns = ['CustomerID', 'Amount', 'Frequency', 'Recency']
rfm.head()

# Rescaling the attributes

rfm_df = rfm[['Amount', 'Frequency', 'Recency']]

# Instantiate
scaler = StandardScaler()

# fit_transform
rfm_df_scaled = scaler.fit_transform(rfm_df)
rfm_df_scaled.shape
rfm_df_scaled = pd.DataFrame(rfm_df_scaled)
rfm_df_scaled.columns = ['Amount', 'Frequency', 'Recency']
```

```
rfm_df_scaled.head()

## Hierarchical Clustering ##
# Single linkage:

mergings = linkage(rfm_df_scaled, method="single", metric='euclidean')
dendrogram(mergings)
plt.show()

# Complete linkage

mergings = linkage(rfm_df_scaled, method="complete", metric='euclidean')
dendrogram(mergings)
plt.show()


# Average linkage

mergings = linkage(rfm_df_scaled, method="average", metric='euclidean')
dendrogram(mergings)
plt.show()
```

4.2.3 Mixture Models

The previously discussed clustering algorithms face some issues in dealing with data points that are not circular in their distribution in space, or the pattern of their distribution is such that the circle-sized K-Means cluster cannot correctly group them. Also, the K-Means algorithm groups data points based on the distance. This may not be applicable to certain datasets. Hence, Gaussian Mixture Models (GMMs) are proposed as a panacea to address the problems with previous clustering methods explained. GMMs group data based on their distribution. The assumption in GMMs is that the data has a certain number of Gaussian distributions, and each represents a cluster.

Figure 4.10 shows a Gaussian distribution curve for different data points. The distributions have a mean (μ) and variance (σ^2). The probability distribution function (PDF) is given by Eq. 4.4, where x is the input vector, μ is the 2-dimensional mean vector, and Σ is the $2x2$ covariance matrix. For multivariate GMM, it would consist of x and μ as vectors with length d, and Σ would be a $d \times d$ covariance matrix. Finally, for a d-feature data, we would need k cluster size Gaussian distributions, each with its mean and variance matrix. And, a technique known as Expectation-Maximization (EM) is used to assign the mean and covariance values. The EM is a statistical algorithm for finding optimum model parameters. The point to note here is that the EM method is used in cases where the data has latent or missing variables.

GMMs assume that the optimum number of clusters is unknown. Because the latent variables are unknown, the EM helps find the optimum values and then the model parameters. The EM does so based on the existing data. The EM follows a two-step approach, the E-step and the M-step. In the E-step, the data is used to guess

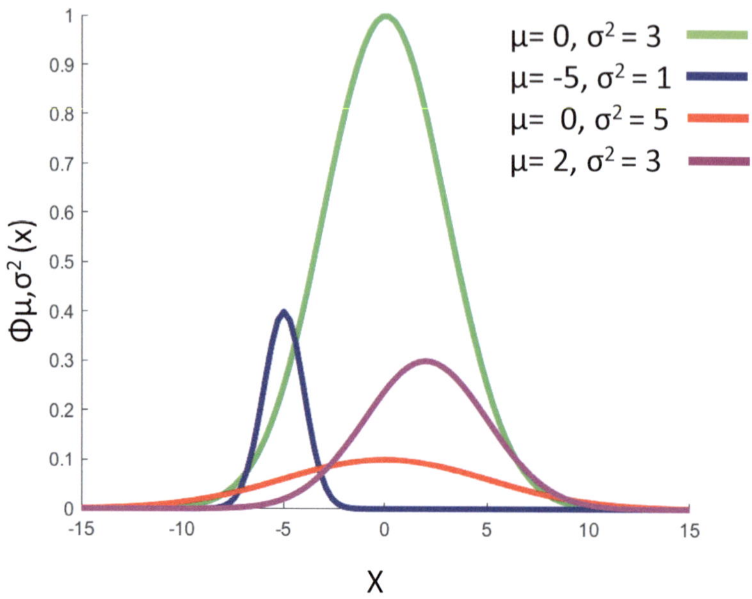

Fig. 4.10 Gaussian (normal) distribution curve with multiple mean and variance

the latent variables' values, while, in the M-step, the complete data is used to update the model parameters. This step uses the values guessed during E-step.

$$f(x|\mu, \Sigma) = \frac{1}{\sqrt{2\pi|\Sigma|}} \exp\left[-\frac{1}{2}(x-\mu)^t(\Sigma)^{-1}(x-\mu)\right]. \tag{4.4}$$

Mixture models are used in speech recognition systems, multiple object tracking, and text and color database retrieval systems. The GMM techniques have faster convergence with flexible cluster assignment compared to K-Means clustering. Furthermore, in contrast, K-Means is a hard algorithm—a data point can belong to only one cluster. Despite the effectiveness, GMM can fail to work for data with very high dimensionality and depends on the user to set the number of mixture models to fit on the data.

Example 4.4 (Gaussian Mixture Model)
Problem: Use the "Clustering_gmm" dataset[a] to test a GMM model.
Solution: To build a GMM model, import the pandas library, the GMM algorithm from sklearn, matplotlib library, load the dataset, and plot the same

(continued)

Example 4.4 (continued)
to visualize the data points. Based on how scattered the data points look, decide the number of components, defined by "n_components" parameter to the `GaussianMixture` function. Fit the model on the data, and generate the predictions. Plot the results, each cluster in a different color.

[a]https://cdn.analyticsvidhya.com/wp-content/uploads/2019/10/Clustering_gmm.csv.

Sample code for a Gaussian Mixture Model is given below.

```
## Importing the libraries and necessary packages

import pandas as pd
from sklearn.mixture import GaussianMixture
import matplotlib.pyplot as plt

## Load the dataset

data = pd.read_csv('Clustering_gmm.csv')

## Visualize the dataset

plt.figure(figsize=(7,7))
plt.scatter(data["Weight"],data["Height"])
plt.xlabel('Weight')
plt.ylabel('Height')
plt.title('Data Distribution')
plt.show()

## Train GMM on the data

gmm = GaussianMixture(n_components=4)
gmm.fit(data)

## Generate predictions from the GMM model

labels = gmm.predict(data)
frame = pd.DataFrame(data)
frame['cluster'] = labels
frame.columns = ['Weight', 'Height', 'cluster']

## Use different colors for each cluster

color=['blue','green','cyan', 'black']
for k in range(0,4):
    data = frame[frame["cluster"]==k]
    plt.scatter(data["Weight"],data["Height"],c=color[k])
plt.show()
```

4.3 Unsupervised Neural Networks

In this section, you will learn how to apply unsupervised learning techniques to identify patterns and structures within datasets.

Unsupervised learning techniques are a valuable set of tools for exploratory analysis. They bring out patterns and structure within datasets, which yield information that may be informative in itself or serve as a guide to further analysis. It is critical to have a solid set of unsupervised learning tools that you can apply to help break up unfamiliar or complex datasets into actionable information.

We will begin by discussing Kohonen's Self-Organizing Map (SOM), a method of topological clustering that enables the projection of complex datasets into two dimensions. Next, we will discuss generative adversarial networks' (GANs) some very popular neural models that can be employed to perform a data-generating process and new samples that can be drawn from it. Then, we will review Deep Belief Networks (DBNs), a very famous generative model that, in an unsupervised scenario, can be employed in order to perform the dimensionality reduction of input dataset X, drawn from a predefined data-generating process.

4.3.1 Self-Organizing Maps

A SOM is a technique to generate topological representations of data in reduced dimensions. It is one of the number of techniques with such applications, with a better-known alternative being PCA. However, SOMs present unique opportunities, both as dimensionality reduction techniques and as a visualization format.

The SOM algorithm involves iteration over many simple operations. When applied at a smaller scale, it behaves similarly to K-Means clustering (as we will see shortly). At a larger scale, SOMs reveal the topology of complex datasets in a powerful way. A SOM is made up of a grid (commonly rectangular or hexagonal) of nodes, where each node contains a weight vector that is of the same dimensionality as the input dataset. The nodes may be initialized randomly, but an initialization that roughly approximates the distribution of the dataset will tend to train faster.

The algorithm iterates as observations are presented as input. Iteration takes the following form: (1) Identifying the winning node in the current configuration—the Best Matching Unit (BMU). The BMU is identified by measuring the Euclidean distance in the data space of all the weight vectors; (2) The BMU is adjusted (moved) toward the input vector; and (3) Neighboring nodes are also adjusted, usually by lesser amounts, with the magnitude of neighboring movement being dictated by a neighborhood function. (Neighborhood functions vary. In this section, we will use a Gaussian neighborhood function.) This process repeats over potentially many iterations, using sampling if appropriate, until the network converges (reaching a position where presenting a new input does not provide an opportunity to minimize loss).

A node in a SOM is not unlike that of a neural network. It typically possesses a weight vector of length equal to the dimensionality of the input dataset. This means that the topology of the input dataset can be preserved and visualized through a lower-dimensional mapping. Let us start working with understanding the implementation of the SOM algorithm in a familiar context.

Implementing SOM As discussed previously, the SOM algorithm is iterative, being based around Euclidean distance comparisons of vectors. This mapping tends to form a fairly readable 2D grid. In the case of the commonly used Iris tutorial dataset, a SOM will map it out pretty cleanly as shown in Figs. 4.11 and 4.12.

In this diagram, the classes have been separated and also ordered spatially. The background coloring in this case is a clustering density measure. The X and Y axis in these figures represent the classes and the shading represents the clustering density. There is some minimal overlap between the blue and green classes, where the SOM performed an imperfect separation. On the Iris dataset, a SOM will tend to approach a converged solution on the order of 100 iterations, with little visible improvement after 1000. For more complex datasets containing less clearly divisible cases, this process can take tens of thousands of iterations.

Unfortunately, there are not implementations of the SOM algorithm within the pre-existing Python packages such as scikit-learn. This makes it necessary for us to use our own implementation.

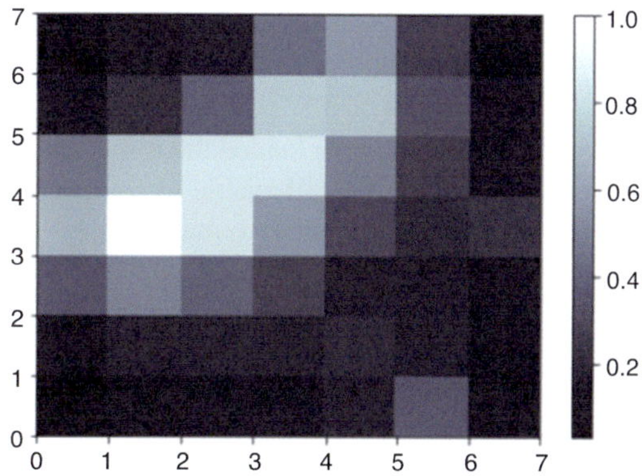

Fig. 4.11 Representation of clusters for Iris dataset using the heatmap

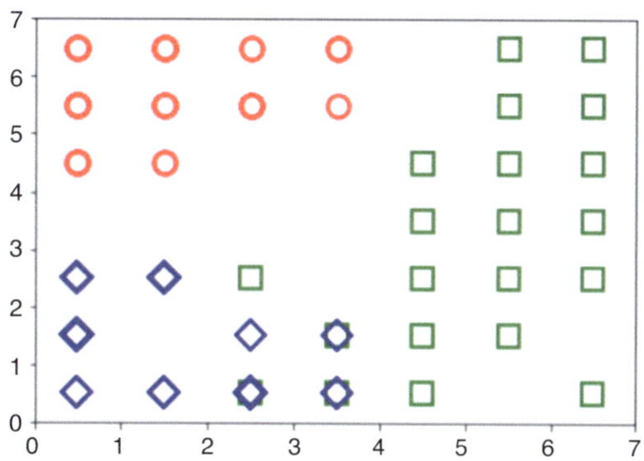

Fig. 4.12 Iris data distribution in clusters Explain legend

Example 4.5 (Self-Organizing Map)
Problem: Employ SOM on digits dataset that is built-in scikit-learn.
Solution: To begin the implementation of SOM, the reader must first import the necessary libraries and dataset. Then import SOM class provided in file Som.py. Set the values for sigma and learning rate. Iterate through each data point and set up label and color assignments for each class, so that one can distinguish classes on the plotted SOM.

For now, let us take a look at the relevant script and get an understanding of how the code works:

```
import numpy as np
from sklearn.datasets import load_digits
from som import Som
from pylab import plot,axis,show,pcolor,colorbar,bone


digits = load_digits()
data = digits.data
labels = digits.target
```

At this point, we have loaded the digits dataset and identified labels as a separate set of data. Doing this will enable us to observe how the SOM algorithm separates classes when assigning them to map:

```
som = Som(16,16,64,sigma=1.0,learning_rate=0.5)
som.random_weights_init(data)
print("Initiating SOM.")
som.train_random(data,10000)
print("\n. SOM Processing Complete")

bone()
pcolor(som.distance_map().T)
colorbar()
```

At this point, we have utilized a Som class that is provided in a separate file, Som.py, in the repository. This class contains the methods required to deliver the SOM algorithm we discussed earlier in the chapter. As arguments to this function, we provide the dimensions of the map (After trialing a range of options, we will start out with 16 x 16 in this case—this grid size gave the feature map enough space to spread out while retaining some overlap between groups.) and the dimensionality of the input data. (This argument determines the length of the weight vector within the SOM's nodes.) We also provide values for sigma and learning rate.

Sigma, in this case, defines the spread of the neighborhood function. As noted previously, we are using a Gaussian neighborhood function. The appropriate value for sigma varies by grid size. For an 8×8 grid, we would typically want to use a value of 1.0 for Sigma, while in this case, we are using 1.3 for a 16×16 grid. It is fairly obvious when one's value for sigma is off; if the value is too small, values tend to cluster near the center of the grid. If the values are too large, the grid typically ends up with several large, empty spaces toward the center.

The learning rate self-explanatorily defines the initial learning rate for the SOM. As the map continues to iterate, the learning rate adjusts according to the following function:

$$learning\ rate(t) = learning\ rate/(1 + t/(0.5 \times t)). \tag{4.5}$$

Here, t is the iteration index. We follow up by first initializing our SOM with random weights. Next, we set up labels and color assignations for each class, so that we can distinguish classes on the plotted SOM. Following this, we iterate through each data point. On each iteration, we plot a class-specific marker for the BMU as calculated by our SOM algorithm. When the SOM finishes iteration, we add a U-Matrix (a colorized matrix of relative observation density) as a monochrome-scaled plot layer:

```
labels[labels == '0'] = 0
labels[labels == '1'] = 1
labels[labels == '2'] = 2
labels[labels == '3'] = 3
labels[labels == '4'] = 4
labels[labels == '5'] = 5
labels[labels == '6'] = 6
labels[labels == '7'] = 7
labels[labels == '8'] = 8
labels[labels == '9'] = 9

markers = ['o', 'v', '1', '3', '8', 's', 'p', 'x', 'D', '*']
colors = ["r", "g", "b", "y", "c", (0,0.1,0.8), (1,0.5,0), (1,1,0.3), "m",
    (0.4,0.6,0)]
for cnt,xx in enumerate(data):
    w = som.winner(xx)
    plot(w[0]+.5,w[1]+.5,markers[labels[cnt]],
    markerfacecolor='None', markeredgecolor=colors[labels[cnt]],
    markersize=12, markeredgewidth=2)
    axis([0,som.weights.shape[0],0,som.weights.shape[1]])
    show()
```

This code delivers a 16 × 16 node SOM plot. As we can see, the map has done a reasonably good job of separating each cluster into topologically distinct areas of the map. Certain classes (particularly the digits five in cyan circles and nine in green stars) have been located over multiple parts of the SOM space. For the most part, though, each class occupies a distinct region, and it is fair to say that the SOM has been reasonably effective. The U-Matrix shows that regions with a high density of points are co-habited by data from multiple classes (Fig. 4.13).

Some advantages of SOM are that the data is easily interpreted and understood. Further, the reduction of dimensionality and grid clustering makes it easy to observe similarities in the data. On the other hand, it does not build a generative model for the data, i.e., the model does not understand how data is created. Also, SOM does not behave so gently when using categorical data, even worse for mixed types of data. The time for preparing SOM is slow, hard to train against slowly evolving data.

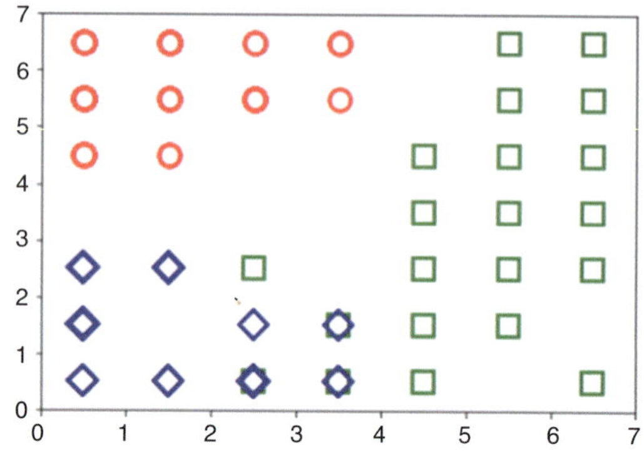

Fig. 4.13 Iris data distribution in clusters

4.3.2 *Generative Adversarial Networks*

Generative adversarial network (GAN) is a class of machine learning frameworks designed by Ian Goodfellow and his colleagues in 2014. Generative adversarial networks, or GANs for short, are an approach to generative modeling using deep learning methods, such as convolutional neural networks.

Generative modeling is an unsupervised learning task in machine learning that involves automatically discovering and learning the regularities or patterns in input data in such a way that the model can be used to generate or output new examples that plausibly could have been drawn from the original dataset. GANs are a clever way of training a generative model by framing the problem as a supervised learning problem with two sub-models: the generator model that we train to generate new examples, and the discriminator model that tries to classify examples as either real (from the domain) or fake (generated) as shown in Fig. 4.14. The two models are trained together in a zero-sum game, adversarial, until the discriminator model is fooled about half the time, meaning the generator model is generating plausible examples.

GANs are an exciting and rapidly changing field, delivering on the promise of generative models in their ability to generate realistic examples across a range of problem domains, most notably in image-to-image translation tasks such as translating photos of summer to winter or day to night, and in generating photorealistic photos of objects, scenes, and people that even humans cannot tell are fake.

GANs are widely used, and some of the applications of GANs are as follows:(1) GAN can be used to detect glaucomatous images helping the early diagnosis that is essential to avoid partial or total loss of vision, (2) GANs that produce photorealistic images can be used to visualize the interior design, industrial design, shoes, bags, and clothing items or items for computer games' scenes. Such networks were reported to be used by Facebook, (3) GANs can be used to age face photographs to show how an individual's appearance might change with age, (4) GANs have been used to visualize the effect that climate change will have on specific houses,

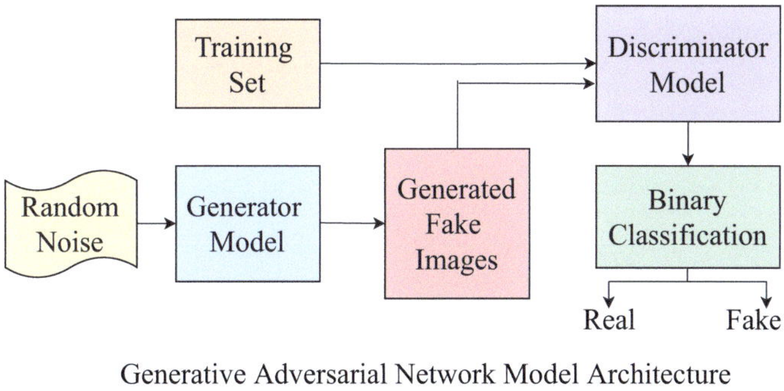

Fig. 4.14 GAN model architecture

and (5) A GAN model called Speech2Face can reconstruct an image of a person's face after listening to their voice.

Example 4.6 (Generative Adversarial Networks)
Problem: Develop a generative adversarial network with deep convolutional networks for generating handwritten digits.
Solution: To begin this, first define and train the standalone discriminator model for learning the difference between real and fake images. Define the standalone generator model and train the composite generator and discriminator model. Then evaluate the performance of the GAN and use the final standalone generator model to generate new images.

```

# example of training a gan on mnist
from numpy import expand_dims
from numpy import zeros
from numpy import ones
from numpy import vstack
from numpy.random import randn
from numpy.random import randint
from keras.datasets.mnist import load_data
from keras.optimizers import Adam
from keras.models import Sequential
from keras.layers import Dense
from keras.layers import Reshape
from keras.layers import Flatten
from keras.layers import Conv2D
from keras.layers import Conv2DTranspose
from keras.layers import LeakyReLU
from keras.layers import Dropout
from matplotlib import pyplot

# define the standalone discriminator model
def define_discriminator(in_shape=(28,28,1)):
    model = Sequential()
```

```
    model.add(Conv2D(64, (3,3), strides=(2, 2), padding='same', input_shape=
       in_shape))
    model.add(LeakyReLU(alpha=0.2))
    model.add(Dropout(0.4))
    model.add(Conv2D(64, (3,3), strides=(2, 2), padding='same'))
    model.add(LeakyReLU(alpha=0.2))
    model.add(Dropout(0.4))
    model.add(Flatten())
    model.add(Dense(1, activation='sigmoid'))
    # compile model
    opt = Adam(lr=0.0002, beta_1=0.5)
    model.compile(loss='binary_crossentropy', optimizer=opt, metrics=['accuracy
       '])
    return model

# define the standalone generator model
def define_generator(latent_dim):
    model = Sequential()
    # foundation for 7x7 image
    n_nodes = 128 * 7 * 7
    model.add(Dense(n_nodes, input_dim=latent_dim))
    model.add(LeakyReLU(alpha=0.2))
    model.add(Reshape((7, 7, 128)))
    # upsample to 14x14
    model.add(Conv2DTranspose(128, (4,4), strides=(2,2), padding='same'))
    model.add(LeakyReLU(alpha=0.2))
    # upsample to 28x28
    model.add(Conv2DTranspose(128, (4,4), strides=(2,2), padding='same'))
    model.add(LeakyReLU(alpha=0.2))
    model.add(Conv2D(1, (7,7), activation='sigmoid', padding='same'))
    return model

# define the combined generator and discriminator model, for updating the
     generator
def define_gan(g_model, d_model):
    # make weights in the discriminator not trainable
    d_model.trainable = False
    # connect them
    model = Sequential()
    # add generator
    model.add(g_model)
    # add the discriminator
    model.add(d_model)
    # compile model
    opt = Adam(lr=0.0002, beta_1=0.5)
    model.compile(loss='binary_crossentropy', optimizer=opt)
    return model

# load and prepare mnist training images
def load_real_samples():
    # load mnist dataset
    (trainX, _), (_, _) = load_data()
    # expand to 3d, e.g. add channels dimension
    X = expand_dims(trainX, axis=-1)
    # convert from unsigned ints to floats
    X = X.astype('float32')
    # scale from [0,255] to [0,1]
    X = X / 255.0
    return X

# select real samples
def generate_real_samples(dataset, n_samples):
    # choose random instances
    ix = randint(0, dataset.shape[0], n_samples)
    # retrieve selected images
    X = dataset[ix]
    # generate 'real' class labels (1)
```

```
  y = ones((n_samples, 1))
  return X, y

# generate points in latent space as input for the generator
def generate_latent_points(latent_dim, n_samples):
  # generate points in the latent space
  x_input = randn(latent_dim * n_samples)
  # reshape into a batch of inputs for the network
  x_input = x_input.reshape(n_samples, latent_dim)
  return x_input

# use the generator to generate n fake examples, with class labels
def generate_fake_samples(g_model, latent_dim, n_samples):
  # generate points in latent space
  x_input = generate_latent_points(latent_dim, n_samples)
  # predict outputs
  X = g_model.predict(x_input)
  # create 'fake' class labels (0)
  y = zeros((n_samples, 1))
  return X, y

# create and save a plot of generated images (reversed grayscale)
def save_plot(examples, epoch, n=10):
  # plot images
  for i in range(n * n):
    # define subplot
    pyplot.subplot(n, n, 1 + i)
    # turn off axis
    pyplot.axis('off')
    # plot raw pixel data
    pyplot.imshow(examples[i, :, :, 0], cmap='gray_r')
  # save plot to file
  filename = 'generated_plot_e%03d.png' % (epoch+1)
  pyplot.savefig(filename)
  pyplot.close()

# evaluate the discriminator, plot generated images, save generator model
def summarize_performance(epoch, g_model, d_model, dataset, latent_dim,
    n_samples=100):
  # prepare real samples
  X_real, y_real = generate_real_samples(dataset, n_samples)
  # evaluate discriminator on real examples
  _, acc_real = d_model.evaluate(X_real, y_real, verbose=0)
  # prepare fake examples
  x_fake, y_fake = generate_fake_samples(g_model, latent_dim, n_samples)
  # evaluate discriminator on fake examples
  _, acc_fake = d_model.evaluate(x_fake, y_fake, verbose=0)
  # summarize discriminator performance
  print('>Accuracy real: %.0f%%, fake: %.0f%%' % (acc_real*100, acc_fake*100)
    )
  # save plot
  save_plot(x_fake, epoch)
  # save the generator model tile file
  filename = 'generator_model_%03d.h5' % (epoch + 1)
  g_model.save(filename)

# train the generator and discriminator
def train(g_model, d_model, gan_model, dataset, latent_dim, n_epochs=100,
    n_batch=256):
  bat_per_epo = int(dataset.shape[0] / n_batch)
  half_batch = int(n_batch / 2)
  # manually enumerate epochs
  for i in range(n_epochs):
    # enumerate batches over the training set
    for j in range(bat_per_epo):
      # get randomly selected 'real' samples
      X_real, y_real = generate_real_samples(dataset, half_batch)
```

```
        # generate 'fake' examples
        X_fake, y_fake = generate_fake_samples(g_model, latent_dim, half_batch)
        # create training set for the discriminator
        X, y = vstack((X_real, X_fake)), vstack((y_real, y_fake))
        # update discriminator model weights
        d_loss, _ = d_model.train_on_batch(X, y)
        # prepare points in latent space as input for the generator
        X_gan = generate_latent_points(latent_dim, n_batch)
        # create inverted labels for the fake samples
        y_gan = ones((n_batch, 1))
        # update the generator via the discriminator's error
        g_loss = gan_model.train_on_batch(X_gan, y_gan)
        # summarize loss on this batch
        print('>%d, %d/%d, d=%.3f, g=%.3f' % (i+1, j+1, bat_per_epo, d_loss,
      g_loss))
    # evaluate the model performance, sometimes
    if (i+1) % 10 == 0:
        summarize_performance(i, g_model, d_model, dataset, latent_dim)

# size of the latent space
latent_dim = 100
# create the discriminator
d_model = define_discriminator()
# create the generator
g_model = define_generator(latent_dim)
# create the gan
gan_model = define_gan(g_model, d_model)
# load image data
dataset = load_real_samples()
# train model
train(g_model, d_model, gan_model, dataset, latent_dim)
```

4.3.3 Deep Belief Nets

In this subsection, we will focus on some more sophisticated techniques, drawing from the area of deep learning. This subsection is dedicated to building an understanding of how to apply the Restricted Boltzmann Machine (RBM) and manage the deep learning architecture one can create by chaining RBMs—the Deep Belief Network (DBN). DBNs are trainable to effectively solve complex problems in the text, image, and sound recognition. They are used by leading companies for object recognition, intelligent image search, and robotic spatial recognition.

A DBN is a graphical model, constructed using multiple stacked RBMs as shown in Fig. 4.15. While the first RBM trains a layer of features based on input from the pixels of the training data, subsequent layers treat the activations of preceding layers as if they were pixels and attempt to learn the features in subsequent hidden layers. This is frequently described as learning the representation of data and is a common theme in deep learning.

From a practical perspective, it is a trade-off between increasing accuracy and increasing computational cost. It is the case that each layer of RBMs will improve the lower bound of the log probability of the training data. In other words; the DBN almost inevitably becomes less bad with each additional layer of features. As far as layer size is concerned, it is generally advantageous to reduce the number of nodes

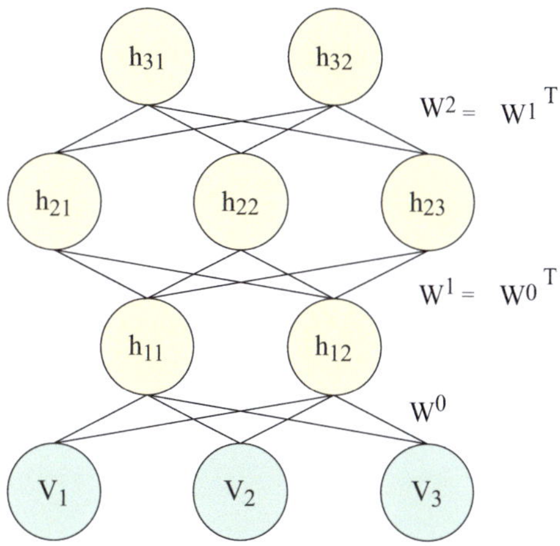

Fig. 4.15 Architecture of deep belief network

in the hidden layers of successive RBMs. One should avoid contexts in which an RBM has at least as many visible units as the RBM preceding it has hidden units.

It can be advantageous (but is by no means necessary) when successive RBMs decrease in layer size until the final RBM has a layer size approximating the dimensionality of variance in the data. Affixing an MLP to the end of a DBN whose layers have too many nodes will harm classification performance; Even an MLP with many neurons may not successfully train in such contexts. On a related note, it has been noted that even if the layers do not contain very many nodes, with enough layers, more or less any function can be modeled.

Determining what the dimensionality of variance in the data is, is not a simple task. One tool that can support this task is PCA; as we saw in the preceding chapter, PCA can enable us to get a reasonable idea as to how many components of meaningful size exist in the input data.

Training a DBN

Training a DBN is typically done greedily, which is to say that it trains to optimize locally at each layer, rather than attempting to reach a global optimum. The steps involved in the learning process are as follows: (1)The first layer of the DBN is trained using the method that we saw in our earlier discussion of RBM learning. As such, the first layer converts its data distribution into a posterior distribution using Gibbs sampling over the hidden units, (2) This distribution is far more conducive for RBM training than the input data itself so the next RBM layer learns that distribution, (3) Successive RBM layers continue to train on the samples output by preceding layers, and (4) All of the parameters within this architecture are tuned using a performance measure.

This performance measure may vary. It may be a log-likelihood proxy used in gradient descent, as discussed earlier in the chapter. In supervised contexts, a classifier (for example, an MLP) can be added as the final layer of the architecture, and prediction accuracy can be used as the performance measure to fine-tune the deep architecture.

Applying the DBN

We will be working in a similar way to the RBM, by walking through a DBN class and connecting the code to the theory, discussing what to expect and how to review the network's performance, before initializing and training our network to see it in action.

Example 4.7 (Deep Belief Networks)
Problem: Employ a DBN in order to find a low-dimensional representation of the MNIST dataset.
Solution: As the complexity of these models can easily grow, we are going to limit the process to 500 random samples. The implementation is based on the deep-belief-network package (https://github.com/albertbup/deep-belief-network), which supports both NumPy and TensorFlow. In the former case, the classes (whose names remain unchanged) must be imported from the dbn package, while in the latter, the package is dbn.tensorflow. In this example, we are going to use the NumPy version, which has fewer requirements.

Let us take a look at the code:

```
import numpy as np

from sklearn.datasets import load_digits
from sklearn.utils import shuffle

nb_samples = 500

digits = load_digits()

X_train = digits['data'] / np.max(digits['data'])
Y_train = digits['target']

X_train, Y_train = shuffle(X_train, Y_train, random_state=1000)
X_train = X_train[0:nb_samples]
Y_train = Y_train[0:nb_samples]
```

We can now instantiate the UnsupervisedDBN class, with the following structure: 64 input neurons (implicitly detected from the dataset) 32 sigmoid neurons 32 sigmoid neurons 16 sigmoid neurons Hence, the last representation is made up of 16 values (one-quarter of the original dimensionality). We are setting a learning rate of $\eta = 0.025$ and 16 samples per batch. The following snippet initializes and trains the model:

```
    from dbn import UnsupervisedDBN

unsupervised_dbn = UnsupervisedDBN(hidden_layers_structure=[32, 32, 16],
                                   learning_rate_rbm=0.025,
                                   n_epochs_rbm=500,
                                   batch_size=16,
                                   activation_function='sigmoid')

X_dbn = unsupervised_dbn.fit_transform(X_train)
```

At the end of the training process, one can analyze the distribution, after projecting it onto a bidimensional space. We employ the t-SNE algorithm, which guarantees to find the most similar low-dimensional distribution:

```
from sklearn.manifold import TSNE

tsne = TSNE(n_components=2, perplexity=10, random_state=1000)
X_tsne = tsne.fit_transform(X_dbn)
```

Finally, we add a logistic regression layer to the end of the DBN so as to form an MLP: t-SNE plot of the unsupervised DBN output representations.

As you can see, most of the blocks are quite cohesive, indicating that the peculiar properties of a digit have been successfully represented in the lower-dimensional space. In some cases, the same digit group is split into more clusters, but in general, the amount of noisy (isolated) points is extremely low. For example, the group containing the digit 2 is indicated with the symbol x. The majority of the samples is in the range $0 < x0 < 30$, $x1 < -40$; however, a subgroup is also located in the range $-10 < x1 < 10$. If we check the neighbors of this small cluster, they are made up of samples representing the digit 8 (represented by a square). It is easy to understand that some malformed twos are very similar to malformed eights, and this justifies the split of the original cluster. From a statistical viewpoint, the explained variance can have a different impact. In some cases, a few components are enough to determine the peculiar features of a class, but this cannot generally be true. When samples belonging to different classes show similarities, a distinction can only be made thanks to the variance of the secondary components. This consideration is very important when working with datasets containing almost (or even partially) overlapping samples. The main task of the data scientist, when performing dimensionality reduction, is not to check the overall explained variance, but rather, to understand whether there are regions that are negatively affected by the dimensionality reduction. In such situations, it is possible to either define multiple detection rules (for example, when a sample, $xi \in R1$ or $xi \in R4 \rightarrow xi$, has the yk label) or to try to avoid models that create this segmentation.

4.3.4 *Method of Moments*

The moment is a term used generally in statistics and data science. Basically, moments are used to measure the shape of data distribution, of a probability density function. Mathematically, we define it as

$$\mu_n = \int_{-\infty}^{\infty} (x - c)^n f(x)dx \tag{4.6}$$

for moment n around value c. We take the difference between each value from some value raised to the nth power, where n is the moment number and integrating across the entire function from negative infinity to infinity.

The first moment works out to just be the mean of the data. The first moment is the mean, the average. The second moment is the variance. The second moment of the dataset is the same thing as the variance value. The variance is really based on the square of the differences from the mean. The third moment is called skew, and it is basically a measure of how lopsided distribution is.

You can see in these two examples above (in Fig. 4.16) that have a longer tail on the left, now then that is a negative skew, and if one has a longer tail on the right then, that is a positive skew. The dotted lines show what the shape of a normal distribution would look like without skew. The dotted line out on the left side ends up with a negative skew, or on the other side, a positive skew in that example.

The fourth moment is called kurtosis. It is a measure of the shape of the data distribution. Kurtosis is how peaked, how squished together the data distribution is.

Example 4.8 (Method of Moments)
Problem: Compute the moments by creating a random data.
Solution: Create a normal distribution of a random data. Center it around zero, with a 0.5 standard deviation and 10,000 data points, and then plot them.

```
import numpy as np
import matplotlib.pyplot as plt

vals = np.random.normal(0, 0.5, 10000)

plt.hist(vals, 50)
plt.show()
```

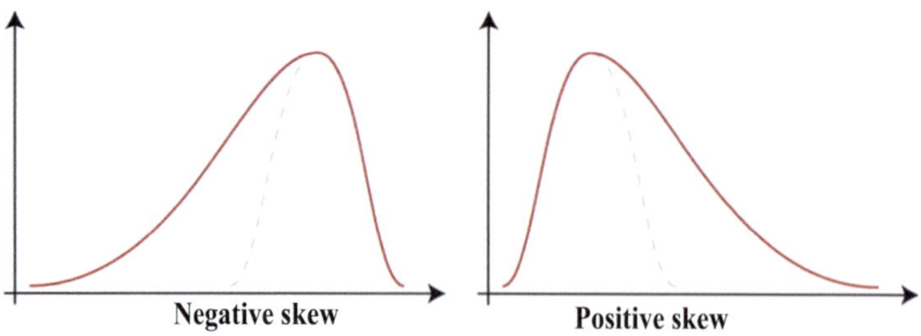

Fig. 4.16 Positive and negative skews

Now, we find the mean and variance. We have done this before; NumPy just gives you a mean and var function to compute that. So, we just call np.mean to find the first moment as shown in the following code:

```
np.mean(vals)
```

Now we find the second moment, which is just another name for variance. We can do that with the following code, as we have seen before:

```
np.var(vals)
```

The third moment is skew, and to do that, we will have to use the SciPy package instead of NumPy. But that again is built into any scientific computing package such as Enthought Canopy or Anaconda. Once we have SciPy, the function call is as simple as our earlier two:

```
import scipy.stats as sp
sp.skew(vals)
```

The fourth moment is kurtosis, which describes the shape of the tail. Again, for a normal distribution that should be about zero. SciPy provides us with another simple function call.

```
sp.kurtosis(vals)
```

In this way, one can compute the moments for any data distribution.

4.4 Feature Selection Techniques

In this subsection, we will introduce and discuss some important techniques that can be implemented to perform both dimensionality reduction and component extraction. In dimensionality reduction, the goal is to transform a high-dimensional dataset into a lower-dimensional one, to try to minimize the amount of information loss. While in the component extraction, it is necessary to find a dictionary of items that can be mixed up so as to build samples.

4.4.1 Principal Component Analysis

Principal Component Analysis (PCA) is a famous feature extraction method in which new independent features are created from old existing features. Furthermore, only the most prominent features for predicting the target output are kept by combining both old and new features. These new features are extracted from old features, and any feature that is less dependent on predicting target output can be dropped.

PCA is by far the most popular dimensionality reduction algorithm, and it is used to interpret and visualize data and to find inter-relation between variables in the data.

PCA is fast, simple to use, and absorbs the global variance of the data. PCA is also used most widely in exploratory data analysis (EDA) and in machine learning for predictive models.

The main limitation of PCA is that it does not consider class separability since it does not take into account the class label of the feature vector. PCA simply performs a coordinate rotation that aligns the transformed axes with the directions of maximum variance. There is no guarantee that the directions of maximum variance will contain good features for discrimination. While PCA is a common approach for linear data, it faces some limitations when dealing with non-linear data. In higher-dimensional space, PCA cannot distinguish non-linear structure from no structure. In other words, PCA does not retain local variance in data. PCA is explained in detail in the steps described below:

Step 1: Standardization—The aim of this step is to standardize the range of the continuous initial variables so that each one of them contributes equally to the analysis. More specifically, the reason why it is critical to perform standardization prior to PCA is that the latter is quite sensitive regarding the variances of the initial variables. That is, if there are large differences between the ranges of initial variables, those variables with larger ranges will dominate over those with small ranges (For example, a variable that ranges between 0 and 100 will dominate over a variable that ranges between 0 and 1.), which will lead to biased results. So, transforming the data to comparable scales can prevent this problem. Mathematically, this can be done by subtracting the mean and dividing by the standard deviation for each value of each variable.

Step 2: Covariance matrix computation—The aim of this step is to understand how the variables of the input dataset are varying from the mean with respect to each other, or in other words, to see if there is any relationship between them, because sometimes variables are highly correlated in such a way that they contain redundant information. So, in order to identify these correlations, we compute the covariance matrix. The covariance matrix is a $p \times p$ symmetric matrix (where p is the number of dimensions) that has as entries the covariances associated with all possible pairs of the initial variables. Since the covariance of a variable with itself is its variance (Cov(a,a)=Var(a)), in the main diagonal (top left to bottom right), we actually have the variances of each initial variable. And since the covariance is commutative (Cov(a,b)=Cov(b,a)), the entries of the covariance matrix are symmetric with respect to the main diagonal, which means that the upper and lower triangular portions are equal.

Step 3: Compute the eigenvectors and eigenvalues of the covariance matrix to identify the principal components—Eigenvectors and eigenvalues are the linear algebra concepts that we need to compute from the covariance matrix in order to determine the principal components of the data. Before getting to the explanation of these concepts, let us first understand what do we mean by principal components. Principal components are new variables that are constructed as linear combinations or mixtures of the initial variables. These combinations are done in such a way that the new variables (i.e., principal components) are uncorrelated, and most of the information within the initial variables is squeezed or compressed into the first components. So, the idea is 10-dimensional data gives you 10 principal components, but PCA tries to put maximum possible information in the first component, then maximum remaining information in the second, and so on. Organizing information in principal components this way will allow you to reduce dimensionality without losing much information, and this by discarding the components with low information and considering the remaining components as your new variables. An important thing to realize here is that the principal components are less interpretable and do not have any real meaning since they are constructed as linear combinations of the initial variables. Geometrically speaking, principal components represent the directions of the data that explain a maximal amount of variance, that is to say, the lines that capture the most information of the data. The relationship between variance and information here is that the larger the variance carried by a line, the larger the dispersion of the data points along with it, and the larger the dispersion along a line, the more information it has. To put all this simply, just think of principal components as new axes that provide the best angle to see and evaluate the data, so that the differences between the observations are better visible.

Step 4: Feature vector—As we saw in the previous step, computing the eigenvectors and ordering them by their eigenvalues in descending order allow us to find the principal components in order of significance. In this step, what we do is to choose whether to keep all these components or discard those of lesser significance (of low eigenvalues) and form with the remaining ones a matrix of vectors that we call feature vector. So, the feature vector is simply a matrix that has as columns the eigenvectors of the components that we decide to keep. This makes it the first step toward dimensionality reduction because if we choose to keep only p eigenvectors (components) out of n, the final dataset will have only p dimensions.

Step 5: Recast the data along the principal component axes—In the previous steps, apart from standardization, you do not make any changes on the data, you just select the principal components and form the feature vector, but the input dataset remains always in terms of the original axes (i.e., in terms of the initial variables). In this step, which is the last one, the aim is to use the feature vector formed using the eigenvectors of the covariance matrix, to reorient the data from the original axes to the ones represented by the principal components (hence, the name Principal Components Analysis). This can be done by multiplying the transpose of the original dataset by the transpose of the feature vector.

Example 4.9 (Principal Component Analysis)
Problem: Perform the Principal Component Analysis (PCA) on the Iris dataset.
Solution: To perform PCA on Iris dataset, one must begin by importing pandas library that helps to load the dataset into dataframe. Then from scikit-learn (sklearn), import StandardScalar that standardizes the features in the dataset. Now import the PCA class implemented in sklearn.decomposition, and then simply pass the number of components to this constructor. Call the fit and then transform methods by passing the feature set to these methods. The transform method returns the specified number of principal components. Plot the results and distribution of data points with respect to the principal components (Fig. 4.17).

```
import pandas as pd

url = "https://archive.ics.uci.edu/ml/machine-learning-databases/iris/iris.data"
# load dataset into Pandas DataFrame
df = pd.read_csv(url, names=['sepal length','sepal width','petal length','petal width','target'])

from sklearn.preprocessing import StandardScaler
features = ['sepal length', 'sepal width', 'petal length', 'petal width']
# Separating out the features
x = df.loc[:, features].values
# Separating out the target
y = df.loc[:,['target']].values
# Standardizing the features
x = StandardScaler().fit_transform(x)

from sklearn.decomposition import PCA
pca = PCA(n_components=2)
principalComponents = pca.fit_transform(x)
principalDf = pd.DataFrame(data = principalComponents
             , columns = ['principal component 1', 'principal component 2'])

finalDf = pd.concat([principalDf, df[['target']]], axis = 1)

fig = plt.figure(figsize = (8,8))
ax = fig.add_subplot(1,1,1)
ax.set_xlabel('Principal Component 1', fontsize = 15)
ax.set_ylabel('Principal Component 2', fontsize = 15)
ax.set_title('2 component PCA', fontsize = 20)
targets = ['Iris-setosa', 'Iris-versicolor', 'Iris-virginica']
colors = ['r', 'g', 'b']
for target, color in zip(targets,colors):
    indicesToKeep = finalDf['target'] == target
    ax.scatter(finalDf.loc[indicesToKeep, 'principal component 1']
               , finalDf.loc[indicesToKeep, 'principal component 2']
               , c = color
               , s = 50)
ax.legend(targets)
ax.grid()
```

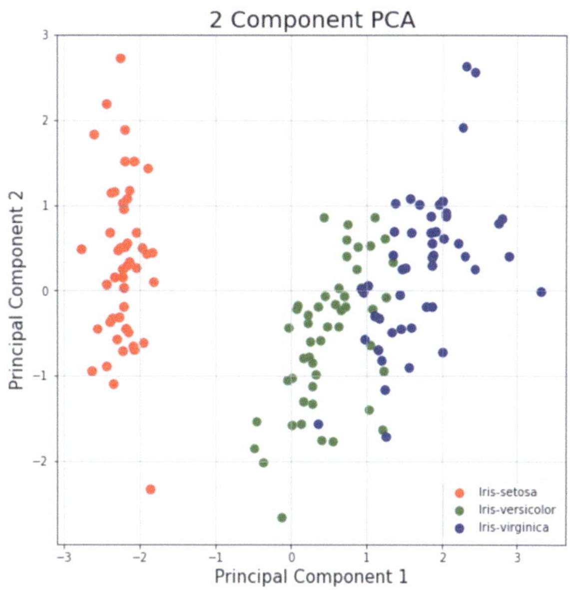

Fig. 4.17 2-component PCA for Iris dataset

4.4.2 T-Distributed Stochastic Neighbor Embedding

Removing redundant information from non-linear/high-dimensional data to obtain useful insights and make data-driven decisions calls for a common approach called t-distributed Stochastic Neighbor Embedding (t-SNE). Unlike PCA that is a famous approach to retain global variance, t-SNE is famous for capturing local and small variances in higher-dimensional data. It converts multi-dimensional data into lower-dimensional data and preserves only small pairwise distances, i.e., neighborhoods. This non-linear algorithm can handle data with a non-linear structure. The t-SNE algorithm tries to preserve neighborhoods from a high-dimensional space by modeling the points in the low-dimensional space to mimic the distance matrix. Similar instances would be modeled by nearby points, and dissimilar instances would be distant points.

Figure 4.18 shows the effectiveness of t-SNE method applied on a non-linear malware dataset. On the left side, this figure shows the result of applying PCA technique, and t-SNE applied on the same dataset displayed on the right side. Due to the non-linearity of this dataset and preserving large distances by PCA, PCA did not produce non-overlapping clusters that would incorrectly preserve the structure of the data. This is because PCA does not preserve local (i.e., small) distances. In contrast with PCA, t-SNE preserves local distances. Hence, it results in clear and non-overlapping clusters used to distinguish malware from benign applications.

The t-SNE algorithm projects high-dimensional data into a low-dimensional space in such a way that the clustering of data in high-dimensional space is

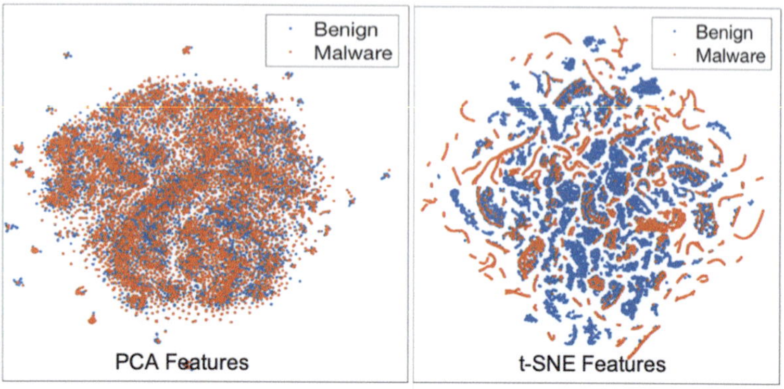

Fig. 4.18 Principal component analysis vs. t-distributed Stochastic neighbor embedding applied on a non-linear malware dataset. While PCA results in polluted features for advanced malware detection, t-SNE provides a clear distinction between malware and benign applications by using local variances

preserved. In the malware dataset example in Fig. 4.18, such clustering of data represents the type of applications in the dataset, i.e., malware or benign.

The t-SNE algorithm has multiple steps. In the first step, it finds the pairwise similarity between similar data points in a high-dimensional space. For every data point x_i in a non-linear dataset, t-SNE considers its Euclidean distance with each of the other data points, i.e., $||x_i - x_j||^2$. Hence, the algorithm would have a distance matrix. The pairwise instances are used to form a conditional probability distribution. The t-SNE algorithm constructs a probability distribution over the pairwise instances such that similar pairs are assigned a higher probability and non-similar pairs get a minuscule probability. The similarity of a data point x_j to x_i is represented as a conditional probability $p_{j|i}$ that is the probability that x_i takes x_j for its neighbor. In other words, the conditional probability value will be proportional to the pairwise similarity.

$$p_{j|i} = \frac{exp(-||x_i - x_j||^2/2\sigma_i^2)}{\sum_{k\neq i} exp(-||x_i - x_k||^2/2\sigma_i^2)} \tag{4.7}$$

In Eq. 4.7, x_i picks the data point x_j as its neighbor according to the proportion of its probability density under a Gaussian distribution centered at x_i. The variance of the Gaussian distribution, i.e., σ_i^2, is centered on data point x_i and is controlled by a parameter called **perplexity** that controls the neighborhood size in t-SNE algorithm by setting the number of effective neighbors that leads to the formation of clusters around data points. In other words, perplexity is a parameter in the t-SNE algorithm that indicates the expected density near a data point.

Next, the conditional probabilities are summarized in high-dimensional space by averaging the two probabilities $p_{j|i}$ and $p_{i|j}$ as shown below to get the final

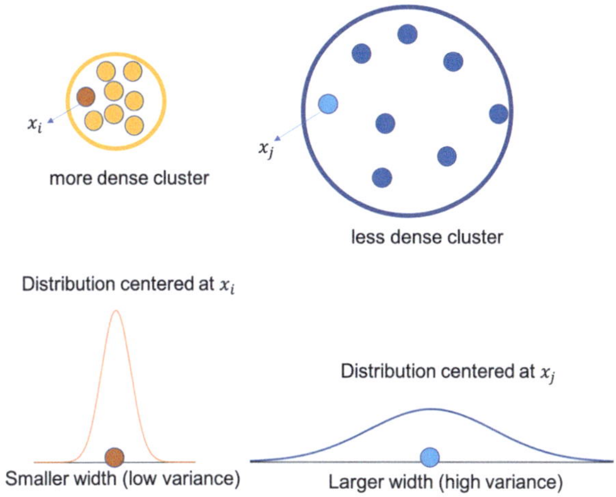

Fig. 4.19 The impact of neighborhood density of data points on the variance of a distribution

similarities, where n refers to the number of data points (we are considering probabilities from two different distributions), and p_{ij} is the probability of instances x_i and x_j being in the same neighborhood. Due to the varying density of clusters (determined by the neighborhood size around a data point), the width of the Gaussian distribution centered at a data point can be different. Essentially, clusters that are relatively less dense have a distribution with higher variance (i.e., larger width), while clusters that are relatively more dense have low variance (i.e., smaller width); this is illustrated in Fig. 4.19. As a result, the distance from x_i to x_j may not be equivalent to the distance from x_j to x_i. Therefore, t-SNE takes the average of the two distances to represent a bidirectional distance between x_i and x_j. In the following equation, p_{ij} is the probability of data points x_i and x_j being in the same neighborhood. For each data point x_i, there would be a set of p_{ij}, referred as P_i. Also, $2n$ refers to the number of data points, since we are considering probabilities from two different distributions. The probabilities represent the similarities between the two data points x_i and x_j.

$$p_{ij} = \frac{p_{j|i} + p_{i|j}}{2n}. \tag{4.8}$$

In the code snippet provided below, **computeDistance(X)** function computes the pairwise distances and generates the distance matrix D. The function **plotGaussian** plots the distance matrix related to the neighborhood of a data point (see Fig. 4.20).

```
import matplotlib.pyplot as plt
from scipy.stats import norm
import itertools

# Standardization
scaler = preprocessing.StandardScaler()
```

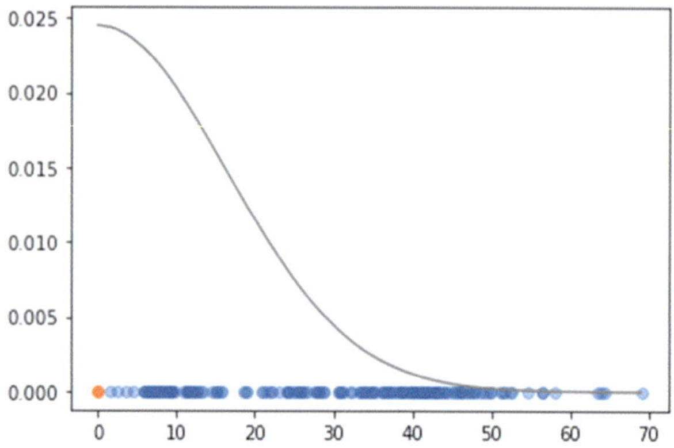

Fig. 4.20 Visualization of the similarity matrix for a sample high-dimensional/non-linear dataset

```
X = scaler.fit_transform(df)

def plotGaussian(points, idx):
    # points[idx] is the datapoint itself, which is 0.
    axis = np.sort(points)
    std = np.std(points) # In the t-SNE implementation, this would also be controlled by PERPLEXITY
    plt.plot(axis, norm.pdf(axis, points[idx] , std), color='tab:grey') # 0 refers to the center, which is the data point itself
    plt.scatter(points[idx], 0, c='tab:orange')
    plt.scatter(points[points != 0], np.zeros(len(points)-1), c='tab:blue', alpha=.4)
    plt.show()

def computeDistance(X):
    N = X.shape[0]
    D = np.zeros((N, N)) # D[i, i] is set to be 0 in t-SNE
    for i, j in itertools.combinations(range(N), 2): # pairwise that like the upper triangular matrix but skip diagonals
        dist = np.linalg.norm(X[i, :] - X[j, :])**2
        D[i, j] = dist
        D[j, i] = dist
    return D

dist_matrix = computeDistance(X)
idx = 0 # You can play with any other instances
plotGaussian(dist_matrix[idx, :], idx)
```

In the second step, each point in the high-dimensional space is mapped to a low-dimensional space according to the pairwise similarities of data points in the high-dimensional space. The data from step-1 is randomly projected into a lower-dimensional space, and then the similarity values for the data points are calculated in the lower-dimensional space.

In the very beginning, we would create random low-dimensional points, $\{y_1, y_2, \ldots, y_n\}$, to represent our original data, $\{x_1, x_2, \ldots, x_n\}$. Then, we would repeat roughly the same structure as what we do in the high-dimensional space, except the probability distribution. Instead of Gaussian distribution, Student's t-distribution is used. That is why this technique is called t-Distributed Stochastic

Neighbor Embedding. The difference between random low-dimensional data points makes dissimilar points more farther apart. Similarly, for each instance y_i, it would also have a set of q_{ij} according to Eq. 4.9, referred to be Q_i.

$$q_{ij} = \frac{(1 + ||y_i - y_j||^2)^{-1}}{\sum_k \sum_{l \neq k}(1 + ||y_k - y_l||^2)^{-1}}. \tag{4.9}$$

Finally, in the third step, the approximation Q is optimized to look like the observations P. To evaluate how two probability distributions are different, Kullback–Leibler divergence (aka relative entropy) is used to minimize their difference. Kullback–Leibler (KL) divergence measures the difference between two probability distributions. Using KL divergence, we can understand how the approximation Q is different from the observations P:

$$KL(P|Q) = \sum_i \sum_j p_{ij} \log \frac{p_{ij}}{q_{ij}}, i \neq j. \tag{4.10}$$

The objective here is to minimize Eq. 4.10 to let Q approximate P. To update y, the low-dimensional representations, gradient descent can be used. Once updated, step 2 is repeated to recompute Q, and each iteration would repeat steps 2 and 3.

The code snippet below shows how t-SNE from sklearn.manifold is applied on a high-dimensional dataset to map the high-dimensional data to a low-dimensional 2-D space. The result of such mapping using t-SNE creates clear non-overlapping clusters (i.e., class 0, class 1, class2), which is illustrated in Fig. 4.21.

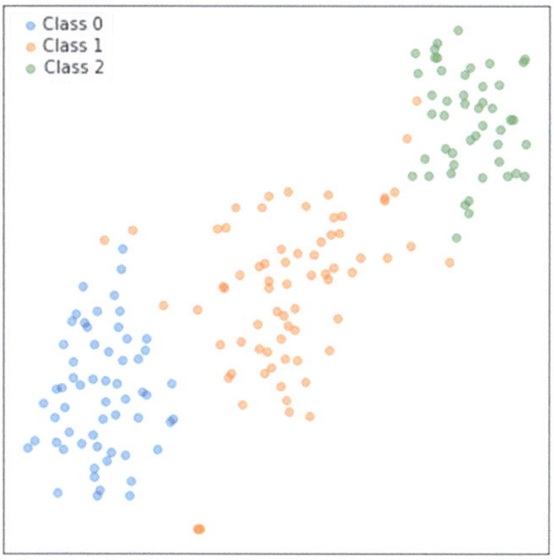

Fig. 4.21 Mapping of a high-dimensional dataset into a 2-D lower-dimensional space using t-SNE algorithm

```
from sklearn import preprocessing
from sklearn.manifold import TSNE

scaler = preprocessing.StandardScaler()
tsne = TSNE(n_components=2)
X = scaler.fit_transform(df)

X_tsne = tsne.fit_transform(X) # Get the projected points

colors = ['tab:blue', 'tab:orange', 'tab:green']
plt.figure(figsize=(7,7))
for c in classes.unique():
    plt.scatter(X_tsne[(classes == c), 0], X_tsne[(classes == c), 1], c=
     colors[c], alpha=0.5, label=f'Class {c}')
plt.legend()
plt.xticks([], [])
plt.yticks([], [])
plt.show()
```

Example 4.10 (T-Distributed Stochastic Neighbor Embedding)
Problem: Perform the T-distributed Stochastic Neighbor Embedding (t-SNE) on the Wine Dataset from UCI (https://archive.ics.uci.edu/ml/datasets/wine). The data is the results of a chemical analysis of wines grown in the same region in Italy. There are 13 different measurements taken from 3 types of wines (cultivators/grapevines). Use this dataset to demonstrate how t-SNE dimensionality reduction method works.
Solution: To perform t-SNE on the Wine Dataset, one must begin by importing pandas library that helps to load the dataset into dataframe. Then from scikit-learn (sklearn), import StandardScalar that standardizes the features in the dataset. Then import the t-SNE class implemented in sklearn.manifold and then simply pass the number of components to this constructor. Call the fit-transform method by passing the feature set to this method. The method returns the specified number of components (here the number of components is 2) in the lower-dimensional space. Plot the results and distribution of data points with respect to the components.

```
import pandas as pd
import numpy as np
from sklearn.datasets import load_wine
df, classes = load_wine(return_X_y=True, as_frame=True)

import matplotlib.pyplot as plt
from scipy.stats import norm
import itertools

# Standardization
scaler = preprocessing.StandardScaler()
X = scaler.fit_transform(df)
print(df.head(5))
```

```

def plotGaussian(points, idx):
    # points[idx] is the datapoint itself, which is 0.
    axis = np.sort(points)
    std = np.std(points) # In the t-SNE implementation, this would also be controlled by PERPLEXITY
    plt.plot(axis, norm.pdf(axis, points[idx] , std), color='tab:grey') # 0 refers to the center, which is the data point itself
    plt.scatter(points[idx], 0, c='tab:orange')
    plt.scatter(points[points != 0], np.zeros(len(points)-1), c='tab:blue', alpha=.4)
    plt.show()

def computeDistance(X):
    N = X.shape[0]
    D = np.zeros((N, N)) # D[i, i] is set to be 0 in t-SNE
    for i, j in itertools.combinations(range(N), 2): # pairwise that like the upper triangular matrix but skip diagonals
        dist = np.linalg.norm(X[i, :] - X[j, :])**2
        D[i, j] = dist
        D[j, i] = dist
    return D

dist_matrix = computeDistance(X)
#dist_matrix = computeDistance(np.asarray(df))
idx = 0 # You can play with any other instances
plotGaussian(dist_matrix[idx, :], idx)

from sklearn import preprocessing
from sklearn.manifold import TSNE

scaler = preprocessing.StandardScaler()
tsne = TSNE(n_components=2)
X = scaler.fit_transform(df)

X_tsne = tsne.fit_transform(X) # Get the projected points

colors = ['tab:blue', 'tab:orange', 'tab:green']
plt.figure(figsize=(7,7))
for c in classes.unique():
    plt.scatter(X_tsne[(classes == c), 0], X_tsne[(classes == c), 1], c=colors[c], alpha=0.5, label=f'Class {c}')
plt.legend()
plt.xticks([], [])
plt.yticks([], [])
plt.show()

```

4.4.3 Pearson Correlation Coefficient

Similar to T-SNE discussed previously, Pearson correlation coefficient (PCC) is another feature selection technique to reduce high dimensionality in the data. For efficient data processing, one needs to preserve maximum information represented by the features; yet, eliminating redundant or less relevant features is important to minimize computational complexity. Correlation is utilized to study the relationship strength between two features. The formula for Pearson correlation coefficient (PCC) is shown in Eq. (4.11). In simple terms, PCC is the covariance of X and Y divided by the square root of standard deviation of X and Y. The PCC ranges from −1 to +1; A negative PCC means the variables' magnitudes follow an inverse trend, while a positive PCC value indicates a proportional trend. For example, when X and Y increase (irrespective of their scale), it indicates a positive PCC, while a negative PCC would mean Y decreases as X increases or vice versa. In regard to a 2D graph, if a straight line can pass through all data points, the PCC is 1. As the data points fall out of the straight line, the PCC is reduced to 0. Thus, the PCC helps to understand the relationship between two features. The user can decide to choose the correlation threshold while selecting features. Referring to the code example below, if the threshold is 0.80, it returns all the features that are correlated to each other by a PCC factor of greater than or equal 0.80. Any one of such correlated features can then be discarded to reduce dimensionality.

$$p_{ij} = \frac{\Sigma(x_i - \bar{x})(y_i - \bar{y})}{\sqrt{\Sigma(x_i - \bar{x})^2 \Sigma(y_i - \bar{y})^2}}. \tag{4.11}$$

```
# Importing libraries
# We use the boston dataset here as an example
from sklearn.datasets import load_boston
import pandas as pd
import matplotlib.pyplot as plt


#Loading the dataset
data = load_boston()
df = pd.DataFrame(data.data, columns = data.feature_names)
df["MEDV"] = data.target

#print the feature titles
data.feature_names

#consider only the feature matrix
X = df.drop("MEDV",axis=1)   #Feature Matrix
y = df["MEDV"]
df.head()
X.head()


# separate dataset into train and test
from sklearn.model_selection import train_test_split
X_train, X_test, y_train, y_test = train_test_split(
    X,
    y,
    test_size=0.3,
    random_state=0)

X_train.shape, X_test.shape

#print the correlation matrix; this will print the correlation between each
     feature to other features
X_train.corr()

# For feature selection, the correlation is observed and then features are
     selected
# Correlation is only applied to train data. Features that are dropped in
     train dataset are also dropped from the test data
# it will remove the first feature that is correlated with any other feature

def correlation(dataset, threshold):
```

```
    col_corr = set()  # Set of all the names of correlated columns
    corr_matrix = dataset.corr()
    for i in range(len(corr_matrix.columns)):
        for j in range(i):
            if abs(corr_matrix.iloc[i, j]) > threshold: # we are interested in absolute coeff value
                colname = corr_matrix.columns[i]  # getting the name of column
                col_corr.add(colname)
    return col_corr


#calculate the correlation of features and select based on the threshold

corr_features = correlation(X_train, 0.8)
len(set(corr_features))
print(corr_features)

# drop highly correlated features from the data
X_train.drop(corr_features,axis=1)
X_test.drop(corr_features,axis=1)
```

4.4.4 Independent Component Analysis

Independent Component Analysis (ICA) is a statistical and computational technique for revealing hidden factors that underlie sets of random variables, measurements, or signals. ICA defines a generative model for the observed multivariate data, which is typically given as a large database of samples. In the model, the data variables are assumed to be linear mixtures of some unknown latent variables, and the mixing system is also unknown.

The latent variables are assumed non-Gaussian and mutually independent, and they are called the independent components of the observed data. These independent components, also called sources or factors, can be found by ICA. ICA is superficially related to Principal Component Analysis and factor analysis. ICA is a much more powerful technique, however, capable of finding the underlying factors or sources when these classic methods fail completely.

ICA has been applied to problems in fields as diverse as speech processing, brain imaging (e.g., fMRI and optical imaging), electrical brain signals (e.g., EEG signals), telecommunications, and stock market prediction. However, because the Independent Component Analysis is an evolving method that is being actively researched around the world, the limits of what ICA may be good for have yet to be fully explored.

ICA is based on the simple, generic, and physically realistic assumption that if different signals are from different physical processes (e.g., different people speaking), then those signals are statistically independent. ICA takes advantage of the fact that the implication of this assumption can be reversed, leading to a new assumption that is logically unwarranted but that works in practice, namely: if statistically independent signals can be extracted from signal mixtures, then these extracted signals must be from different physical processes (e.g., different people speaking).

Accordingly, ICA separates signal mixtures into statistically independent signals. If the assumption of statistical independence is valid, then each of the signals extracted by independent component analysis will have been generated by a different physical process and will therefore be the desired signal.

Having identified the prominent factors, it would then be possible to estimate the extent to which each individual neuron depended on each factor, so that neurons could be classified as coding for luminance or edge orientation. In every case, it is these factors or source signals that are of primary interest, but they are buried within a large set of measured signals or signal mixtures. ICA can be used to extract the source signals underlying a set of measured signal mixtures.

In every case, it is these factors or source signals that are of primary interest, but they are buried within a large set of measured signals or signal mixtures. ICA can be used to extract the source signals underlying a set of measured signal mixtures.

FastICA is a way of separating signals that have multivariate data into their additive subcomponents using statistical methods to separate a single voice signal from a mixture of sounds like other voices and background noise. The FastICA algorithm is a highly efficient computational method for performing ICA estimation. It uses a fixed-point iteration scheme that is 10–100 times faster than conventional gradient descent methods for ICA in the independent experiments. Another advantage of the FastICA algorithm is that it can also be used for projection pursuit, thus providing a general-purpose data analysis method that can be used both in an exploratory manner and for the estimation of independent components (or sources).

Compared to the existing ICA methods, the FastICA algorithm has several desirable properties. The FastICA has most of the benefits of neural algorithms: it is parallel, distributed, computationally simple, and requires very little memory space. Independent components can be estimated one by one, which is roughly equivalent to the pursuance of projection. This is useful in exploratory data analysis and reduces the method's computational load in circumstances where it is only necessary to estimate some of the independent components. This algorithm finds directly independent components of (practically) any non-Gaussian distribution by using non-linearity. The performance of the method can be optimized by selecting the appropriate non-linearity.

Example 4.11 (Independent Component Analysis)
Problem: Apply PCA and ICA on the randomly generated data and compare the results.
Solution: To compare the PCA and ICA, first let us generate some sample data. Let us generate three signals sinusoidal signal, square signal, and a sawtooth signal. Then some noise is added followed by standardizing the data. Mix the matrices and general observations. Finally, compute ICA and PCA. Plot the results to compare the ICA and PCA on the randomly generated signals (Fig. 4.22).

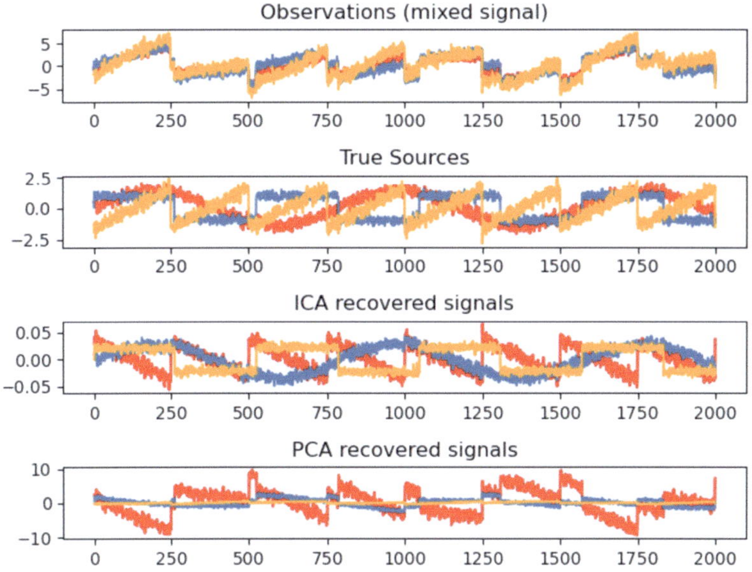

Fig. 4.22 Estimating source signals from noisy data

```
import numpy as np
import matplotlib.pyplot as plt
from scipy import signal

from sklearn.decomposition import FastICA, PCA

###########################################################################

# Generate sample data
np.random.seed(0)
n_samples = 2000
time = np.linspace(0, 8, n_samples)

s1 = np.sin(2 * time)  # Signal 1 : sinusoidal signal
s2 = np.sign(np.sin(3 * time))  # Signal 2 : square signal
s3 = signal.sawtooth(2 * np.pi * time)  # Signal 3: saw tooth signal

S = np.c_[s1, s2, s3]
S += 0.2 * np.random.normal(size=S.shape)  # Add noise

S /= S.std(axis=0)  # Standardize data
# Mix data
A = np.array([[1, 1, 1], [0.5, 2, 1.0], [1.5, 1.0, 2.0]])  # Mixing matrix
X = np.dot(S, A.T)  # Generate observations

# Compute ICA
ica = FastICA(n_components=3)
S_ = ica.fit_transform(X)  # Reconstruct signals
A_ = ica.mixing_  # Get estimated mixing matrix

# We can 'prove' that the ICA model applies by reverting the unmixing.
assert np.allclose(X, np.dot(S_, A_.T) + ica.mean_)

# For comparison, compute PCA
pca = PCA(n_components=3)
```

```
H = pca.fit_transform(X)  # Reconstruct signals based on orthogonal
    components

###########################################################################

# Plot results

plt.figure()

models = [X, S, S_, H]
names = ['Observations (mixed signal)',
         'True Sources',
         'ICA recovered signals',
         'PCA recovered signals']
colors = ['red', 'steelblue', 'orange']

for ii, (model, name) in enumerate(zip(models, names), 1):
    plt.subplot(4, 1, ii)
    plt.title(name)
    for sig, color in zip(model.T, colors):
        plt.plot(sig, color=color)

plt.tight_layout()
plt.show()
```

4.4.5 Non-negative Matrix Factorization (NMF)

NMF (Non-negative Matrix Factorization) has a wide range of uses, from topic modeling to signal processing. NMF is a matrix factorization method where we constrain the matrices to be non-negative. In order to understand NMF, we should clarify the underlying intuition between matrix factorization.

Suppose we factorize a matrix X into two matrices W and H so that X $\approx$ W $\times$ H. There is no guarantee that we can recover the original matrix, so we will approximate it as best as we can. Now, suppose that X is composed of m rows x_1, x_2, . . . x_m , W is composed of k rows w_1, w_2, . . . w_k, and H is composed of m rows h_1, h_2, . . . h_m. Each row in X can be considered a data point. For instance, in the case of decomposing images, each row in X is a single image, and each column represents some feature. Consider the i-th row in X, x_i can be written as

Basically, we can interpret x_i to be a weighted sum of some components (or bases if you are more familiar with linear algebra), where each row in H is a component, and each row in W contains the weights of each component (Fig. 4.23).

Note that in this explanation, we treat each row of the input matrix X to be a single data point. In other explanations, each column might be considered a data point, in which case each column in W becomes a component, and each column in H becomes a set of weights.

In practice, we introduce various conditions on the components, so that they can be interpreted in a meaningful manner. In the case of NMF, we constrict the underlying components and weights to be non-negative. Essentially, NMF decomposes each data point into an overlay of certain components.

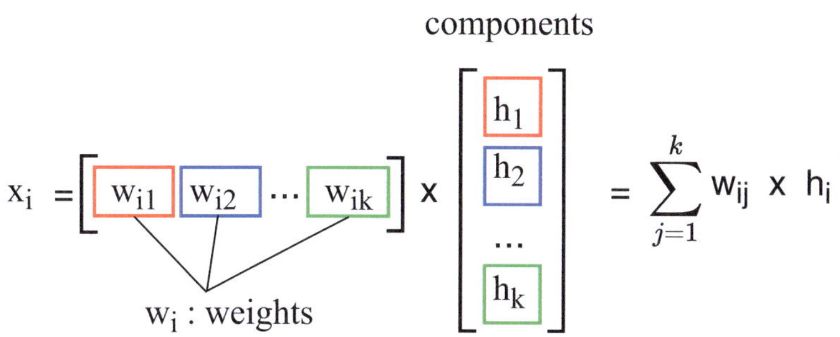

Fig. 4.23 Visualizing non-negative matrix factorization

The reason why NMF has become so popular is because of its ability to automatically extract sparse and easily interpretable factors. Suppose we take a gray-level image of a face containing p pixels and squash the data into a single vector such that the ith entry represents the value of the ith pixel. Let the rows of $\mathbf{X} \in \mathbb{R}^{p \times n}$ represent the p pixels, and the n columns each represent one image. NMF will produce two matrices W and H. The columns of W can be interpreted as images (the basis images), and H tells us how to sum up the basis images in order to reconstruct an approximation to a given face. In the case of facial images, the basis images are features such as eyes, noses, moustaches, and lips, while the columns of H indicate which feature is present in which image.

NMF can be applied for recommender systems, for collaborative filtering for topic modeling, and for dimensionality reduction. In the code below, we will check out how NMF can be used for dimensional reduction. We apply NMF on eurovision dataset. In Python, it can work with a sparse matrix where the only restriction is that the values should be non-negative. The logic for dimensionality reduction is to take our m $\times$ n data and to decompose it into two matrices of m $\times$ features and features $\times$ n, respectively. The features will be the reduced dimensions, and the code to implement NMF for dimensionality reduction is presented below.

Example 4.12 (Non-negative Matrix Factorization)

Problem: Use the NMF technique to prepare the food recommendations for people shown in Fig. 4.24.

Solution: To process with this, put clients as columns and products/ratings as rows of an array (let us call it V). As values, you should put adequate statistics like the number of purchases or ratings. Then perform segmentation to obtain W and H segment-defining arrays. Finally, by multiplying W and H, we obtain an initial V matrix approximation.

Fig. 4.24 Reconstructed value for Example 4.12

	John	Alice	Mary	Greg	Peter	Jennifer
Vegetables	0.00	0.98	0.04	0.99	2.11	1.89
Fruits	2.01	2.84	1.23	0.87	2.08	2.07
Sweets	1.01	1.30	0.53	0.25	0.86	0.87
Bread	0.00	2.01	2.98	4.02	0.99	1.00
Coffee	0.00	0.21	0.00	0.20	0.45	0.41

```
import pandas as pd
import numpy as np
from sklearn.decomposition import NMF

V = np.array([[0,1,0,1,2,2],
              [2,3,1,1,2,2],
              [1,1,1,0,1,1],
              [0,2,3,4,1,1],
              [0,0,0,0,1,0]])

V = pd.DataFrame(V, columns=['John', 'Alice', 'Mary', 'Greg', 'Peter', '
    Jennifer'])
V.index = ['Vegetables', 'Fruits', 'Sweets', 'Bread', 'Coffee']

nmf = NMF(3)
nmf.fit(V)

H = pd.DataFrame(np.round(nmf.components_,2), columns=V.columns)
H.index = ['Fruits pickers', 'Bread eaters',  'Veggies']

W = pd.DataFrame(np.round(nmf.transform(V),2), columns=H.index)
W.index = V.index

reconstructed = pd.DataFrame(np.round(np.dot(W,H),2), columns=V.columns)
reconstructed.index = V.index
# Print the extracted values
print(reconstructed)
print(H)
print(W)
print(V)
```

The reconstructed matrix serves as a basis for the recommendation. The process of assigning values for previously unknown values (zeros in this case) is called collaborative filtering. One can find attraction weight toward certain products in columns of the matrix. By sorting the values in descending order, one can determine which products should be proposed to the customer to match their preferences. For example, Mary should be offered products in the following order Bread, Fruits, and Sweets. Recommendation order for Alice: Fruits, Bread, Sweets, Vegetables, and Coffee.

In H matrix, the higher the weight value, the more the person belongs to the specific segment. Some people like John can be assigned in 100% to one cluster, and some people like Peter belong to all the segments with some weights (Fig. 4.25).

The W matrix can be called a segment-defining array. By observing the values (weights—note that they do not sum up to 1) in each column. The higher the weight, the more "determined" the column (segment) is by the variable in the row in Fig. 4.26.

Fig. 4.25 H value for Example 4.12

	John	Alice	Mary	Greg	Peter	Jennifer
Fruits pickers	1.04	1.34	0.55	0.26	0.89	0.90
Bread eaters	0.00	0.60	1.12	1.36	0.03	0.07
Veggies	0.00	0.35	0.00	0.34	0.77	0.69

Fig. 4.26 W value for Example 4.12

	Fruits pickers	Bread eaters	Veggies
Vegetables	0.00	0.04	2.74
Fruits	1.93	0.15	0.47
Sweets	0.97	0.00	0.00
Bread	0.00	2.66	1.18
Coffee	0.00	0.00	0.59

Fig. 4.27 V value for Example 4.12

	John	Alice	Mary	Greg	Peter	Jennifer
Vegetables	0	1	0	1	2	2
Fruits	2	3	1	1	2	2
Sweets	1	1	1	0	1	1
Bread	0	2	3	4	1	1
Coffee	0	0	0	0	1	0

For example, one segment is named "Bread eaters," because it is almost entirely driven by bread consumption. "Fruit pikers" are driven by two product categories—Fruits and Sweets. The third one is a mixed segment with the leading Vegetable category (Fig. 4.27).

4.5 Multi-Dimensional Scaling

With the proliferation of data-driven technologies, the world is using more data than we ever did previously. This has led to the massive amounts of data. This massive data is usually high-dimensional. But, with the proliferation of such high-dimensional data, the need for its interpretation and analysis also arose. Various dimensionality reduction techniques for accurate data representation and analysis are introduced to address this challenge, with one such technique being the multi-dimensional scaling or MDS. Dimensionality scaling techniques are useful in extracting features or properties of data for simpler analysis. In other words, dimensionality reduction is the process of scaling down the number of variables by extracting a set of principal variables.

The classical-MDS technique is explained here. MDS is also known as Principal Coordinate Analysis (PCoA). MDS is a complex technique similar to Principal Component Analysis (PCA); hence, MDS is discussed here with a real-world example for better understanding. Table 4.1 shows the distance between different

Table 4.1 Distance between US cities—A sample data for MDS [74]

	Atlanta	Chicago	Denver	Houston	Los Angeles	Miami	New York	San Francisco	Seattle	Washington, DC
Atlanta	0	587	1212	701	1936	604	748	2139	2182	543
Chicago	587	0	920	940	1745	1188	713	1858	1737	597
Denver	1212	920	0	879	831	1726	1631	949	1021	1494
Houston	701	940	879	0	1374	968	1420	1645	1891	1220
Los Angeles	1936	1745	831	1374	0	2339	2451	347	959	2300
Miami	604	1188	1726	968	2339	0	1092	2594	2734	923
New York	748	713	1631	1420	2451	1092	0	2571	2408	205
San Francisco	2139	1858	949	1645	347	2594	2571	0	678	2442
Seattle	2182	1737	1021	1891	959	2734	2408	678	0	2329
Washington, DC	543	597	1494	1220	2300	923	205	2442	2329	0

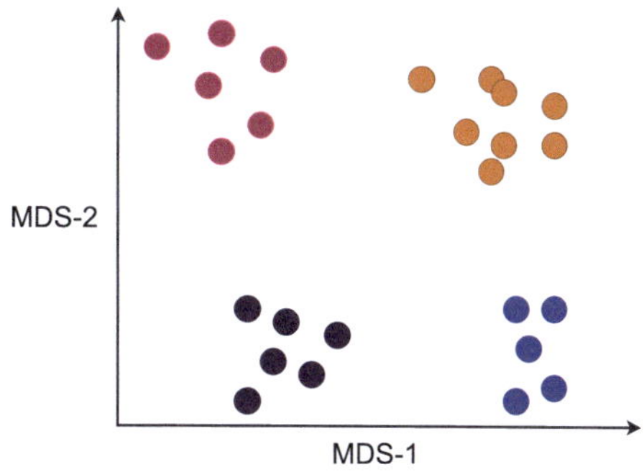

Fig. 4.28 MDS output showing multi-dimensional data plotted on a 2D plane for visualization

cities [74] in the USA as a matrix. As seen from the Table, the data is high-dimensional. To represent and analyze such data, MDS can be employed. The columns are represented by X, while the rows are represented by Y. The MDS calculates the distance between the points on X_1 and X_2. The entire column is considered, meaning all the points on the Y-axis. The distance method could be Euclidean, Manhattan, log-fold change, hamming distance, etc. MDS calculates distances between each column with all other columns, basically, a combination of all the possible mix of any two columns. These values are then plotted on a 2D plot for analysis and representation as shown in Fig. 4.28. In conclusion, an n-dimension variable space is reduced to a 2D space (typically) representation. The next example demonstrates on how MDS can be employed to visualize a high-dimensionality data on a 2D plot.

MDS algorithm also has its own advantages and disadvantages:

- With MDS, the user can assess how closely related different data points are.
- MDS makes plotting of large multi-dimensional features to low dimensions, better for visualization and assessment.
- The downside is that MDS does not deal with real numbers, as it is based on the relative relationships among dimensions.
- It is difficult to evaluate the depth of the relation between features.

Example 4.13 (Multi-Dimensional Scaling)
Problem: Use the MDS algorithm on the Iris dataset and generate output plots.
Solution: The Iris dataset is available in the sklearn package under datasets. We import all the necessary libraries and functions we need for MDS. After loading the dataset, the min-max scaling is done to normalize the data. The imported `MDS` is used as a model to fit the data. The MDS algorithm is then employed to scale the data to two dimensions, which is easier to plot. Different colors are used for ease of visibility. The MDS implementation code is shown in listing below. The generated 2D plot is shown in Fig. 4.29.

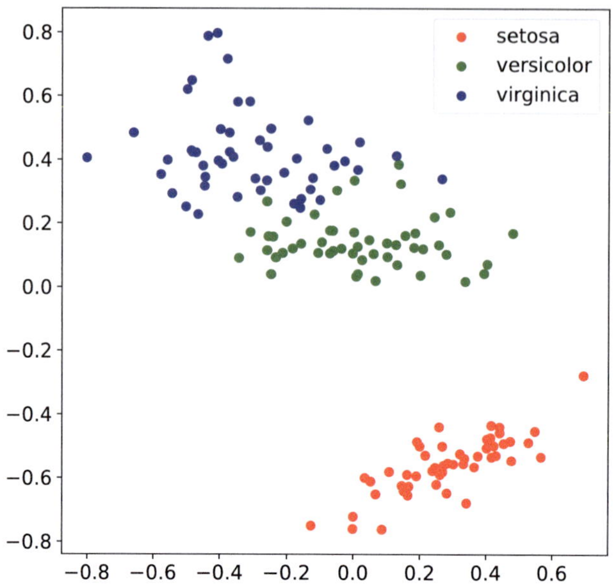

Fig. 4.29 MDS output for the Iris dataset

```
import numpy as np
from sklearn.datasets import load_iris
import matplotlib.pyplot as plt
from sklearn.manifold import MDS
from sklearn.preprocessing import MinMaxScaler

# Loading the dataset
data = load_iris()
X = data.data

#Min-max scaling
scaler = MinMaxScaler()
X_scaled = scaler.fit_transform(X)

mds = MDS(2,random_state=0)
X_2d = mds.fit_transform(X_scaled)

#Plotting the scaled dataset
colors = ['red','green','blue']
plt.rcParams['figure.figsize'] = [7, 7]
plt.rc('font', size=14)
for i in np.unique(data.target):
  subset = X_2d[data.target == i]

  x = [row[0] for row in subset]
  y = [row[1] for row in subset]
  plt.scatter(x,y,c=colors[i],label=data.target_names[i])
plt.legend()
plt.show()
```

4.6 Google Page Ranking Algorithm

It is very important for a search engine to display the websites that are most relevant to what we search. There must be some factor that must decide the ranking or listing of these relevant websites. PageRank (PR) is very important and one of the factors which a search engine like Google takes into account when it decides which results to show at the top of its search engine listings—where they can be easily seen. (In fact, PageRank is a Google trade mark—but other search engines use similar techniques.) Google search ranks web pages in their search engine using PageRank algorithm. The algorithm is named after Larry Page, who is one of the founders of Google. Importance of website pages is measured using PageRank algorithm. PageRank determines a rough estimate of how important a website is by counting the number and quality of links to a page.

The output of PageRank algorithm is a probability distribution used to represent the likelihood that a person randomly clicking on links will arrive at any particular page. PageRank can be calculated for groups or collections of documents of any size. It is assumed in several research papers that the distribution is evenly divided among all documents in the collection at the beginning of the computational process. The PageRank computations require several passes, called "iterations," through the collection to adjust approximate PageRank values to more closely reflect the theoretical true value.

Example 4.14 (Page Rank)
Problem: Implement Page Rank using random walk method.
Solution: To implement Page Rank using the random walk method, one has to first select a random graph from python library and then initialize all the nodes to a rank value "0." Then randomly pick a source node, create a list to store neighbors of the source node, and pick a node from the list randomly and increment its rank. Remember to check if the node is a sink node, i.e., node having no outgoing edges. If yes, pick a node from a set of nodes randomly and increment its rank. Otherwise, select a node from the list obtained randomly and increment its rank. This process is repeated until the resulting vector, with rank for every single page, converges.

```

import networkx as nx
import random
import operator

g = nx.gnp_random_graph(10, 0.5, directed = True)

# nx.draw(g, with_labels = True)

# plt.show()
```

```
# pick source node randomly

x = random.choice([i for i in range(g.number_of_nodes())])

# initialize weight/count of each node as 0

dict_count = {}

for i in range(g.number_of_nodes()):
    dict_count[i]= 0
dict_count[x]= dict_count[x]+1

# Look at neighbors of x

for i in range(100):
    # create list to store neighbors
    list_n = list(g.neighbors(x))
    if (len(list_n)==0):
        x = random.choice([i for i in range(g.number_of_nodes())])
        dict_count[x]= dict_count[x]+1

    else:
        x = random.choice(list_n)
        dict_count[x]= dict_count[x] +1


# verify page rank values

p = nx.pagerank(g)

#sort values

sort_p = sorted(p.items(), key = operator.itemgetter(1))
sort_rw = sorted(dict_count.items(), key = operator.itemgetter(1))

print(sort_p)
print(sort_rw)
```

```
# Output
>> [(3, 0.0604892548254l0406), (9, 0.06859674242363363), (4,
    0.08950883219300496), (5, 0.09677783028897), (6, 0.10328132399446883),
    (8, 0.10449415902756215), (2, 0.10568727592939682), (1,
    0.11425507072221733), (0, 0.12470785722731952), (7, 0.13220165336801626)
    ]
>> [(3, 4), (1, 6), (9, 8), (2, 9), (5, 9), (6, 10), (4, 13), (8, 13), (0,
    14), (7, 15)]
```

Thus, we can conclude that after some iterations page rank obtained from the random walk method discussed in the above example matches with the values obtained from built-in functions in python.

4.7 Putting It All Together

Unsupervised learning is a type of machine learning approach that is used when the data is not labeled. Thus, data processing for classification or other purposes can be complicated, as learning algorithms must leverage the underlying link in order to classify (say). As a result, the computational complexity is comparable to that of some supervised learning techniques.

Clustering is the least difficult of the unsupervised learning strategies mentioned in this chapter. However, complexity can increase in direct proportion to the distribution of the underlying data and the number of classes (clusters). Unsupervised neural networks are effective for unsupervised learning of large datasets with complicated relationships between variables that are difficult to represent using simple rules. The feature selection strategies can aid in the creation of a more effective clustering algorithm.

4.8 Exercise Problems

Problem 4.1 The Online Retail Dataset[1] contains transactions for an online retail business. The objective of this assignment is to find the best set of customers which the company should target. Perform the following tasks on the dataset:

1. Preprocess the dataset to address any NaN (not a number) cells.
2. Perform EDA analysis on the dataset to visualize the dataset.
3. Plot multiple graphs based on your intuition that suit the dataset.
4. Use K-Means clustering on the dataset to create clusters with similar characteristics.
5. Use hierarchical clustering (HC) to cluster the dataset and generate a dendrogram to determine the optimum number of clusters in the dataset. Plot Elbow graph as well.

Problem 4.2 Perform K-Means clustering on the Mall Customer dataset[2] with a cluster size of 1 through 10. Find the optimum number of clusters using the Elbow method, and perform K-Means on different cluster sizes. Support your answers with an Elbow method plot and results for all the cluster sizes.

Problem 4.3 Perform K-Means clustering for segmentation using the Annual Income and Spending Score features. Use the Mall Customer dataset. Determine the optimum number of clusters, plot the results, and comment on the performance of the K-Means classifier.

Problem 4.4 Perform K-Means clustering for segmentation using the Age, Annual Income, and Spending Score features. Use the Mall Customer dataset. Determine the optimum number of clusters, plot the results, and comment on the performance of the K-Means classifier. Note: The final plot will be in 3-Dimensions.

---

[1] https://www.kaggle.com/hellbuoy/online-retail-customer-clustering.

[2] https://www.kaggle.com/vjchoudhary7/customer-segmentation-tutorial-in-python.

Problem 4.5 Consider any suitable dataset of your choice. The dataset must have more than 8 features. Visualize the dataset with the MDS algorithm into 2, 4, and 6 dimensions. Use the new, reduced dataset and perform K-Means clustering.

Problem 4.6 Perform K-Means clustering and GMM-based clustering on the ClusterGMM dataset.[3] Compare the results and provide a reason for the differences in the results. Comment on which method, K-Means or GMM, is suitable for the given data.

Problem 4.7 Consider a dataset with at least 16 dimensions (features). Use Principal Component Analysis (PCA) algorithm to reduce the dimensionality of the dataset. Evaluate the performance of a neural network (NN) with original dimensions and with reduced dimensions. Comment on your results.

Problem 4.8 Compare different dimensionality reduction techniques. Comment on specific advantages of each.

Problem 4.9 During PCA on a high-dimensional dataset, it produces a number of PCs or Principal Components. Are the new features the same as the original ones? If not, what other dimensionality techniques can you employ to keep the original features in the dataset as is?

Problem 4.10 Import a Wine recognition dataset from the scikit-learn library and perform dimensionality reduction using Self-Organizing Maps (SOM). Report the features pre- and post-performing dimensionality reduction.

Problem 4.11 Understand the functioning of Deep Belief Networks (DBNs) and apply it to classify the breast cancer dataset available in scikit-learn.

Problem 4.12 Perform dimensionality reduction using PCA (Principal Component Analysis) and ICA (Independent Component Analysis) on the wild face recognition dataset from scikit-learn.

Problem 4.13 Understand the functioning of each module of GAN (generator, discriminator) and develop a generative adversarial network with deep convolutional networks for evaluating the difference between real and fake fashion MNIST images (please use fashion MNIST dataset that is built in python library).

Problem 4.14 Use any built-in graph from networkx library and calculate the page rank using the random walk method. Compare the values of page rank with built-in page rank values.

Problem 4.15 Compute the moments of the randomly generated data. The data must be a normal distribution of random data. Center it around zero, with a 0.7 standard deviation and 15,000 data points.

---

[3] https://cdn.analyticsvidhya.com/wp-content/uploads/2019/10/Clustering_gmm.csv.

Chapter 5
Reinforcement Learning

5.1 Introduction

Reinforcement learning is a kind of machine learning technique that mimics one of the most common learning styles in natural life, which is to learn to achieve a goal by trial-and-error interaction with a dynamic/uncertain environment [75, 76]. The interactions between the learning agent and the environment are generally modeled using a finite state space **S** (corresponding to environment inputs), a set of available actions **A** (corresponding to control/optimization knobs used by the agent), and a reward function $\mathbf{R} : \mathbf{S} \times \mathbf{A} \rightarrow \mathbf{R}$ (used to decide which action to take for a given state). The ultimate goal of reinforcement learning is to figure out a policy $\pi(s) = a$, which chooses action $a \in \mathbf{A}$ in each state $s \in \mathbf{S}$ (i.e., a mapping between the states and the actions), to optimize a reward function (i.e., to maximize the cumulative rewards over a potentially infinite time span).

Decision epochs are a sequence of points in time $\{t_0, t_1, t_2, \ldots, t_k, \ldots\}$ at which an action is chosen and a state transition may appear. At time t_k, when the system just transitioned to state $s_k \in \mathbf{S}$, the agent selects an action $a_k \in \mathbf{A}$. This action will lead to an instant reward rate $r_{(s_k,a_k)}(t)$ in regard to state–action pair (s_k, a_k). In the next decision epoch (i.e., at time t_{k+1}), the system switches to state s_{k+1}.

An important issue in reinforcement learning is exploration vs. exploitation. A reinforcement learning agent must exploit the best action known so far in order to gain rewards while exploring all possible actions such that it can find a potentially better choice. The risk is thus always choosing the action with the temporary highest reward, as this can lead to reaching a local maximum and getting stuck in a sub-optimal solution.

S. Rafatirad et al., *Machine Learning for Computer Scientists and Data Analysts*,
https://doi.org/10.1007/978-3-030-96756-7_5

5.2 Q-Learning

Q-learning is one of the most popular algorithms used to perform reinforcement learning [75, 76]. In Q-learning, a Q-value is associated with every state–action pair (s, a), denoted as $Q(s, a)$. The value of $Q(s, a)$ approximates the expected long-term cumulative reward of taking action a starting from state s [77]. In this way, the agent decides which action should be taken in the current state in order to achieve the maximum long-term rewards based on this value function $Q(s, a)$. Namely, at decision epoch t_k, when the system has just transitioned to state $s_k \in \mathbf{S}$, the action a_k with the highest Q-value will be chosen. Furthermore, given that it is a model-free learning algorithm, it is not necessary for the Q-learning agent to have any prior system information, such as the transition probability from one state to another. Therefore, it is a highly adaptive and flexible algorithm.

The fundamental aspect of the Q-learning algorithm is a value iteration update of the Q-value function. Particularly, the Q-value for each state–action pair is initially chosen by the designer. However, these values are updated every time an action is issued and a reward is received. That is, at decision epoch t_{k+1}, the Q-value $Q(s_k, a_k)$ is updated according to the received reward as shown in the following expression:

$$Q(s_k, a_k) \leftarrow \underbrace{Q(s_k, a_k)}_{\text{old value}} + \underbrace{\beta_k(s_k, a_k)}_{\text{learning rate}} \cdot \left[\overbrace{\underbrace{r_{k+1}}_{\text{reward}} + \underbrace{\gamma}_{\text{discount factor}} \cdot \underbrace{\max_{a \in \mathbf{A}} Q(s_{k+1}, a)}_{\text{max future value}}}^{\text{expected discounted reward}} - \overbrace{Q(s_k, a_k)}^{\text{old value}} \right], \tag{5.1}$$

where r_{k+1} is the reward measured at time t_{k+1} for having taken action a_k at time t_k, value $\gamma \in (0, 1)$ is the discount factor, and $\beta_k(s_k, a_k) \in (0, 1)$ is the learning rate at time t_k for state–action pair (s_k, a_k) (which may or may not be equal for all pairs and which may be constant or variable in time). The next time state s is visited, the action with the maximum Q-value will be chosen, i.e., $\pi(s) = \max_{a \in \mathbf{A}} Q(s, a)$, and given that the Q-value was updated, it might be a different action from the one taking the last time state s was visited. In order to choose the action, an ϵ-greedy policy can be employed. In this case, the ϵ-greedy policy chooses the action with a probability of $1 - \epsilon$ that leads to a high Q-value and chooses a random action with ϵ probability.

As an example, a state diagram depicting the change of states by the basic Q-learning algorithm is shown in Fig. 5.1. We consider four states and five actions. When the system is in state s_1 (s_2) and action a_1 (a_2) is selected, then the system changes to state s_2 (s_1), whereas when action a_3 is chosen, then the state changes to s_3 (s_4) and remains in the same state when action a_5 is chosen. Similarly, other state transitions also happen. One needs to note that the state transition entirely depends

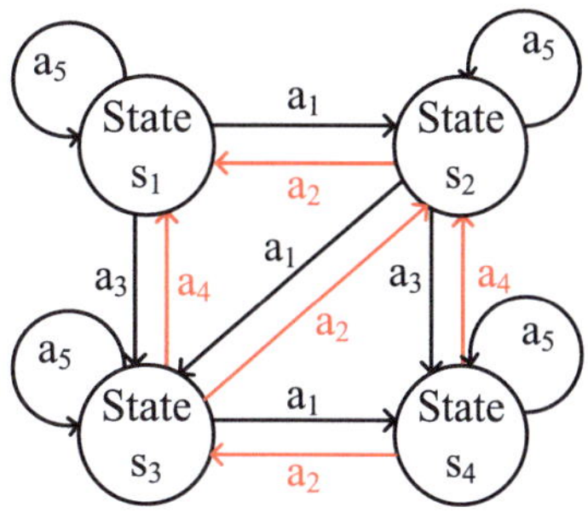

Fig. 5.1 State transition based on basic Q-learning

on the action chosen. This is a mere illustration of Q-learning. The number of states, actions, transitions, and rewards highly depends on the problem settings.

Example 5.1 (Q-Learning)
Problem: Create a policy for reinforcement learning based on discussed greedy policy.
Solution: In order to create an ϵ-greedy policy, we first create a module based on the aforementioned policy update, presented in (5.1). Furthermore, a Q-learning module can be defined with an iterative loop to keep the policy updated based on the ϵ-greedy policy. The code snippet showcasing this procedure is shown below:

```
def createEpsilonGreedyPolicy(Q, epsilon, num_actions):
  """
  Creates an epsilon-greedy policy-based
  on a given Q-function and epsilon.

  Returns a function that takes the state
  as an input and returns the probabilities
  for each action in the form of a numpy array
  of length of the action space(set of possible actions).
  """
  def policyFunction(state):

    Action_probabilities = np.ones(num_actions,
        dtype = float) * epsilon / num_actions

    best_action = np.argmax(Q[state])
    Action_probabilities[best_action] += (1.0 - epsilon)
    return Action_probabilities

  return policyFunction
```

```python
def qLearning(env, num_episodes, discount_factor = 1.0,
                alpha = 0.6, epsilon = 0.1):
  """
  Q-Learning algorithm: Off-policy TD control.
  Finds the optimal greedy policy while improving
  following an epsilon-greedy policy"""

  # Action value function
  # A nested dictionary that maps
  # state -> (action -> action-value).
  Q = defaultdict(lambda: np.zeros(env.action_space.n))

  # Keeps track of useful statistics
  stats = plotting.EpisodeStats(
    episode_lengths = np.zeros(num_episodes),
    episode_rewards = np.zeros(num_episodes))

  # Create an epsilon greedy policy function
  # appropriately for environment action space
  policy = createEpsilonGreedyPolicy(Q, epsilon, env.action_space.n)

  # For every episode
  for ith_episode in range(num_episodes):

    # Reset the environment and pick the first action
    state = env.reset()

    for t in itertools.count():

      # get probabilities of all actions from current state
      action_probabilities = policy(state)

      # choose action according to
      # the probability distribution
      action = np.random.choice(np.arange(
          len(action_probabilities)),
          p = action_probabilities)

      # take action and get reward, transit to next state
      next_state, reward, done, _ = env.step(action)

      # Update statistics
      stats.episode_rewards[ith_episode] += reward
      stats.episode_lengths[ith_episode] = t

      # TD Update
      best_next_action = np.argmax(Q[next_state])
      td_target = reward + discount_factor * Q[next_state][best_next_action]
      td_delta = td_target - Q[state][action]
      Q[state][action] += alpha * td_delta

      # done is True if episode terminated
      if done:
        break

      state = next_state

  return Q, stats

Q, stats = qLearning(env, 1000)
```

The above code snippet shows the process of Q-learning with ϵ-greedy approach. First, the greedy policy is created, followed by Q-learning model, as described in Eq. (5.1).

5.2.1 Accelerated Q-learning by Environment Exploration

The traditional Q-learning algorithm [78] converges to the optimal after unlimited iterations that may be too slow for convergence [79]. To overcome this convergence issue, we propose the use of reinforcement Q-learning for adaptive tuning of output-voltage swing to achieve low power and faster convergence. Here, we will first present the modeling of a Markov decision process (MDP) and then introduce a reinforcement Q-learning algorithm for the adaptive tuning.

Accelerated Q-learning [80] can be utilized to find the optimum with a faster convergence based on the predicted next state and the according transition probability with an initialized random action at first few states. Accelerated Q-learning has two transition rules. Random actions make the system explore environment faster and more easily find optimal states. Optimal states are not found only based on Q-value but also by the random selection.

Similar to the Q-learning algorithm, the set of states and actions is known, but the reward for each action is unknown. The reward function is calculated. To achieve faster convergence, the transition probability is utilized to select the action instead of directly selecting the next state.

To find the optimal MDP, reinforcement Q-learning algorithm can be utilized to evaluate the pair of state and action as the Q-value. In addition to the Q-learning algorithm discussed in the previous section, we utilize the reinforcement Q-learning to find the optimal for the modeled MDP.

The first phase is initialization to form a look-up-table with states and corresponding actions. In addition, the transition probability P for all the states is set as 1 and the reward is set to a maximum value L. This process of initialization is presented as $Init()$ of Algorithm 1.

Prediction of the next state (say voltage swing level) is performed using the auto-regression technique as discussed in the earlier chapters to obtain the corresponding action. In the action selection phase, given by $Selection()$, the Q-value for the state and action pair is found iteratively, where the Q-value is defined as the weighted sum of the reward and its past values by

$$Q'(s_i,\ a_k) = (1-\alpha) * Q(s_i,\ a_k) + \alpha * delta \tag{5.2a}$$

$$delta = R(s_i, a_k,\ s_{i+1}) + \gamma * \min_{a \in A}(Q(s_{i+1},\ a_k)). \tag{5.2b}$$

$Q'(s_i, a_k)$ shows the updated Q-value after taking the action a_k to the next state s_{i+1}.

In each iteration, the action is selected either based on the transition probability or based on the maximum Q-value (or policy). If the transition probability is larger than the threshold, a random action is selected; otherwise, the policy action with the minimum Q-value is selected. The random action will happen at the first few rounds to explore the design space. As the learning process continues, the policy action with the calculated Q-value will dominate and become more accurate to use.

Algorithm 1 Reinforcement Q-learning algorithm

Input: Communication power Pw, BER feedback
Output: Output-voltage
function Init()
  $1 \rightarrow P(s_i, a_k, s_{i+1})$
  Reward $R(s_i, a_k, s_{i+1}) = L$
  $v_{predict} \rightarrow V_{s_i}$
  Selection()
end function

function Selection()
  for $k = 1 : n$
    $V_{s_i}, BER_i \rightarrow s_i \in S$
    $Q'(s_i, a_k) \leftarrow (1 - \alpha) * Q(s_i, a_k) +$
    $\alpha * (R(s_i, a_k, s_{i+1}) + \gamma * \min(Q(s_{i+1}, a_k))$
    If $P(s_i, a_k, s_{i+1}) > rand(0, 1)$
      $a_k \leftarrow rand(A)$
    else
      $a_k \leftarrow \min(Q(s_{i+1}, a_k))$
    end if
    Update()
  end for
end function

function Update()
  $Reward$: $R(s_i, a_k, s_{i+1}) = b_1 \Delta V_s(P_i) + b_2 \Delta V_s(BER_i)$
  Update Policy (s_i, a_i), based on new Q
  $\forall\, s_i \in S$ {
    $a_k \leftarrow rand(A)$
    $Q'(s_i, a_k) = Q(s_i, a_k)$
    $P(s_i, a_k, s_{i+1}) = 1 - \frac{1}{\log(N_{s_i}+2)}$
  }
end function

As such, a higher probability exists that the action a_k with the minimum Q-value. The policy action with the maximum Q-value (5.2) can be described as below:

$$a_k \leftarrow \min(Q(s_{i+1},\ a_k)). \tag{5.3}$$

Lastly, the phase of $Update()$ is activated at the end of each iteration of $Selection()$ function. The reward is defined as the weighted value of parameters say bit error rate (BER) and power consumption (Pw) and updated. At the end of $Update$, each state will be randomly visited and Q-value (5.2) will be updated accordingly. The transition probability $P(s_i,\ a_k,\ s_{i+1})$ is also updated as N_{s_i} (the number of visits to state s_i) will increase after each iteration. Note that with

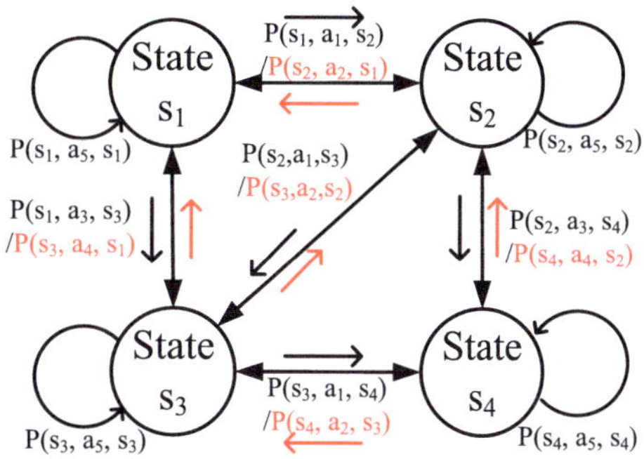

Fig. 5.2 State transition based on reinforcement Q-learning

the prediction of states s_i as in function $Init()$ and $Update()$ and the transition probability, the convergence to the optimal solution is accelerated [81, 82]. This is done at the end of each round with the random action a_k to visit the state s_i.

Example 5.2 (Accelerated Q-Learning)
Problem: Create an accelerated Q-learning based on the discussion and Eqs. (5.2a) and (5.2b).
Solution: One example with 4-state is shown in Fig. 5.2. For state s_1, action a_1 can change its state to state s_2 with probability $P(s_1,\ a_1,\ s_2)$, and for state s_2, action a_2 can change its state to state s_1 with probability $P(s_2,\ a_2,\ s_1)$. Whereas action a_5 causes no change in state, whose probability is given as $P(s_1,\ a_5,\ s_2)$. The state transition probability P is given by a decaying function. The probability under the decaying function is given by $P = 1/(\log(N_{s_i} + 2)$ with N_{s_i} denoting the number of visits to state s_i. The probability-based action will ensure the visit to all states at starting period. This will calculate Q-value to every available state accordingly. After this, Q-value-based action will dominate and the optimal action with the largest Q-value will be selected.

5.3 TD(λ)-Learning

In some of the real-time problems, the system may not have a predefined policy or knowledge regarding the state transitions. In such cases, the system has to learn the policy as well as make the decision in parallel. The TD(λ)-learning methods can be applied to learn the policy and perform the decision-making.

For every state s_k visited at epoch t_k, the TD(λ) algorithm chooses an action either with a maximum Q-value, i.e., $\max_{a \subset \mathbf{A}} Q(s_k, a)$ for different possible actions

a, or by using the semi-greedy policies given in [83]. The estimated Q-value is updated in the next epoch based on the action chosen a_k and the next state s_{k+1}. The Q-value update is similar to that of traditional Q-learning algorithm, but with different estimated Q-value and error terms, particularly,

$$\forall(s,a) \in \mathbf{S} \times \mathbf{A}: \quad Q(s,a) \leftarrow Q(s,a) + \beta \cdot \varepsilon_k(s,a) \cdot$$
$$\left[\frac{1-e^{-\gamma\tau_k}}{\gamma} r(s_k,a_k) + e^{-\gamma\tau_k} \max_{a' \in \mathbf{A}} Q(s_{k+1},a') - Q(s_k,a_k)\right],$$

where the amount of time that system remains in state s_k is given by $\tau_k = t_{k+1} - t_k$, $\beta \in (0,1)$ is the learning rate, $\frac{1-e^{-\gamma\tau_k}}{\gamma} r(s_k,a_k)$ is the sample discounted reward received in τ_k time units, and $Q(s_{k+1},a')$ is the estimated value of the state–action pair (s_{k+1},a') with s_{k+1} being the next state. The term $\varepsilon_k(s,a)$ represents the eligibility for each state-action pair, updated as

$$\varepsilon_k(s,a) = \lambda \cdot e^{-\gamma\tau_{k-1}} \cdot \varepsilon_{k-1}(s,a) + \delta((s,a),(s_k,a_k)),$$

where $\delta((s,a),(s_k,a_k))$ is the delta-Kronecker function.

5.4 SARSA Learning

SARSA learning algorithm is on-policy learning and an extension of the TD-learning algorithm. In contrast to the traditional Q-learning, in SARSA learning, the maximum reward for the future state is not necessarily used in the Q-update Eq. (5.1). Instead, a new action using the same policy as original action could be chosen leading to a different award. The name SARSA is derived from the quadruple $Q(s,a,r,s',a')$, i.e., the current state and the action taken with a given reward r in the next state, with s' and a' being the next state and action, respectively. In the case of SARSA learning, the Q is chosen as follows:

$$Q(s,a) \leftarrow Q(s,a) + \alpha[r + \gamma \max_{\alpha} Q(s',a') = Q(s,a)]. \tag{5.4}$$

This process is repeated iteratively until convergence. Also, the ϵ-greedy algorithm can be deployed for convergence.

Traditional Q-learning and SARSA learning techniques have some similarities and differences. For instance, both the techniques follow ϵ-greedy algorithm for exploration and perform the Q-value update based on the rewards with state transition(s). The way the Q-update happens in Q-learning and SARSA learning are different as noted from their respective equations. Furthermore, the SARSA algorithm is on-policy learning as the new action a' is based on the same policy used to determine action a, i.e., the one that leads to state s'.

Example 5.3 (SARSA Learning)
Problem: Create a reinforcement learning based on the discussed SARSA learning policy.
Solution: The SARSA learning technique primarily deviates from traditional Q-learning primarily in terms of the update and the action selection policies. The code snippet that depicts this variation is shown below. One can integrate this with the Q-learning example discussed in the earlier part of this chapter for full implementation.

```
#Function to choose the next action
def choose_action(state):
  action=0
  if np.random.uniform(0, 1) < epsilon:
    action = env.action_space.sample()
  else:
    action = np.argmax(Q[state, :])
  return action

#Function to learn the Q-value
def update(state, state2, reward, action, action2):
  predict = Q[state, action]
  target = reward + gamma * Q[state2, action2]
  Q[state, action] = Q[state, action] + alpha * (target - predict)
```

The above code snippet shows the function that defines the Q-update in the SARSA algorithm. The update function is different compared to the traditional Q-learning algorithm. The initialization and other components can be similar to the traditional Q-learning technique, though not exactly the same.

5.5 Deep Q-Learning

As observed from the previous sections, Q-learning algorithms are effective and can solve a wide range of complex problems. Thus, they are adopted in products like Deepmind's AlphaGo, gaming applications and drone navigation, and similar use cases. However, the complexity of the reinforcement or Q-learning algorithms are extremely large. To address this challenge, deep Q-learning is introduced.

As seen in previous chapters that the neural networks function efficiently for estimation, prediction, and forecasting applications. Thus, in deep Q-learning, neural networks are used as function approximators for target optimization, mapping of action-state pairs with maximizing rewards. Deep Q-learning is adopted in Deepmind's AlphaGo.

In deep Q-learning, the neural network can be initialized randomly or stochastically, and depending on the obtained feedback from the Q-learning algorithm, the neural network coefficients or model parameters will be updated in an interactive manner. Similar to traditional neural networks, the networks used in deep Q-learning will also utilize the coefficients to approximate the functions that map the inputs to the outputs. The model learning happens in a gradient descent manner.

In addition to traditional or deep neural networks, CNNs can as well be used for deep Q-learning. In this case, the CNN can be used to detect the state of the system, i.e., say based on the image captured from a gaming window, the CNN captures the state information instead of traditional classification, and this can be processed by a neural network for further action–state mapping that maximizes the reward.

5.6 Policy Optimization

5.6.1 *Stochastic Policy Gradient*

Consider optimizing $J(\pi)$, the value of policy π under some initial state distribution μ. Consider a class of parametric stochastic policies $\{\pi_\theta : \theta \in \Theta\}$, such that $\pi_\theta(a|s)$ is differentiable with respect to θ. Thus, the problem of interest is

$$\max_{\theta \in \Theta} J(\pi_\theta).$$

If we can compute (the stochastic estimate of) $\nabla_\theta J(\pi_\theta)$, we can optimize objective $J(\pi_\theta)$ with gradient-based methods (e.g., SGD, natural gradient descent).

5.6.2 *REINFORCE*

Let $\tau = (s_0, a_0, r_0, s_1, a_1, r_1, \ldots)$ denote the state–action–reward trajectory. The probability distribution over trajectories under policy π can be expressed as

$$P^\pi(\tau) = \mu(s_0)\pi(a_0|s_0)R(r_0|s_0, a_0)P(s_1|s_0, a_0)\pi(a_1|s_1)\ldots,$$

where μ is the initial state distribution and (P, R) are the transition and reward functions of the MDP. Note this expression is obtained by first decomposing the joint distribution into conditional distributions (the chain rule of probability) and then using the conditional independence of Markov transition and Markov policy. The discounted total reward of trajectory τ is

$$R(\tau) := \sum_{t=0}^{H-1} \gamma^t r_t,$$

where horizon H can be finite or infinite.

Per-trajectory REINFORCE For per-trajectory REINFORCE, we start by decomposing the objective into trajectories, i.e., $J(\pi) = \sum_\tau [R(\tau)P^\pi(\tau)]$. Then, the gradient can be computed as (dropping the θ in the subscript of ∇_θ)

$$\begin{aligned}
\nabla J(\pi_\theta) &= \nabla \sum_\tau \left[R(\tau)P^{\pi_\theta}(\tau)\right] \\
&= \sum_\tau \left[R(\tau)\nabla P^{\pi_\theta}(\tau)\right] \\
&= \sum_\tau \left[R(\tau)P^{\pi_\theta}(\tau)\nabla \log P^{\pi_\theta}(\tau)\right] \\
&= E_{\tau\sim P^\pi(\tau)}\left[R(\tau)\nabla \log P^{\pi_\theta}(\tau)\right] \quad \text{(likelihood ratio trick)} \\
&= E_{\tau\sim P^\pi(\tau)}\left[R(\tau)\nabla \log(\mu(s_0)\pi_\theta(a_0|s_0)R(r_0|s_0,a_0)P(s_1|s_0,a_0)\pi_\theta(a_1|s_1)\ldots)\right] \\
&= E_{\tau\sim P^\pi(\tau)}\left[R(\tau)\sum_{t=0}^{H-1}\nabla \log \pi_\theta(a_t|s_t)\right].
\end{aligned}$$

This means that one can compute stochastic gradient of $\nabla J(\pi_\theta)$ by (1) generating a trajectory under π_θ and (2) computing $R(\tau)\sum_{t=0}^{H-1}\nabla \log \pi_\theta(a_t|s_t)$.

Per-step REINFORCE We can obtain a lower variance version of REINFORCE by decomposing the total discounted rewards into steps

$$\nabla J(\pi_\theta) = \nabla E_{\tau\sim P^{\pi_\theta}(\tau)}\left[\sum_{t=0}^{H-1}\gamma^t r_t\right] = \sum_{t=0}^{H-1}\gamma^t \nabla E_{\tau\sim P^{\pi_\theta}(\tau)}[r_t],$$

where

$$\begin{aligned}
\nabla E_{\tau\sim P^{\pi_\theta}(\tau)}[r_t] &= \nabla E_{\tau_{0:t}\sim P^{\pi_\theta}(\tau_{0:t})}[r_t] \qquad (\tau_{0:t} \text{ is } \tau \text{ truncated by } t) \\
&= E_{\tau_{0:t}\sim P^{\pi_\theta}(\tau_{0:t})}\left[r_t\sum_{t'=0}^{t}\nabla \log \pi_\theta(a_{t'}|s_{t'})\right] \qquad \text{(likelihood ratio trick)} \\
&= E_{\tau\sim P^{\pi_\theta}(\tau)}\left[r_t\sum_{t'=0}^{t}\nabla \log \pi_\theta(a_{t'}|s_{t'})\right] \qquad \text{(back to untruncated } \tau\text{)}.
\end{aligned}$$

Plugging this into the previous step, we have

$$\nabla J(\pi_\theta) = \sum_{t=0}^{H-1}\gamma^t \nabla E_{\tau\sim P^{\pi_\theta}(\tau)}[r_t]$$

$$= \sum_{t=0}^{H-1} \gamma^t E_{\tau \sim P^{\pi_\theta}(\tau)} \left[r_t \sum_{t'=0}^{t} \nabla \log \pi_\theta(a_{t'}|s_{t'}) \right]$$

$$= E_{\tau \sim P^{\pi_\theta}(\tau)} \left[\sum_{t=0}^{H-1} \gamma^t r_t \sum_{t'=0}^{t} \nabla \log \pi_\theta(a_{t'}|s_{t'}) \right]$$

$$= E_{\tau \sim P^{\pi_\theta}(\tau)} \left[\sum_{t=0}^{H-1} \gamma^t \left(\sum_{t'=t}^{H-1} \gamma^{t'-t} r_{t'} \right) \nabla \log \pi_\theta(a_t|s_t) \right].$$

This gives us another unbiased estimator of $\nabla J(\pi_\theta)$: (1) generate a trajectory under π_θ, (2) pick a random timestep t with probability $\propto \gamma^t$, and (3) compute $\left(\sum_{t'=t}^{H-1} \gamma^{t'-t} r_{t'}\right) \nabla \log \pi_\theta(a_t|s_t)$. This per-step estimator has lower variance than the per-trajectory estimator.

Action Value Expression

The estimator obtained by per-step REINFORCE suggests the following policy gradient expression:

$$\nabla J(\pi_\theta) = E_{\tau \sim P^{\pi_\theta}(\tau)} \left[\sum_{t=0}^{H-1} \gamma^t Q^{\pi_\theta}(s_t, a_t) \nabla \log \pi_\theta(a_t|s_t) \right]$$

because $\left(\sum_{t'=t}^{H-1} \gamma^{t'-t} r_{t'}\right) \nabla \log \pi_\theta(a_t|s_t)$ is an unbiased estimator of $Q^{\pi_\theta}(s_t, a_t)\nabla \log \pi_\theta(a_t|s_t)$. Note that this expression is equivalent to

$$\nabla J(\pi_\theta) = \frac{1}{1-\gamma} E_{s \sim d^{\pi_\theta}} E_{a \sim \pi_\theta(\cdot|s)} \left[Q^{\pi_\theta}(s, a) \nabla \log \pi_\theta(a|s) \right],$$

where d^π is the normalized state occupancy with initial distribution μ:

$$\frac{1}{1-\gamma} d^\pi(s) = \sum_{t=0}^{H-1} \gamma^t \Pr(s_t = s | s_0 \sim \mu, \pi),$$

which makes $E_{\tau \sim P^\pi(\tau)} \left[\sum_{t=0}^{H-1} \gamma^t f(s_t, a_t)\right] = \frac{1}{1-\gamma} E_{s \sim d^\pi} E_{a \sim \pi(\cdot|s)} [f(s, a)]$ for any f. We now prove this action value expression of policy gradient, assuming the horizon is infinite, $H = \infty$, and dropping subscript θ in π_θ. The proof starts with the fact that $V^\pi(s) = \sum_a \pi(a|s) Q^\pi(s, a)$. Differentiate both sides:

$$\nabla V^\pi(s) = \nabla \sum_a \pi(a|s) Q^\pi(s, a)$$

$$
\begin{aligned}
&= \sum_a \nabla\pi(a|s) \cdot Q^\pi(s,a) + \pi(a|s) \cdot \nabla Q^\pi(s,a) \\
&= \sum_a \nabla\pi(a|s) \cdot Q^\pi(s,a) + \pi(a|s) \cdot \nabla \left(r(s,a) + \gamma E_{s'\sim P(\cdot|s,a)}[V^\pi(s')]\right) \\
&= \sum_a \pi(a|s)\nabla \log\pi(a|s) \cdot Q^\pi(s,a) + \pi(a|s)\gamma E_{s'\sim P(\cdot|s,a)}[\nabla V^\pi(s')] \\
&= E_{a\sim\pi(\cdot|s)}\left[Q^\pi(s,a)\nabla \log\pi(a|s) + \gamma E_{s'\sim P(\cdot|s,a)}[\nabla V^\pi(s')]\right].
\end{aligned}
$$

Note that ∇V^π appears on both sides, and now we apply this recursion. We let $d_t^\pi(s) = \Pr(s_t = s|s_0 \sim \mu, \pi)$ be the state distribution at time t under policy π with initial state distribution μ. Note $d_0^\pi = \mu$ and thus $J(\pi) = E_{s\sim d_0^\pi}[V^\pi(s)]$. Now, by applying the recursion, we have

$$
\begin{aligned}
\nabla J(\pi) &= \nabla E_{s\sim d_0^\pi}[V^\pi(s)] = E_{s\sim d_0^\pi}[\nabla V^\pi(s)] \\
&= E_{s\sim d_0^\pi}\left[E_{a\sim\pi(\cdot|s)}\left[Q^\pi(s,a)\nabla \log\pi(a|s) + \gamma E_{s'\sim P(\cdot|s,a)}[\nabla V^\pi(s')]\right]\right] \\
&= E_{s\sim d_0^\pi, a\sim\pi(\cdot|s)}[Q^\pi(s,a)\nabla \log\pi(a|s)] + \gamma E_{s\sim d_1^\pi}[\nabla V^\pi(s)] \\
&= E_{s\sim d_0^\pi, a\sim\pi(\cdot|s)}[Q^\pi(s,a)\nabla \log\pi(a|s)] \\
&\quad + \gamma E_{s\sim d_1^\pi, a\sim\pi(\cdot|s)}[Q^\pi(s,a)\nabla \log\pi(a|s)] + \gamma^2 E_{s\sim d_2^\pi}[\nabla V^\pi(s)] \\
&= \ldots = \sum_{t=0}^{\infty} \gamma^t E_{s\sim d_t^\pi, a\sim\pi(\cdot|s)}[Q^\pi(s,a)\nabla \log\pi(a|s)] \\
&= \frac{1}{1-\gamma} E_{s\sim d^\pi} E_{a\sim\pi(\cdot|s)}\left[Q^\pi(s,a)\nabla \log\pi(a|s)\right].
\end{aligned}
$$

This concludes the proof.

Variance Reduction by Baseline A useful fact is that, for any fixed s, we have

$$
E_{a\sim\pi(\cdot|s)}[\nabla \log\pi(a|s)] = \sum_a \nabla\pi(a|s) = \nabla \sum_a \pi(a|s) = \nabla 1 = 0.
$$

Therefore, adding any $b : \mathcal{S} \to \mathbb{R}$ to the policy gradient estimator will not affect its unbiasedness:

$$
\nabla J(\pi) = \frac{1}{1-\gamma} E_{s\sim d^\pi} E_{a\sim\pi(\cdot|s)}\left[\nabla \log\pi(a|s)(Q^\pi(s,a) - b(s))\right].
$$

Adding a function $b : \mathcal{S} \to \mathbb{R}$, often called the baseline, does not introduce bias but does affect the variance. A popular choice is $b(s) = V^\pi(s)$, which can reduce the variance, with $Q^\pi(s,a) - b(s) = Q^\pi(s,a) - V^\pi(s) = A^\pi(s,a)$ becoming the advantage function.

Other Expressions

There are many other (unbiased) expressions and estimators for the policy gradient. For example, TD residual $r(s, a) + V^{\pi}(s) - \gamma V^{\pi}(s')$ can replace the advantage function.

5.7 Gradient-Based Policy Optimization

The previous section explained how we can obtain an unbiased estimator of $\nabla J(\pi_t heta)$ from a sampled trajectory under $\pi_t heta$. With such a gradient estimator, we can use gradient-based methods to optimize objective $J(\pi_t heta)$. Below we give such an algorithm using SGD, with the gradient estimated from per-step REINFORCE with a learned baseline. Methods more sophisticated than SGD include TRUST region methods, natural gradient, etc.

- Input: differentiable policy $\pi_\theta(a|s)$, differentiable baseline function $b_w(s)$.
- Initialize: $\theta(a|s)$, w, step sizes α^θ, α^w
- Loop forever:

  Generate a trajectory on policy π_θ $(s_0, a_0, r_0, s_1, a_1, r_1, \ldots, s_H)$
  Loop for each step $t = 0, 1, \ldots, H - 1$:
    $G \leftarrow \sum_{k=t}^{H-1} \gamma^{k-t} R_k$
    $A \leftarrow G - b_w(s_t)$
    $w \leftarrow w + \alpha^w A \nabla_w b_w(s)$
    $\theta \leftarrow \theta + \alpha^\theta \gamma^t A \nabla_\theta \log \pi_\theta(a_t|s_t)$.

5.8 Putting It All Together

Reinforcement learning is a human learning inspired technique that learns through experience and does not require large amounts of labeled data. Reinforcement learning technique is an iterative learning process that requires exploration to better understand the environment and the impact of the considered actions. Unlike supervised learning, reinforcement learning can better adapt to the varying operating conditions.

Reinforcement learning is widely used in applications such as games, navigation, and other similar applications. For simpler applications with a smaller number of actions, traditional reinforcement learning can be sufficient. However, as the number of states and actions increases, the convergence can be a concern. For such scenarios, techniques such as SARSA and TD(λ) can be efficient. In addition, the techniques such as reinforce can be utilized for well-defined data and the corresponding state and action pairs.

5.9 Exercise Problems

Problem 5.1 Build a simple reinforcement learning based problem using Q-function containing:

Rules: The agent (yellow box) has to reach one of the goals to end the game (green or red cell). Rewards: Each step gives a negative reward of -0.04. The red cell gives a negative reward of -1. The green one gives a positive reward of $+1$. States: Each cell is a state the agent can be. Actions: There are only 4 actions. Up, Down, Right, and Left.

Problem 5.2 The task is simply to reach point G by starting from point S in the map. However, there is a cliff we should avoid so that a -100 reward incurs if a transition into the cliff is made. We will use two approaches to solve the problem as it can be seen from the picture.

Problem 5.3 Make use of Policy Iteration and Value Iteration algorithms to solve a simple MDP problem. The environment is a simple MDP problem formulated as States—Happy, Sad; Actions—Studying, Drinking; and Rewards—ranges between -10 and 40 depending on the transition.

Problem 5.4 Implement an explicit policy for the mountain car environment without using any learning algorithm. Explain in detail your reasoning behind your policy and run several test episodes to measure its performance.

Problem 5.5 Given a maze as a cube, try to use a simple reinforcement learning algorithm to solve the problem.

Problem 5.6 In this game, the user can choose how many rounds the AI will be trained on and then access its performance after training. With this tool, we can examine if the algorithm implemented (Q-learning) performs worse or better with more training.

Problem 5.7 Build a simple implementation and comparison of three ϵ-greedy bandits with a single state.

Problem 5.8 Consider the per-trajectory REINFORCE expression for policy gradient with a constant baseline:

$$\nabla J(\pi_\theta) = E_{\tau \sim P^{\pi_\theta}(\tau)} \left[\nabla \log \pi_\theta(\tau)(R(\tau) - b) \right],$$

where $\nabla \log \pi_\theta(\tau) := \sum_{t=0}^{H-1} \nabla \log \pi_\theta(a_t|s_t)$ and $b \in \mathbb{R}$ is the constant baseline. For simplicity, let us consider the 1D case where $\theta \in \mathbb{R}$. We now look at the variance of the gradient estimator $\widehat{g} := \nabla \log \pi_\theta(\tau)(R(\tau) - b)$. What is the optimal baseline that minimizes the variance of $\widehat{g}$?

Problem 5.9 Suppose for any state–action pair (s, a), we have a feature mapping $\phi_{s,a} \in \mathbb{R}^d$. The log-linear policy parameterization is of the form:

$$\pi_\theta(a|s) = \frac{\exp(\theta \cdot \phi_{s,a})}{\sum_{a'} \exp(\theta \cdot \phi_{s,a'})}$$

with $\theta \in \mathbb{R}^d$. Compute $\nabla \log \pi_\theta(a|s)$ for the log-linear policy parameterization.

Part II
Advanced Machine Learning

Chapter 6
Online Learning

6.1 Introduction

Offline or batch learning is a traditional machine learning paradigm in which a model is learned from the complete dataset at once via batches. When dealing with new training data, such a learning style incurs high retraining costs. Data expands and evolves quickly in the era of big data, making classic batch learning methods difficult to scale for real-world applications.

Online learning is a sub-field of machine learning that deals with data entering in a sequential sequence, as opposed to batch machine learning methods. At each step, the online learning model seeks to learn and update the best predictor for the new data. Online learning approaches address the drawbacks of offline learning in handling streaming data by successfully handling streaming data. If we want to create a real-time stock price prediction model, we should use an online model that updates the model in real time rather than an offline model that requires all of the data to be retrained when new stock features are added.

We introduce online learning approaches in both supervised and unsupervised environments in this chapter. In particular, supervised online learning requires that the ground truth labels are readily available online. Online classification is one of the most popular problems in supervised online learning, and it seeks to predict the categories for a new data instance based on historical training data and fresh streaming data observations. For example, in a spam email detection system, we can use the online classification approach to categorize each email as "spam" or "benign." Unsupervised online learning, on the other hand, presupposes that the new data is unlabeled. Online clustering, for example, is a technique of grouping data instances into groups in which data instances in the same group are more similar than data instances in other groups.

Finally, to help readers comprehend the function of online learning in the actual world, we provide several instances of online learning applications.

S. Rafatirad et al., *Machine Learning for Computer Scientists and Data Analysts*,
https://doi.org/10.1007/978-3-030-96756-7_6

6.2 Online Supervised Learning

In this section, we introduce the fundamental approaches and principles for online learning methodologies toward supervised learning tasks [84, 85]. Here, we discuss linear online learning methods, which are widely used in real-world applications. Concretely, consider an input domain $\mathcal{X}$ and an output domain $\mathcal{Y}$ for a learning task; we aim to learn a hypothesis $f : \mathcal{X} \rightarrow \mathcal{Y}$ where the target model f is a linear function. For example, consider a typical linear binary classification task; our goal is to learn a linear classifier $f : \mathcal{X} \rightarrow \{+1, -1\}$ as follows: $f(x_t; w) = sgn(w \cdot x_t)$, where $\mathcal{X}$ is typically a d-dimensional vector space $\mathbb{R}^d$, $w \in \mathcal{X}$ is a weight vector specified for the classifier to be learned, and $sgn(z)$ is an indicator function that outputs +1 when $z > 0$ and -1 otherwise. We review two major types of online learning algorithms: first-/second-order online learning and online learning with regularization.

6.2.1 *First-/Second-Order Online Learning*

In the following, we discuss two important algorithms for first-order linear online learning and one for second-order online learning. The first-order linear online learning exploits the first-order information of the model during the learning process. In the contrast, second-order online learning algorithms exploit both first-order and second-order information in order to accelerate optimization convergence. Despite the better learning performance, second-order online learning algorithms often fall short in higher computational complexity.

Passive Aggressive Online Learning (PA)

Passive aggressive algorithm is a popular family of first-order online learning algorithms which generally follows the principle of margin-based learning [86]. Specifically, given an instance x_t at round t, PA formulates the updating optimization as follows:

$$w_{t+1} = \arg\min_{w \in \mathbb{R}^d} \frac{1}{2}\|w - w_t\|^2 \qquad s.t. \ell_t(w) = 0, \tag{6.1}$$

where $\ell_t(w) = \max(0, 1 - y_t w \cdot x_t)$ is the hinge loss. The above resulting update is passive whenever the hinge loss is zero, i.e., $w_{t+1} = w_t$ when $\ell = 0$. In contrast, when the loss is not zero, the approach will force w_{t+1} aggressively to satisfy the constraint regardless of any step size. Intuitively, PA algorithm aims to keep the updated classifier w_{t+1} stay close to the previous classifier and ensure every incoming instance to be classified correctly by the updated classifier. The regular

PA algorithm assumes training data is always separable, which may not be true for noisy training data in real-world applications. To tackle these problems, two variants of PA algorithm relax the assumption as the following:

$$\begin{aligned} &\text{PA-I}: w_{t+1} = \arg\min_{w\in\mathbb{R}^d} \frac{1}{2}\|w - w_t\|^2 + C\xi \\ &\text{subject to } \ell_t(w) \leq \xi \text{ and } \xi \geq 0 \\ &\text{PA-II}: w_{t+1} = \arg\min_{w\in\mathbb{R}^d} \frac{1}{2}\|w - w_t\|^2 + C\xi^2 \\ &\text{subject to } \ell_t(w) \leq \xi, \end{aligned} \tag{6.2}$$

where C is a positive parameter to balance the trade-off between first regularization term and second slack variable term. By solving the three optimization tasks, we can derive the closed-form updating rules of three PA algorithms:

$$w_{t+1} = w_t + \tau_t y_t x_t, \qquad \tau_t = \begin{cases} \ell_t/\|x_t\|^2 & \text{(PA)} \\ \min\{C, \ell_t/\|x_t\|^2\} & \text{(PA-I)} \\ \frac{\ell_t}{\|x_t\|^2 + \frac{1}{2C}} & \text{(PA-II).} \end{cases} \tag{6.3}$$

In the following example, we demonstrate the sample code of passive aggressive regressor.

```
from sklearn.linear_model import PassiveAggressiveRegressor
from sklearn.datasets import make_regression

X, Y = make_regression(n_features=4, random_state=0)
regr = PassiveAggressiveRegressor(random_state=0)
regr.fit(X, Y)
```

Example 6.1 (Iris Classification with Passive Aggressive Algorithm)
Problem: Perform classification with passive aggressive algorithm on Iris dataset.
Solution: In order to perform the passive aggressive algorithm-based prediction on Iris dataset, one needs to first create a passive aggressive classifier using the `PassiveAggressiveClassifier()` and train the model with the loaded Iris dataset. The code snippet for Iris dataset classification with passive aggressive algorithm is shown below:

```
# Importing modules
from sklearn.datasets import load_iris
from sklearn.linear_model import PassiveAggressiveClassifier
from sklearn.metrics import classification_report, accuracy_score
```

```python
from sklearn.model_selection import train_test_split

# Loading
dataset = load_iris()
X, y = dataset.data, dataset.target

# Splitting
X_train, X_test, y_train, y_test = train_test_split(X, y)

# Creating model
model = PassiveAggressiveClassifier()

# Fitting
model.fit(X_train, y_train)

# Prediction
test_pred = model.predict(X_test)

# Evaluation
print("Accuracy : {accuracy_score(y_test, test_pred)}")
```

Online Gradient Descent (OGD)

Many online learning problems can be formulated as an online convex optimization task, which can be solved by applying the OGD algorithm. Consider the online binary classification as an example, where we use the hinge loss function, i.e., $\ell_t(w) = \max(0, 1 - y_t w \cdot x_t)$. By applying the OGD algorithm, we can derive the updating rule as follows:

$$w_{t+1} = w_t + \eta_t y_t x_t, \tag{6.4}$$

where η_t is the learning rate (or step size) parameter. The OGD algorithm is outlined in Algorithm 2, where any generic convex loss function can be used. $\prod_S$ is the projection function to constrain the updated model to lie in the feasible domain.

Algorithm 2 Online gradient descent

Input: w_1, convex set S, step size η_t

for $t = 1, 2, \ldots, T$ **do**
  Receive $x_t \in \mathbb{R}^d$, predict $\hat{y}_t$ using w_t
  Suffer loss $\ell_t(w_t)$
  Update $w_{t+1} = \prod_S(w_t - \eta_t \nabla \ell_t(w_t))$
end for

Both OGD and PA algorithms share similar updating rules but differ in that OGD method usually employs some predefined learning rate scheme while PA algorithm chooses the optimal learning rate τ_t at each round (but subject to a predefined cost parameter C). Recently, different OGD variants have been proposed to improve either theoretical bounds or practical issues, such as adaptive OGD [87] and mini-batch OGD [88], among others.

Example 6.2 (House Price Prediction with OGD)
Problem: Predict the Boston house prices with OGD.
Solution: We will need to load the dataset first by `load_boston()` from `sklearn.datasets` and then perform `SGDRegressor()` to the training set. The code snippet for OGD is shown below:

```
from sklearn.datasets import load_boston
from sklearn.model_selection import train_test_split
from sklearn.linear_model import SGDRegressor
from sklearn.preprocessing import StandardScaler

boston = load_boston()
X_train, X_test, Y_train, Y_test = train_test_split(boston.data, boston.
    target, test_size=0.2, random_state=42)
regr = SGDRegressor(loss='huber', penalty='l2', alpha=0.0001, fit_intercept=
    False, n_iter=5, shuffle=True, verbose=1, epsilon=0.1, random_state=42,
    learning_rate='invscaling', eta0=0.01, power_t=0.5)

sc_boston = StandardScaler()
X_train = sc_boston.fit_transform(X_train)
X_test = sc_boston.transform(X_test)

regr.fit(X_train, Y_train)
```

Second-Order Perceptron (SOP)

Second-order Perceptron algorithm [89] aims to exploit certain geometrical properties of the data which are ignored by the first-order algorithms. Indeed, SOP can be viewed as an online variant of the whitened Perceptron algorithm. Assuming that the instances $x_1, \ldots, x_T$ are preliminarily available, we can get the correlation matrix $M = \sum_{t=1}^{T} x_t x_t^T$. The whitened Perceptron algorithm is simply the standard Perceptron run on the transformed sequence $(M^{-1/2}x_1, y_1), \ldots, (M^{-1/2}x_T, y_T)$. By reducing the correlation matrix of the transformed instances, the whitened Perceptron algorithm can achieve significantly better mistake bound. In online setting, the correlation matrix M can be approximated by the previously seen instances. SOP is outlined in Algorithm 3.

6.2.2 *Online Learning with Regularization*

Traditional online learning methods learn a classifier $w \in \mathbb{R}^d$ where the magnitude of each element $|w^j|$ weights the importance of each feature, which are often nonzero. When dealing with high-dimensional data, traditional online learning methods suffer from expensive computational time and space costs. This drawback is often

Algorithm 3 Second-order perceptron

Input: $w_1 = 0$, $X_0 = []$, $v_0 = 0$, $k = 1$
1: **for** $t = 1, 2, \ldots, T$ **do**
2: Given an incoming instance x_t, set $S_t = [X_{k-1}, x_t]$
3: predict $\hat{y}_t = f_t(x_t) = sign(w_t \cdot x_t)$, where $w_t = (aI_n + S_t S_t^T)^{-1} v_{k-1}$
4: Receive the true class label $y_t \in \{+1, -1\}$;
5: **if** $\hat{y}_t \neq y_t$ **then**
6: $v_k = v_{k-1} + y_t x_t$, $X_k = S_t$,$k = k + 1$
7: **end if**
8: **end for**

addressed using regularization by performing sparse online learning, which aims to exploit the sparsity property with real-world high-dimensional data. Specifically, a batch sparse learning problem can be formalized as

$$P(w) = \frac{1}{n} \sum_{i=1}^{n} \ell_t(w) + \phi_s(w), \tag{6.5}$$

where ϕ_s is a sparsity-inducing regularizer. For example, when choosing $\phi_s = \lambda\|w\|_0$, it is equivalent to imposing a hard constraint on the number of non-zero elements in w. Instead of choosing ℓ_0-norm which is hard to be optimized, a more commonly used regularizer is ℓ_1-norm, i.e., $\phi_s = \lambda\|w\|_1$, which can induce sparsity of the weight vector but does not explicitly constrain the number of non-zero elements. The following reviews some popular sparse online learning methods.

Truncated Gradient Descent

A straightforward idea to sparse online learning is to modify Online Gradient Descent and round small coefficients of the weight vector to 0 after every K iterations:

$$w_{t+1} = T_0(w_t - \eta \nabla \ell_t(w_t), \theta), \tag{6.6}$$

where the function $T_0(v, \theta)$ performs an element-wise rounding on the input vector: if the j-th element v^j is smaller than the threshold θ, set $v^j = 0$. Despite its simplicity, this method struggles to provide satisfactory performance because the aggressive rounding strategy may ignore many useful weights which may be very small due to the low frequency of appearance. Motivated by addressing the above limitation, the Truncated Gradient Descent (TGD) method [90] explores a less aggressive version of the truncation function:

$$w_{t+1} = T_1(w_t - \eta \nabla \ell_t(w_t), \eta g_i, \theta) \tag{6.7}$$

$$\text{where } T_1(v^j, \alpha, \theta) = \begin{cases} \max(0, v^j - \alpha) & \text{if } v^j \in [0, \theta] \\ \min(0, v^j + \alpha) & \text{if } v^j \in [-\theta, 0] \\ v^j & \text{otherwise,} \end{cases} \tag{6.8}$$

where $g_i > 0$ is a parameter that controls the level of aggressiveness of the truncation. By exploiting sparsity, TGD achieves efficient time and space complexity that is linear with respect to the number of non-zero features and independent of the dimensionality d. In addition, it is proven to enjoy a regret bound of $O(\sqrt{T})$ for convex loss functions when setting $\eta = O(1/\sqrt{T})$.

Forward-Looking Subgradients (FOBOS)

Consider the objective function in the t-th iteration of a sparse online learning task as $\ell_t(w) + r(w)$, FOBOS [91] assumes f_t is a convex loss function (differentiable), and r is a sparsity-inducing regularizer (non-differentiable). FOBOS updates the classifier in the following two steps:

(1) Perform Online Gradient Descent:

$$w_{t+\frac{1}{2}} = w_t - \eta_t \nabla \ell_t(w_t). \tag{6.9}$$

(2) Project the solution in (i) such that the projection stays close to the interim vector $w_{t+\frac{1}{2}}$ and (ii) has a low complexity due to r:

$$w_{t+1} = \arg\min_{w}\{\frac{1}{2}\|w - w_{\frac{1}{2}}\|^2 + \eta_{t+\frac{1}{2}} r(w)\}. \tag{6.10}$$

When choosing ℓ_1-norm as the regularizer, the above optimization can be solved with the closed-form solution for each coordinate:

$$w^j_{t+1} = \text{sgn}(w^j_{t+\frac{1}{2}})[|w^j_{t+\frac{1}{2}}| - \eta_{t+\frac{1}{2}}]_+. \tag{6.11}$$

The FOBOS algorithm with ℓ_1-norm regularizer can be viewed as a special case of TGD, where the truncation threshold $\theta = \infty$ and the truncation frequency $K = 1$. When $\eta_{t+\frac{1}{2}} = \eta_{t+1}$ and $\eta_t = O(1/\sqrt{t})$, this algorithm also achieves $O(\sqrt{T})$ regret bound.

Regularized Dual Averaging (RDA)

Motivated by the theory of dual averaging techniques [92], the RDA algorithm [93] updates the classifier by

$$w_{t+1} = \arg\min_{w}\{\bar{g}_t w + \Psi(w) + \frac{\beta_t}{t}h(w)\}, \tag{6.12}$$

where $\Psi(w)$ is the original sparsity-inducing regularizer, i.e., $\Psi(w) = \lambda\|w\|_1$, $h(w) = \frac{1}{2}\|w\|^2$ is an auxiliary strongly convex function, and $\bar{g}_t$ is the averaged gradients of all previous iterations, i.e., $\bar{g} = \frac{1}{t}\sum_{\tau=1}^{t} \nabla\ell_\tau(w_\tau)$. Setting the step size $\beta_t = \gamma\sqrt{t}$, one can derive the closed-form solution:

$$w_{t+1}^{j} = \begin{cases} 0 & \text{if } |\bar{g}_t^{j}| < \lambda \\ -\frac{\sqrt{t}}{\lambda}(\bar{g}_t^{j} - \lambda \text{sgn}(\bar{g}_t^{j})) & \text{otherwise.} \end{cases} \tag{6.13}$$

To further pinpoint the differences between RDA and FOBOS, we rewrite FOBOS in the same notation as RDA:

$$w_{t+1} = \arg\min_{w}\{{g_t}^T w + \Psi(w) + \frac{1}{2\alpha_t}\|w - w_t\|_2^2\}. \tag{6.14}$$

Specifically, RDA differs from FOBOS in several aspects. First, RDA uses the averaged gradient instead of the current gradient. Second, $h(w)$ is a global proximal function instead of its local Bregman divergence. Third, the coefficient for $h(w)$ is $\beta_t/t = \gamma/\sqrt{t}$ which is $1/\alpha_t = O(\sqrt{t})$ in FOBOS. Fourth, the truncation of RDA is a constant λ, while the truncation in FOBOS $\eta_{t+\frac{1}{2}}$ decreases with a factor $\sqrt{t}$. Clearly, RDA uses a more aggressive truncation threshold and thus usually generates significantly more sparse solutions. RDA also ensures the $O(\sqrt{T})$ regret bound.

Follow-the-Regularized-Leader-Proximal (FTRL-Proximal)

As we mentioned before, the aim of the online convex optimization task is to optimize the regret. Traditional approaches (termed as Follow the Leader (FTL)) can be unstable, leading to high regret (e.g., linear regret) in the worst case [94]. This motivates the need to stabilize the approaches through regularization. Here we discuss Follow-the-Regularized-Leader-Proximal (FTRL-Proximal), which is widely used in industry programs. The FTRL-Proximal algorithm can be seen as a hybrid of FOBOS and RDA algorithms and significantly outperforms both on a large real-world dataset [95]. The idea is to solve the following optimization problem in each iteration:

$$w_{t+1} = \arg\min_{w}\{\bar{g}_t w + \Psi(w) + \frac{1}{2\alpha_t}\|w - w_t\|_2^2\}. \tag{6.15}$$

As indicated by Eq. 6.15, when the non-smooth term Ψ is omitted, FTRL-Proximal algorithm is in fact identical to FOBOS. On the other hand, its update is essentially the same as that of dual averaging, except that additional strong convexity is centered at the current feasible point. The FTRL-Proximal algorithm is outlined in Algorithm 4.

Algorithm 4 Per-coordinate FTRL-proximal with L_1 and L_2 regularization for logistic regression

---

Input: parameters $\alpha, \beta, \lambda_1, \lambda_2, \forall i \in \{1, \ldots, d\}$, initialize $z_i = 0$ and $n_i = 0$

for $t = 1$ to T **do**

  Receive feature vector x_t and let $I = \{i | x_i \neq 0\}$

  For $i \in I$ compute

$$w_{t,i} = \begin{cases} 0 & \text{if } |z_i| \leq \lambda_1 \\ -(\frac{\beta + \sqrt{n_i}}{\alpha} + \lambda_2)^{-1}(z_i - \text{sgn}(z_i)\lambda_1) & \text{otherwise.} \end{cases}$$

  Predict $p_t = \sigma(x_t \cdot w)$ using the $w_{t,i}$ computed above

  Observe label $y_t \in \{0, 1\}$

  for all $i \in I$ **do**

    $g_i = (p_t - y_t)x_i$ # gradient of loss w.r.t w_i

    $\sigma_i = \frac{1}{\alpha}(\sqrt{n_i + g_i^2} - \sqrt{n_i})$ # equals $\frac{1}{\eta_{t,i}} - \frac{1}{\eta_{t-1,i}}$

    $z_i \leftarrow z_i + g_i - \sigma_i w_{t,i}$

    $n_i \leftarrow n_i + g_i^2$

  end for

end for

6.3 Online Unsupervised Learning

In this section, we briefly introduce the online learning methods worked in the literature of unsupervised learning, where models are learned from unlabeled data streams. Due to the vast number of unsupervised learning methods in online settings that have been explored, it is almost impossible to give a comprehensive introduction in all the related areas. Instead, we will focus on the most important tasks, online clustering, and give a brief introduction to other unsupervised online learning methods such as dimension reduction, online density estimation, and online anomaly detection.

6.3.1 Online Clustering

Clustering is an unsupervised learning process that groups a set of data instances in a way that instances in the same group (called a cluster) are more similar than those in other groups. For batch learning settings, clustering methods group all the data instances entirely. In contrast, online clustering handles the streaming data that arrive continuously. In the following, we introduce the online learning approaches for clustering on streaming data especially for the partition-based and density-based clustering approaches.

Partition-Based Online Clustering

Traditional partition-based clustering methods such as K-Means [96] split the data instances into partitions (called clusters) based on some distance measures such as Euclidean distance. The number of clusters is usually predefined as prior knowledge. The most representative clustering method, the K-Means algorithm, identifies k centroids by minimizing the sum of square errors between each instance to their corresponding centroids. The online algorithms based on K-Means [96] clustering usually try to break the stream of instances into chunks whose size is decided by the memory budget. After that, the batch-based clustering algorithm can be directly applied to each chunk. The STREAM [97] algorithm utilizes the idea and achieves a constant factor approximation in a single pass.

To understand the idea of STREAM algorithm, we first introduce the *Small-Space* algorithm that shows clustering can be conducted in small spaces. Basically, Small-Space is a divide-and-conquer algorithm that divides the data into small pieces and clusters each data piece. Given a data stream D with multiple chunks $D = \{D_1, D_2, \ldots, D_n\}$ where n is the number of chunks. The steps of *the Small-Space* method are summarized in Algorithm 5.

Algorithm 5 Small-space algorithm

Input: Data stream D, cluster number k
1: Divide data stream D into n disjoint pieces $D_1, \ldots, D_n$
2: **for** $i = 1$ to n **do**
3: Find $O(k)$ centers in D_i
4: Assign each point in D_i to its closest center
5: **end for**
6: Let D' be the $O(nk)$ centers obtained from each data partition D_i
7: Cluster D' to find k centers.

The issue with the *Small-Space* algorithm is the number of subsets n is limited since it has to store all the intermediate medians in memory. Therefore, if M is the size of memory, then we need to fit weighted nk centers into the memory and make sure $nk < M$. However, such a chunk number n may not always exist.

The STREAM algorithm solves the problem of storing intermediate medians by the following steps:

1. For the first m data instances, use a bi-criterion algorithm [97] to reduce them to $O(k)$ instances.
2. Repeat the above step until $m^2/(2k)$ of the original data instances are seen. At this point we have m intermediate medians.

3. Cluster these m first-level medians into $2k$ second-level medians.
4. Maintain at most m level-i medians, and, on seeing m, generate $2k$ level-$(i + 1)$ medians, with the weight of a new median as the sum of the weights of the intermediate medians assigned to it.
5. When all the original data points are observed, we cluster all the intermediate medians into k final medians.

The STREAM algorithm can solve the k-Median problem on a data stream in a single pass with a constant factor approximation using a small space. Besides the STREAM algorithm, there are also some sampling methods [98] designed for the extremely large data streams. For example, StreamKM++ algorithm [99] use an adaptive non-uniform sampling approach to obtain small coresets from the data stream, which can significantly improve the efficiency.

Similar to the idea of STREAM algorithm, let's introduce another popular partition-based online clustering method, called Mini-Batch K-Means, which is an online version of K-Means algorithm. Mini-Batch K-Means uses mini-batches to reduce the amount of computation required to optimize the same objective of original K-Means algorithm and produces results that are generally the same as the K-Means.

Example 6.3 (Mini-Batch K-Means)
Problem: Make a synthetic dataset and perform Mini-Batch K-Means.
Solution: In two-dimensional space, initialize the three center points as (0, 0), $(-1, -1)$, and $(-1, -1)$. $(-1, 1)$. Afterward, create 10 thousand dots all around the three focal locations. Finally, run small batch K-Means on the data. Readers may also experiment with non-batch K-Means to see how well they do in terms of efficiency.

```
import numpy as np
from sklearn.cluster import MiniBatchKMeans
from sklearn.datasets.samples_generator import make_blobs

np.random.seed(0)
centers = [[0, 0], [-1, -1], [-1, 1]]
X, labels_true = make_blobs(n_samples=10000, centers=centers, cluster_std
    =0.7)

mbk = MiniBatchKMeans(init='k-means++', n_clusters=3, batch_size=45, n_init
    =10, max_no_improvement=10, verbose=0)
mbk.fit(X)
```

Furthermore, the results of K-Means and Mini-Batch K-Means are shown in Fig. 6.1. The results show that the difference between the both methods is quite small in practice.

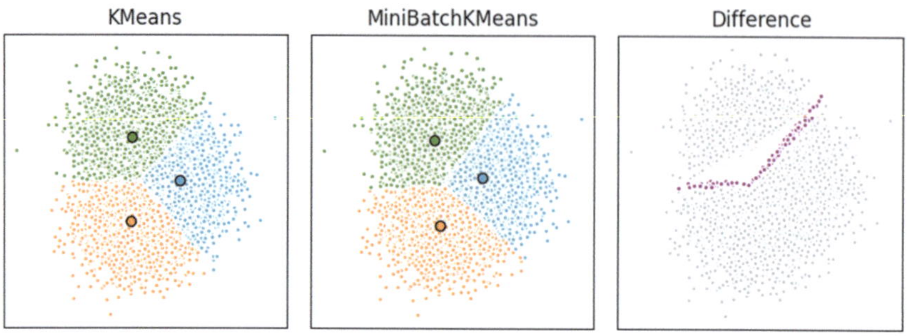

Fig. 6.1 Comparison between K-means and mini-batch K-means

Density-Based Online Clustering

Density-based clustering algorithms are usually designed to handle the two limitations of partition-based approaches: (1) infeasible to handle arbitrary cluster shapes but spherical clusters and (2) require prior knowledge to determine the number of clusters. The density-based approaches such as DBSCAN [100] cluster dense regions separated by sparse regions, in which a cluster can be arbitrary shapes without any prior knowledge of the cluster numbers. However, the traditional density-based clustering methods still suffer from several challenges including dynamic cluster evolution and memory limitation. The online density-based clustering methods are designed to handle these challenges, which can be categorized into micro-clustering and grid-based clustering.

The micro-clustering algorithms [101, 102] summarize data instances in an online manner and cluster the instances based on the summaries. Grid-based algorithm divides the data space into grids, where each new data instance is assigned into one grid. Then clustering method is performed based on the density of the grids and independent of the number of data instances. Figure 6.2 shows an example of grid online clustering. The data points in the grids are categorized into three clusters based on their positions in the grids.

6.3.2 *Other Unsupervised Tasks*

In addition to online clustering tasks, we will briefly introduce the other three unsupervised online learning tasks in this section: Online Dimension Reduction, Online Density Estimation, and Online Anomaly Detection.

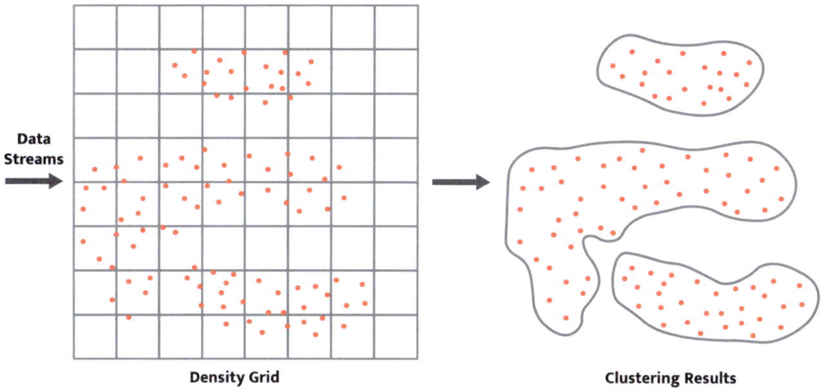

Fig. 6.2 Example of grid-based online clustering

Online Dimension Reduction

Dimension reduction techniques can be used to transform high-dimensional data into a low-dimensional space when feature dimensions are extremely large. The transformed low-dimensional representation is necessary to retain some meaningful properties of the original data but improves learning efficiency and makes it applicable to real-world applications. Formally, consider a data instance $x \in \mathbb{R}^d$; the goal of dimension reduction is to learn a new low-dimensional representation $x' \in \mathbb{R}^l$, where the reduced dimension l is much less than its original dimension d. The studies in this area can be mainly categorized into two groups: subspace learning and manifold learning.

Subspace learning aims to find an optimal linear mapping of input data from high-dimensional space to low-dimensional space. Popular linear subspace methods include Principal Component Analysis (PCA) [103] and Independent Component Analysis (ICA) [104]. The manifold learning method assumes that data lie on an embedded non-linear manifold in the high-dimensional space. These methods aim to find a low-dimensional representation in the preservation of some manifold properties. For example, the multi-dimensional scaling (MDS) [105] and IsoMap [106] preserve global properties, while Locally Linear Embedding (LLE) [107] focuses on preserving local properties.

For the online dimension reduction tasks, most of the efforts have been focused on addressing how to extend the existing methods to streaming data settings. For example, incremental PCA [108] is used as replacement of PCA when the dataset decomposed is too large to fit in memory. Usually, incremental PCA builds a low-rank approximation for the input data using memory which is independent on the number of input data instances. Last, the similar online learning extensions are applied in the other dimension reduction tasks such as Independent Component Analysis [109], IsoMap [110], and Locally Linear Embedding [111].

Online Density Estimation

Online density estimation aims to estimate an underlying unobserved probability density function based on observed data streams. Kernel density estimation (KDE) [112], as one of the most extensively explored topics in density estimation, is a nonparametric way to estimate the probability density function of a target random variable. Specifically, given a sequence of instances $\mathcal{D} = \{x_1, \ldots, x_N\}$, where each instance $x_i \in \mathbb{R}^d$, KDE tries to estimate the density at a point $\mathbf{x}$ as follows:

$$f(x) = \frac{1}{N} \sum_{i=1}^{N} \kappa(x, x_i) = \frac{1}{Nh} \sum_{i=1}^{N} \kappa\left(\frac{x - x_i}{h}\right), \tag{6.16}$$

where the kernel $\kappa(x, x_i)$ is a radially symmetric unimodal function and h is a smoothing parameter. The same as the online learning with kernel, the KDE problem also suffers from the curse of kernelization issue, in which the estimation of the density at any point x requires to compute the kernel function with respect to all the data instances in the data stream.

The attempts to overcome this issue can be grouped into merging and sampling approaches. Merging approaches [113, 114] rely on a prespecified budget on the usages of instances or kernels stored in memory, which guarantees newly arriving samples can be stored in memory as kernel unless the budget is exceeded. One of the approaches [115] performs clustering using self-organizing maps and then performs kernel merging with the clustering results. Sampling approaches [116] randomly select points to be kept in memory and attempt to maintain a certain level of accuracy. The proposed online density estimation methods can be applied in many real-world applications such as real-time visual tracking [117].

Online Anomaly Detection

Anomaly detection is a task to detect abnormal behavior in the data. Although the notion of "anomaly" varies from domains, the anomaly detection is well studied due to its wide applications such as intrusion detection and fraud detection. Distance-based outlier detection algorithms are widely applied because they detect outliers without any underlying data distribution assumption. The online anomaly detection methods usually assume the normal behavior is changing through time. The model is required to update the normal behavior profile with the data records that are probably normal (e.g., have low anomaly score).

A widely used online anomaly detection method is the incremental local outlier factor (ILOF) algorithm [118], which detects outliers in the streaming data. The approach achieves equivalent detection performance as the iterated static LOF algorithm when every time a new point is inserted into the dataset. Also, the ILOF algorithm shows the number of updates each insertion/deletion is independent of the current number of data instances in the dataset to ensure the time complexity of the incremental LOF algorithm is comparable to the static LOF algorithm. However, the ILOF algorithm requires the whole dataset to detect outliers in a data stream and cannot handle new types of outlier. To solve the issues, new approaches such as density summarizing ILOF (DILOF) algorithms [119] were proposed to detect outliers by summarizing the old data with a skipping scheme of outlier sequence.

Example 6.4 (Local Outlier Factor)
Problem: Perform LOF on four data points: -1.2, 0.3, 100.2, and 0.4.
Solution: Obviously, 100.2 is the outlier, as it is far greater than the rest of the data. In the LOF algorithm, the number of neighbors is fixed at two. Then you may go ahead and perform LOF to find the outlier.

```
import numpy as np
from sklearn.neighbors import LocalOutlierFactor
X = [[-1.2], [0.3], [100.2], [0.4]]
clf = LocalOutlierFactor(n_neighbors=2)
clf.fit_predict(X)
# array([ 1,  1, -1,  1])
```

6.4 Application and Resources

Since much data naturally arrives in a streaming fashion, online learning is a critical technique to reduce retraining costs and ensure the model can be updated in a timely fashion. In this section, we briefly introduce several online learning framework applications. At the end of this subsection, we introduce several popular open source toolboxes.

6.4.1 Time Series Prediction

Time series data records any real-value observations over time. It widely exists in research domains such as meteorology, finance, and astronomy. Time series prediction is a fundamental task in time series-related applications. The task aims at training a prediction model based on historical data. The trained model can be used to predict future observations. Since time series data typically arrives in sequential order, online learning plays an important role to ensure a prediction model can be updated efficiently.

AutoRegressive Moving Average (ARMA) model is a fundamental time series prediction model for short-term prediction. The model can be described via the following equation:

$$\bar{x}_{t+1} = \sum_{k=1}^{p} \alpha_k x_{t-k+1} + \sum_{k=1}^{q} \beta_k \epsilon_{t-k+1}, \tag{6.17}$$

where historical observations are denoted as $x_1, x_2, \ldots, x_t$. $\epsilon_{t-k+1} \sim \mathcal{N}(0, 1)$ is a noise observation generated from $\mathcal{N}(0, 1)$ at time step $t-k+1$. $\bar{x}_{t+1}$ is the prediction value generated via model. Given all historic data, the model can be trained via root mean square error (RMSE).

Since time series data often contains a large number of observations recorded over a large time span, retraining the model via all observations when a new observation arrived is not practical. Anava et al. introduce an Online Gradient Descend framework based ARMA model [120] to efficiently update the prediction model.

AutoRegressive Integrated Moving Average (ARIMA) is another popular time series prediction. ARIMA model is much more powerful than the ARMA model and has been applied to predict the nonstationary time series. Different from the ARMA model, ARIMA assumes the difference between two consecutive data follows equation 1.17 (the difference operator can apply more than once). Similar to the online ARMA model, an Online Gradient Descend framework based ARIMA framework is introduced by Liu et al. [121]. The authors use the online version ARIMA model to predict multiple real-world time series such as Dow Jones Industrial Average Index [121].

6.4.2 *Information Retrieval*

Information retrieval (IR) is an important task in information system research. Typically, an IR task aims at finding relevant items given a user query in a large volume of data. There are many real-world applications related to IR. For example, large-scale ad click-through rate (CTR) prediction is a popular application in the IR research field. The task aims at predicting whether an ad on the website will be clicked. Since CTR is an important metric for evaluating ad performance, predicting CTR accurately can greatly help the industry correctly understand users' demand. Ad click-through rate prediction tasks can significantly benefit from online learning as the CTR prediction applications often are time-sensitive.

McMahan et al. [122] described a CTR prediction system used at Google. The system uses FTRL-Proximal online learning approach described in Sect. 1.2.2.4 for the task. To conquer the challenge faced in this real-world application, the authors further conduct a series of engineering-based improvements. First, the authors use a rolling set of counting Bloom filters to adaptively change the set of features. The authors found that the Bloom filter-based approach outperforms traditional feature engineering solutions (e.g., remove fewer information features). In addition, the bit size of the variable can greatly affect the speed performance. Thus all values in the system are stored in *q2.13* fixed-point format. In this format, two bits are used to encode the left of the binary decimal point and thirteen bits are used to encode the value on the right of the binary decimal. To convert a float format w into *q2.13* fixed-point format, the system first converts w into the desired accuracy level via

$$w' = 2^{-13}(2^{13}w + R), \tag{6.18}$$

where R is a random variable generated from uniform distribution between interval [0, 1]. And then the newly generated value w' is stored in *q2.13* format. This new encoding format saves 75% RAM cost compared with classical 64-bit floating-point values without changing model performance. There are many additional engineering solutions mentioned in the work. The readers interested in this topic may refer to the article for detail [122].

6.4.3 Online Portfolio Selection

Online Portfolio Selection (OLPS) [123] is an essential problem in financial and business management. Intuitively, OLPS asks a solution model to make a series of decisions to maximize a utility function. For example, given N stock prices over a time span, the OLPS problem asks a model to adaptively select a stock from these N stocks in each time step to maximize the profit.

OLPS is known as a special type of online learning problem. Das et al. [124] show that Online Gradient Descend can be used to solve this problem. Readers interested in this topic may refer to the article for detail.

6.4.4 Other Applications: Combined with Deep Learning

Since online learning can easily be combined with the offline models, online learning can also integrate with deep learning models. Therefore, online learning can be applied to a wide range of applications that deep learning models are heavily used. These applications include:

- Object Detection [125]
- Natural Language Dependency Parsing [126]
- Image Retrieval [127]
- Graph Representation Learning [128]

The readers interested in these applications may refer to the articles for detail.

6.4.5 Resources

Some popular open source toolboxes of online learning include:

- **Vowpal Wabbit**[1]: Vowpal Wabbit is a machine learning toolbox sponsored by Microsoft Research. The toolbox is specially optimized for fast training and testing. The toolbox contains various types of machine learning models programmed under online, active, and interactive learning frameworks. Online learning algorithms such as Truncated Gradient Descend and Sparse Gradient Descend can be found in the toolbox.
- **SOL, ODL, KOL, and LIBOL**[2]: Hoi and Sahoo from Singapore Management University developed a set of online learning libraries. LIBOL consists of the most popular online learning algorithms. SOL focuses on implementing scalable online learning. ODL and KOL contain online learning frameworks specially designed for deep learning and kernel learning, respectively.
- **Application-Driven Toolbox**: Some applications such as OLPS have a specially designed open source toolbox. For example, OLPS toolbox[3] provided by the research group from SMU University contains a set of models designed for solving OLPS problems.

We next use an example to show how to use Vowpal Wabbit to solve a linear regression problem. The other two source codes are written on Matlab and are well documented. The reader can refer to their GitHub page to learn more about the tools.

We first show how to use Vowpal Wabbit in the Python environment through their Python drivers. A different operating system has different options to install the package. The reader can follow the installation instruction on their website[4] to get to know the installation process.

Example 6.5 (Using Vowpal Wabbit)
Problem: Suppose the following stock price time series is observed:

$$p = 0.21, 0.17, 0.23, 0.18, 0.53, 0.3; \tag{6.19}$$

build a linear regression model with $p = 3$ via Vowpal Wabbit.
Solution: The following is an example of Python code that makes use of Vowpal Wabbit. Lines 1–2 of the sample code include the fundamental code

(continued)

[1] https://vowpalwabbit.org/.
[2] https://github.com/LIBOL.
[3] https://github.com/OLPS/OLPS.
[4] https://vowpalwabbit.org/start.html.

Example 6.5 (continued)
for initializing a Vowpal Wabbit model. Using Eq. (6.19), lines 4–8 generated training data based on the equation referenced above. Finally, lines 10–11 are a basic implementation of Vowpal Wabbit, which is used to train the machine learning model. Lines 13–15 can be used to anticipate a future data point, which is useful for forecasting.

```
1  from vowpalwabbit import pyvw
2  model = pyvw.vw()
3
4  train_examples = [
5    ".18 | t0:.23 t1:.17 t2:.21",
6    "0.53 | t0:.18  t1:.23 t2:.17",
7    "0.3 | t0:.53 t1:.18  t2:.23",
8  ]
9
10 for example in train_examples:
11     model.learn(example)
12
13 observation = "| t0:0.3 t1:.53 t2:.18"
14 prediction = model.predict(observation)
15 print(prediction)
```

In addition, Vowpal Wabbit is well known for its original C++ command-line software. After generating a dataset based on VW format (the format used in lines 4–8 in the previous code), the users can run the following command line for training the model:

```
vw [training_dataset_name] -cache_file cache_train -f [model_name]
```

To predict the newly observed data, the users can use the following command:

```
vw -t [testing_dataset_name] --cache_file cache_test -i [model_name] -p [
    output_file]
```

6.5 Putting It All Together

In the era of big data, massive amounts of data are collected in a real-time fashion. Online learning becomes a popular machine learning tool in real-world applications since it provides an efficient way to process massive data and fulfill the real-time feedback demands. In this chapter, we discussed two types of online learning models including supervised and unsupervised approaches. The summary of the methods is shown in Table 6.1, in which the supervised methods are categorized by the three properties: first- or second-order gradient and data sparsity.

For supervised methods, we discuss a family of first-/second-order linear online learning algorithms and sparse online learning with regularization algorithms.

Table 6.1 Characteristics of Online Algorithms

Algorithm	Supervised			Unsupervised
	First-order gradient	Second-order gradient	Data sparsity	
PA	√			
OGD	√			
TGD		√		
FOBOS		√		
RDA	√		√	
FTRL-proximal		√	√	
STREAM				√
Mini-batch K-means				√

First-/second-order linear online learning algorithms are efficient in the task with non-sparse/low-dimensional data. For example, online binary classification (e.g., spam email filtering) only involves two categories ("spam" vs "benign" emails). Compared with the first-order linear online learning algorithms, the second-order online learning algorithms accelerate the optimization convergence with higher computational complexity. Online learning with regularization methods aims to exploit the sparsity property with real-world high-dimensional data. For example, online learning with regularization methods is naturally applied in financial portfolio management where an online learner aims to find a good (e.g., profitable and low risk) strategy for making a sequence of decisions for portfolio selection.

For unsupervised online learning tasks, we introduce four categories of machine learning methods to handle unlabeled data streams: online clustering, online dimension reduction, online density estimation, and online anomaly detection. Each type of method solves one of the problems without label supervision. Online clustering methods partition the streaming data into different groups, in which the data within one group share more close properties than data samples in different groups. For instance, the online clustering method can be applied in face clustering in long and real-world videos, in which the faces are grouped in their scale, pose, illumination, and expressions. The online clustering methods can help to handle the video streams when the complete video data may not be available at the same time or the data distribution itself may exhibit significant variation over time. Online dimension reduction methods aim to transform data streams from a high-dimensional space to a low-dimensional one without sacrificing the important properties of the original data. Since the high-dimensional space can usually cause the computational and curse of dimensionality issues, online dimension reduction methods are always used in on-demand services to provide a real-time response for huge data. For example, the recommendation system in e-commerce websites can use the online dimension reduction method to reduce the feature dimension for more efficient recommendations. Online density estimation approaches are usually applied in the informal investigation of the streaming data property. Online density estimation helps to yield the valuable indication of data features such as skewness or multimodality. Online anomaly detection is a useful tool to identify rare items, events, or observations that have a significant difference from the majority of data. The anomalous items usually translate to some problems. For example, the unusual bank transactions are always triggered by bank fraud; the unexpected bursts in network flow are caused by the network intrusion behaviors.

6.6 Exercise Problems

Problem 6.1 Which of the following statements about online learning are True? Check all that apply.

(a) One of the disadvantages of online learning is that it requires a large amount of computer memory/disk space to store all the training examples we have seen.
(b) One of the advantages of online learning is that if the function we are modeling changes over time (such as if we are modeling the probability of users clicking on different URLs, and the user tastes/preferences are changing over time), the online learning algorithm will automatically adapt to these changes.
(c) One of the advantages of online learning is that there is no need to pick a learning rate α.
(d) When using online learning, in each step, we get a new example (x, y), perform one step of (essentially stochastic gradient descent) learning on that example, and then discard that example and move on to the next.

Problem 6.2 Which of the following statements about online learning are False? Check all that apply.

(a) Online learning algorithms are most appropriate when we have a fixed training set of size m that we want to train on.
(b) When using online learning, you must save every new training example you get, as you will need to reuse past examples to retrain the model even after you get new training examples in the future.
(c) Online learning algorithms are usually best suited to problems where we have a continuous/non-stop stream of data that we want to learn from.

Problem 6.3 Use the Perceptron algorithm to classify the Digits dataset.

Problem 6.4 What is the major difference between PA and Perceptron algorithms?

Problem 6.5 Suppose you use mini-batch gradient descent on a training set of size m, and you use a mini-batch size of b. Under what situation does the algorithm become the same as batch gradient descent?

Problem 6.6 Suppose you are facing a supervised learning problem and have a very large dataset. How can you tell if using all of the data is likely to perform much better than using a small subset of the data?

Problem 6.7 Compare the different classification performance between Passive Aggressive I and Passive Aggressive II on the Digits dataset.

Problem 6.8 What is the issue of Small-Space algorithm in online clustering?

Problem 6.9 Use the incremental PCA to reduce the data dimension of Iris dataset to 2 and plot the results.

Problem 6.10 Suppose $x = 1.32148$, suppose we stored it in $q2.13$ format, and $R = 0.6$, what is the encoded value? Compute the difference between x and the $q2.13$ format value and briefly discuss the benefit of using $q2.13$ format.

Problem 6.11 In what situations are online learning methods used?

Problem 6.12 What is the curse of kernelization issue of KDE?

Problem 6.13 What is the main difference between traditional anomaly detection method and online anomaly detection?

Problem 6.14 Suppose we have an ARMA model that can be written as follows:

$$x_t = 0.8x_{t-1} + 0.2x_{t-2} + 0.5\epsilon_{t-1} + 0.2\epsilon_{t-2},$$

where $\epsilon \sim \mathcal{N}(0, 1)$. What is the distribution of the output x_t, given input $x_{t-1} = 1$ and $x_{t-2} = 2$?

Problem 6.15 Consider a spam email detection task in which the following data are observed:

Type	Email length	Title length	Received time
Spam	10	3	20
Spam	15	8	21
Normal	20	5	11
Normal	15	8	12
Normal	30	8	20

Write a Python program that uses the VowpalWabbit package to identify the spam email. What are the outputs of your program for the following records?

Email length	Title length	Received time
3	3	22
30	12	11

Chapter 7
Recommender Learning

7.1 Introduction

Extensive studies have made significant progress on recommendation techniques in the past decade. In general, two main classes of techniques are content-based approaches and collaborative filtering [129]. For instance, when recommending a restaurant to a user, these two approaches make recommendations based on the following intuitions, respectively:

- Content-based approaches: Does the menu of the restaurant satisfy the user's taste?
- Collaborative filtering: How did the other users with the same taste rate the restaurant?

This chapter will start with a rating prediction problem and introduce the concept of content-based approach and collaborative filtering, which are the fundamental techniques for all recent recommendation methods. After that, several popular models in industrial recommender systems, including the factorization machine and deep learning models, are introduced. Last, this chapter summarizes different recommendation applications.

7.2 The Recommendation Problem

Rating Matrix

A typical recommendation scene is product recommendation. In e-commerce platforms such as Amazon, the users often leave ratings of 1–5 stars on items purchased. The rating data is often represented as a *rating matrix*, where each column represents an item and each row represents a user. Each entry in the matrix

S. Rafatirad et al., *Machine Learning for Computer Scientists and Data Analysts*,
https://doi.org/10.1007/978-3-030-96756-7_7

Table 7.1 An example rating matrix

	p_1	p_2	p_3	p_4	p_5	p_6
u_1	?	5	?	4	?	3
u_2	?	2	4	2	4	?
u_3	3	?	4	?	?	2
u_4	?	4	1	?	1	?

denotes a user's rating on an item. Table 7.1 shows an example rating matrix with four users ($u_1, \ldots, u_4$) and six items ($p_1, \ldots, p_6$).

The question marks in the matrix denote "unknown" ratings. Note that in practice, most of the ratings should be unknown because each user could explore and purchase only a few items but not all of them. In other words, the rating matrix is often sparse.

Content Information
In many recommendation applications, each item may have some features in addition to the ratings. For instance, a book may have attributes such as authors and types. These features are regarded as *content information*. The content information enables a recommender system to quantify the user's preference on different item attributes. For example, if a user often browses and purchases fiction books, a simple guess on a user's preference is fiction.

The Recommendation Problem
For each target user, the task of recommendation is to explore the rating matrix and the content information of items, discover the user's preferences, and suggest a list of N items that best satisfy the user's preferences but *have not yet been rated* by the user.

7.3 Content-Based Approach

As mentioned at the beginning of this chapter, the intuition behind content-based approaches is to find out items that best match the preferences of the target user. Remind that content information of items can be used to quantify user preferences. As such, we can reconsider the recommendation problem as a *profile matching problem*. Because each user can be profiled using the attributes of items purchased by the user (e.g., fiction books), the user profiles well align with the item profiles. Then, we can recommend items with the most similar profiles to the target user.

Profiling Items
Profiling an item with its attributes such as categories is straightforward. Suppose there are P possible item attributes. The profile of an item is a vector where each entry in the vector denotes the value of an attribute. Table 7.2 lists the binary attributes ($a_1, \ldots, a_5$) of the six items in Table 7.1, where ticks denote the

Table 7.2 Attributes of items in Table 7.1

	a_1	a_2	a_3	a_4	a_5
p_1		✓			✓
p_2		✓	✓		
p_3	✓			✓	
p_4		✓	✓		
p_5	✓			✓	
p_6	✓				✓

corresponding item has the attributes. For instance, the profile of item p_1 will be a vector of length five, i.e., $\mathbf{p}_1 = (0, 1, 0, 0, 1)$.

There are some other recommendation scenarios where item attributes may be very limited or may not even be available. For instance, in news article recommender systems, only the text of an article is available. Given the text content of an item, the item profile should reflect the semantic meaning of the text. Since the text content may contain various words that are not important for measuring the semantic such as stop words, we should find representative words from the text content to profile the items. **T**erm **F**requency **I**nverse **T**erm **F**requency (TF-IDF [130]) is a commonly used technique to quantify the importance of a word in a document. The TF-IDF of a word w in a document $d \in D$ is defined as

$$\text{TF-IDF}(w, d) = \underbrace{\text{TF}(w, d)}_{\text{frequency of word } w \text{ in } d} \cdot \log \underbrace{\frac{|D|}{|\{d \in D : w \in d\}|}}_{\text{inverse document frequency of word } w}. \tag{7.1}$$

The denominator in the inverse document frequency is the number of documents that contain the word w. The intuition behind TF-IDF score is that a word should be representative of a document if it is used frequently in the current document but merely appears in other documents.

Using TF-IDF, we can select a set of words with the highest TF-IDF scores as representative words to construct the item profiles. The profile vector of an item has each component corresponding to a word in a predefined vocabulary. If a word is a representative word of the item, its corresponding component in the profile vector takes the value of 1, otherwise 0.

Note that, no matter what content information is used, each item can be converted to a real-value vector $\mathbf{p} \in \mathbb{R}^P$, where P denotes the number of features (attributes or words).

Profiling Users

User preferences can be observed from the user's historical ratings on items. If a user often gives high ratings to items that contain some attribute a, the user's profile should contain that attribute a. In other words, the user profile should aggregate the profiles of the items that have been rated by the user. Intuitively, items with a higher rating should be considered as more important to the user's profile in the aggregation process. Thus, a typical aggregation method is a weighted sum—summing vectors from all rated items weighted by the rating. Let the $r_{u,v}$ be the rating of a user u to item v, R_u be historical ratings of user u, and $\mathbf{q}_v$ be the profile vector of item v. The profile of user u can be computed as

$$\mathbf{q}_u = \frac{\sum_{v \in \{v' | r_{u,v'} \in R_u\}} r_{u,v} \cdot \mathbf{p}_v}{\sum_{v \in \{v' | r_{u,v'} \in R_u\}} r_{u,v}}. \tag{7.2}$$

Example 7.1 (Profiling Users)
Problem: Profile user u_1 in Table 7.1.
Solution: $\mathbf{q}_1 = \frac{5}{12} \cdot \mathbf{p}_2 + \frac{4}{12} \cdot \mathbf{p}_4 + \frac{3}{12} \cdot \mathbf{p}_6$. Since $\mathbf{p}_2 = \mathbf{p}_4 = (0, 1, 1, 0, 0)$ and $\mathbf{p}_6 = (1, 0, 0, 0, 1)$, $\mathbf{q}_1 = (\frac{1}{4}, \frac{3}{4}, \frac{3}{4}, 0, \frac{1}{4})$.

Matching User and Item Profiles
Once the user and item profiles are obtained, the recommendation problem can be solved by calculating the profile similarity between each item and the target user. A common choice of similarity metric is cosine similarity, which calculates the profile similarity as the cosine of the angle between the two profile vectors:

$$sim(u, v) = \frac{\mathbf{q}_u \cdot \mathbf{p}_v}{||\mathbf{q}_u|| \cdot ||\mathbf{p}_v||}. \tag{7.3}$$

The profile similarity is used for ranking the items not rated by the target user. The recommender system will then pick the top-N items with the highest profile similarity to the user.

Content-based approaches work well for new items or unpopular items. Even when the items are rarely or not rated by any user, the item profiles can be constructed from item descriptions and attributes. However, this type of approach often recommends similar items to users without diversity. It may not work for users with diverse interests.

7.4 Collaborative Filtering

In content-based approaches, we directly match the item content to the user preferences. Now, we will explore completely different manners for the recommendation. Consider a real-life recommendation scenario—restaurant recommendations. If you visit a restaurant for the first time with your friends who have visited the place several times, you would consider what your friends ordered before. Collaborative Filtering (CF) is based on the above idea, which generates a list of recommended items for a user by aggregating the ratings from other users who share similar tastes.

CF techniques are often divided into two rough groups [131]: (1) memory-based CF and (2) model-based CF. In this section, we will introduce the memory-based approach, followed by the latent factor model, one of the typical model-based approaches.

7.4.1 Memory-Based Collaborative Filtering

If the recommender system *remembers* what items have been rated by each user, the idea collaborative filtering can be achieved by two steps: (1) find users with similar tastes and (2) aggregate the ratings from similar users.

User Similarity

To simplify the problem, assume we only have the rating matrix in Table 7.1. The historical ratings often imply user preferences. As such, a straightforward way to measure the similarity between two users is to calculate the similarity between their historical ratings on items. Formally, each user can be represented by a rating vector where each component corresponds to an item with a value equal to the rating. For instance, u_1 in Table 7.1 is represented by the rating vector $(0, 5, 0, 4, 0, 3)$.

Given the rating vectors of two users, the user similarity can be naturally computed using cosine similarity. However, unknown ratings are treated as "negative" ratings and the rating bias of different users is not considered. A critical user with an average rating of 2 stars gave a 3-star rating to an item may be a positive rating, while a user with an average rating of 4 stars gave a 3-star rating to an item may indicate a negative rating. To address the above limitations of cosine similarity, Resnick et al. [132] proposed to use the Pearson correlation coefficient. The idea is to measure the offset of a rating to the user's average rating and only consider items commonly rated by the two users. Let S_{ik} be the set of items that are commonly rated by users i and k, and $\bar{r}_i$ and $\bar{r}_k$ denote the average rating of users i and k, respectively. The Pearson correlation coefficient computes the user similarity as follows:

$$sim_{\text{pearson}}(i, k) = \frac{\sum_{j \in S_{ik}} (r_{i,j} - \bar{r}_i)(r_{k,j} - \bar{r}_k)}{\sqrt{\sum_{j \in S_{ik}} (r_{i,j} - \bar{r}_i)^2} \sqrt{\sum_{j \in S_{ik}} (r_{k,j} - \bar{r}_k)^2}}. \tag{7.4}$$

Example 7.2 (User Similarity)
Problem: Based on Table 7.1, calculate the average ratings, the Pearson correlation coefficients between u_4 and u_1, u_3?

(continued)

Example 7.2 (continued)
Solution: The average ratings of users u_1, u_3, and u_4 are $\bar{r}_1 = 4, \bar{r}_3 = 3$, and $\bar{r}_4 = 2$, respectively. The Pearson correlation coefficients between u_4 and u_1, u_3 are computed as

$$sim_{\text{pearson}}(u_1, u_4) = \frac{(5-4)(4-2)}{\sqrt{(5-4)^2}\sqrt{(4-2)^2}} = 1,$$

$$sim_{\text{pearson}}(u_3, u_4) = \frac{(4-3)(1-2)}{\sqrt{(4-3)^2}\sqrt{(1-2)^2}} = -1.$$

Obviously, the Pearson correlation coefficient ranges from -1 to 1, and it enables negative impacts on the similarity—if two users rated reversely on the same items, they should have a negative similarity. For instance, u_4 placed a low rating on p_3, while u_3 rated p_3 highly. The Pearson correlation coefficient between u_3 and u_4 is -1.

The negative similarity enables the recommender systems to consider the reverse behaviors of users. If u_3 rated on an item negatively, u_4 may rate it positively. In the extreme case that two users behave totally differently, they can be considered as reversely similar.

Rating Aggregation

To predict the rating of a user i to an item j, we shall find the set of top-k similar users $U_{i,j}$, who have rated on item j. Given a target user i and a similarity measure (either cosine similarity or Pearson correlation coefficient), the set of similar users $U_{i,j}$ can be obtained by two steps: (1) ranking all users who have rated item j by the absolute value of similarity to user i and (2) picking the top-k users as $U_{i,j}$.

The next stage of collaborative filtering is to aggregate the ratings from similar users. One simple way is to use weighted average:

$$\hat{r}_{i,j} = \frac{\sum_{k \in U_{i,j}} sim(i,k) r_{k,j}}{\sum_{k \in U_{i,j}} |sim(i,k)|}. \tag{7.5}$$

The weighted average approach has two main drawbacks. First, it cannot recommend items to users who have an empty set of similar users. Second, the bias of users and the bias of ratings on items are not explored. Inspired by the fact that both users and items may have a bias on the ratings, the weighted average approach can be extended by incorporating the rating bias.

In general, rating bias can be decomposed into three factors:

- **Global bias** b^{global} is the average rating of all historical ratings in the rating matrix regardless of individual user or item.
- **User offset** b_i^{user} measures the offset between the average rating of the particular user i and the global bias, i.e., $\bar{r}_i - b^{global}$.
- **Item offset** b_j^{item} measures the offset between the average rating of the particular item j and the global bias, i.e., $\bar{r}_j - b^{global}$.

The rating bias of a user i on an item j is the sum of the above three factors: $b_{i,j} = b^{global} + b_i^{user} + b_j^{item}$. The intuition of rating bias is to provide a baseline on the rating based on the historical average. Then, the ratings from similar users can be used to refine the baseline for accurate rating prediction. The extended weighted average approach predicts the rating of user i on item j as

$$\hat{r}_{i,j} = b_{i,j} + \frac{\sum_{k \in U_{i,j}} sim(i,k)(r_{k,j} - b_{k,j})}{\sum_{k \in U_{i,j}} |sim(i,k)|}. \tag{7.6}$$

Example 7.3 (Rating Aggregation)
Problem: Consider predicting the rating of u_4 on p_6 in Table 7.1.
Solution: Only u_1 and u_3 have rated the item p_6. So, we only need to compute the similarities between u_4 and u_1, u_3.

From Example 7.2, we have known $sim(u_1, u_4) = 1$ and $sim(u_3, u_4) = -1$. Suppose we pick u_1 and u_3 as similar users, i.e., $U_{4,6} = \{u_1, u_3\}$. We calculate the three types of bias:

$$b^{global} = 3, b_1^{user} = 1, b_3^{user} = 0, b_4^{user} = -1, b_6^{item} = -0.5$$

and the bias rating of each user on the item

$$b_{1,6} = 3 + 1 - 0.5 = 3.5$$

$$b_{3,6} = 3 + 0 - 0.5 = 2.5$$

$$b_{4,6} = 3 - 1 - 0.5 = 1.5.$$

Then, the prediction is

$$\hat{r}_{4,6} = 1.5 + \frac{(3 - 3.5) * 1 + (2 - 2.5) * (-1)}{|1| + |-1|} = 1.5.$$

From the above example, we can observe that user u_3 contributes a positive offset $(2-2.5)*(-1) = 0.5$, because u_4 always rates reversely to u_3 and u_3 rated negatively on item p_6. Without using rating bias, u_3 will contribute a negative rating (e.g., -1×2). Another observation is that, even if the set of similar users $U_{4,6}$ is empty, the bias rating $b_{4,6}$ can be used for prediction.

The method we discuss so far is called memory-based collaborative filtering, because it requires storing the whole rating matrix in memory and looking up the corresponding rows of similar users for rating aggregation. Although memory-based collaborative filtering is simple and works well with a large rating matrix without any content information, the computation of user similarities requires overlaps on the historical ratings, which could be difficult on a highly sparse rating matrix. In practice, the rating matrix is very sparse, where many of the users may not have commonly rated items.

7.4.2 *Latent Factor Model*

In memory-based collaborative filtering, a user is modeled by the rating vector, which has a length equal to the number of items. Obviously, the rating vector is of high dimension, and it is sparse. If we know the types of items such as digital devices, games, and groceries, we can group the items by their types to reduce the dimensionality. In practice, the type of items may not be available. The latent factor model is a method to learn low-dimensional vector representations for users and items from the rating matrix. Each component in the low-dimensional vector denotes a latent factor, which can be regarded as the type of items.

The idea of the latent factor model comes from *matrix factorization* [133]. The objective of matrix factorization is to approximate the $M \times N$ rating matrix $\mathbf{R}$ with the multiplication of two matrices $\mathbf{Q} \in \mathbb{R}^{M\times d}$ and $\mathbf{P} \in \mathbb{R}^{N\times d}$, i.e., $\mathbf{R} \approx \mathbf{Q}\mathbf{P}^T$. Each row of the matrix $\mathbf{Q}$ corresponds to the d-dimensional vector representation of a user, while each row of the matrix $\mathbf{P}$ denotes the d-dimensional vector representation of an item.

One typical method for matrix factorization is singular value decomposition (SVD), which decomposes the rating matrix into the multiplication of three matrices $\mathbf{U}$, $\boldsymbol{\Lambda}$, and $\mathbf{V}$:

$$\underbrace{\begin{bmatrix} r_{11} & \dots & r_{1N} \\ \vdots & \ddots & \vdots \\ r_{M1} & \dots & r_{MN} \end{bmatrix}}_{\mathbf{R}\in\mathbb{R}^{M\times N}} \approx \underbrace{\begin{bmatrix} u_{11} & \dots & u_{1d} \\ \vdots & \ddots & \vdots \\ u_{M1} & \dots & u_{Md} \end{bmatrix}}_{\mathbf{U}\in\mathbb{R}^{M\times d}} \underbrace{\begin{bmatrix} \lambda_{11} & & \\ & \ddots & \\ & & \lambda_{dd} \end{bmatrix}}_{\boldsymbol{\Lambda}\in\mathbb{R}^{d\times d}} \underbrace{\begin{bmatrix} v_{11} & \dots & v_{1N} \\ \vdots & \ddots & \vdots \\ v_{d1} & \dots & v_{dN} \end{bmatrix}}_{\mathbf{V}^T\in\mathbb{R}^{d\times N}}. \tag{7.7}$$

From the three matrices, we can get the user and item representations by setting $\mathbf{Q} = \mathbf{U}$ and $\mathbf{P}^T = \boldsymbol{\Lambda}\mathbf{V}^T$. Note that both $\mathbf{U}$ and $\mathbf{V}$ are orthogonal matrices. To estimate the values of the above three matrices, we shall minimize the sum square errors on all entries of the rating matrix $\mathbf{R}$:

$$\min_{\mathbf{U},\boldsymbol{\Lambda},\mathbf{V}} \sum_{i\in[1,M],j\in[1,N]} (r_{i,j} - [\mathbf{U}\boldsymbol{\Lambda}\mathbf{V}]_{i,j})^2. \tag{7.8}$$

Matrix Factorization as Collaborative Filtering

It is not difficult to show that SVD preserves collaborative information from similar users. Note that the dot product of the rating matrix and its transpose $\mathbf{RR}^T \in \mathbb{R}^{M\times M}$ is the unnormalized cosine similarity matrix for users:

$$\mathbf{RR}^T = \begin{bmatrix} \sum_j r_{1j}r_{1j} & \cdots & \sum_j r_{1j}r_{Mj} \\ \vdots & \ddots & \vdots \\ \sum_j r_{Mj}r_{1j} & \cdots & \sum_j r_{Mj}r_{Mj} \end{bmatrix}.$$

Leveraging the orthogonal property of $\mathbf{V}$, the similarity matrix $\mathbf{RR}^T$ can be decomposed as

$$\mathbf{RR}^T \approx \mathbf{U}\boldsymbol{\Lambda}\mathbf{V}^T\mathbf{V}\boldsymbol{\Lambda}\mathbf{U}^T = \mathbf{U}\boldsymbol{\Lambda}^2\mathbf{U}^T.$$

The above equation gives an eigenvalue decomposition of $\mathbf{RR}^T$, where each column of $\mathbf{U}$ is an eigenvector of the matrix. In other words, the columns of $\mathbf{U}$ define a coordinate system that preserves the unnormalized cosine similarity between users, with each axis of the coordinate system as a latent factor.

From Matrix Factorization to Latent Factor Model

Matrix factorization techniques such as SVD optimize the sum of square errors on all entries, including unknown ratings with zero values. The objective of matrix factorization is to approximate the original matrix. In an extreme case of perfect decomposition with zero-sum square errors, all unknown ratings will be predicted as zeros. This counters the objective of recommender systems, which tries to make predictions for unknown ratings. Therefore, we should make the optimization only on existing ratings, which gives us the latent factor model:

$$\min_{\mathbf{Q},\mathbf{P}} \sum_{r_{i,j}\neq 0} (r_{i,j} - \mathbf{q}_i \cdot \mathbf{p}_j)^2, \tag{7.9}$$

where $\mathbf{q}_i$ denotes the i-th row of $\mathbf{Q}$ and $\mathbf{p}_j$ denotes the j-th row of $\mathbf{P}$.

Incorporating Rating Bias

Similar to memory-based collaborative filtering, it turns out to be more robust for the latent factor model to incorporate rating bias. We can think of the dot product $\mathbf{q}_i \cdot \mathbf{p}_j$ as an approximation to the offset to the baseline rating $b_{i,j} = b^{global} + b_i^{user} + b_j^{item}$ and optimize the following instead:

$$\min_{\mathbf{\Phi}} \sum_{r_{i,j} \neq 0} (r_{i,j} - b_{i,j} - \mathbf{q}_i \cdot \mathbf{p}_j)^2, \tag{7.10}$$

where $\mathbf{\Phi} = (\mathbf{Q}, \mathbf{P}, b^{global}, b^{user}, b^{item})$. Note that the rating bias b^{global}, b^{user}, and b^{item} can be learned directly from data.

Regularization

Learning the latent factor model by optimizing Eq. (7.10) may cause overfitting. Regularization is a technique that reduces the risk of overfitting. In particular, we can add L2 regularization on each parameters to the objective function:

$$\begin{aligned} \min_{\mathbf{\Phi}} \sum_{r_{i,j} \neq 0} (r_{i,j} - b_{i,j} - \mathbf{q}_i \cdot \mathbf{p}_j)^2 + \lambda_1 \sum_{i \in [1,M]} ||\mathbf{q}_i||^2 + \lambda_2 \sum_{j \in [1,N]} ||\mathbf{p}_i||^2 \\ + \lambda_3 \sum_{i \in [1,M]} ||b_i^{user}||^2 + \lambda_4 \sum_{j \in [1,N]} ||b_j^{item}||^2 + \lambda_5 ||b^{global}||^2, \end{aligned} \tag{7.11}$$

where $\lambda_{1:5}$ are tunning parameters that control the importance of each regularization term. The optimization problem in Eq. (7.11) can be solved by Stochastic Gradient Descent (SGD).

7.5 Factorization Machine

In this section, we will briefly introduce the factorization machine (**FM**) approach [134], which is another powerful and flexible framework for recommender systems. The FM model has two main advantages.

- It is able to handle extremely sparse interaction data.
- It can be computed in linear time and scaled to large datasets.

In previous sections, we have presented two types of collaborative filtering methods, which have been widely used in real recommender systems. Given user historical rating matrix, conventional recommendation methods directly learn the personalized rating scores. Typically, memory-based collaborative filtering approaches attempt to calculate the user similarity (user-based CF) or item similarity

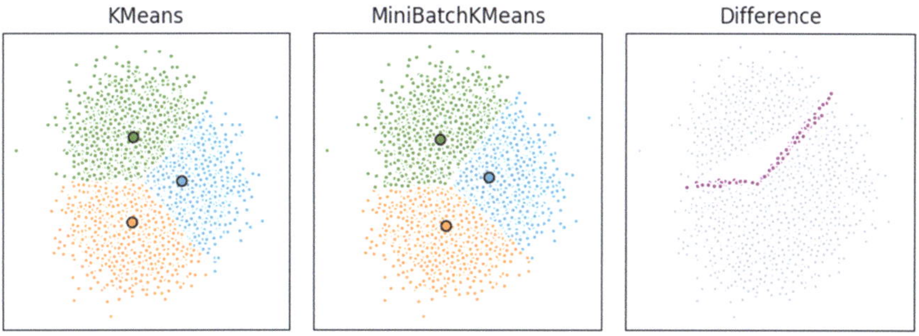

Fig. 7.1 An example of a factorization machine

(item-based CF), while latent factor methods directly decompose the user–item interactions into the product of two lower-dimensional vectors. Different from them, the FM approach represents user–item interactions as real-valued feature vectors, just like the data format commonly used in standard classification or regression models. After that, these constructed features are used as the input feature vector x to predict the target output y.

Let us consider a toy example illustrated in Fig. 7.1. There are three users $\{u_1, u_2, u_3\}$ and four items $\{i_1, i_2, i_3, i_4\}$. The historical data contains the interactions between users and items. For instance, user u_1 has two rating records: user u_1 rates item i_1 with a score of 3 and rates item i_3 with a score of 5.

In addition, auxiliary features can provide supplementary information for learning rating scores. In this example, we consider three types of auxiliary features. The first one is user profile information, e.g., user gender information represented by g_1 and g_2. The second type is related to item feature, such as item attribute, indicated by c_1 and c_2, respectively. The last type provides side information of interactions, such as rating days in a week, which reflects the temporal information of each rating action. For instance, $\{u_1, i_1, 2\}$ means that user u_1 rates item i_1 on day 2.

Given the rating matrix and auxiliary features, we can construct feature vectors as the input of FM model. Totally, there are three types of features: users, items, and auxiliary features. First, there are $|U| = 3$ binary indicator variables (marked as blue) that indicate the active user of a rating action. The following $|I| = 4$ variables represent active items of rating actions. The last $|A| = 5$ variables embody auxiliary features of user gender, item category, and rating day.

The recommendation task is modeled as a regression task, i.e., estimating the score $\hat{y}$ given the input feature vector x. One intuitive idea is to predict the rating score by a linear model.

$$\hat{y} = b + \sum_{i=1}^{n} w_i \cdot x_i + \sum_{i=1}^{n} \sum_{j=i+1}^{n} w_{ij} \cdot (x_i x_j). \tag{7.12}$$

Here, $n = |U|+|I|+|A|$ is the size of the input feature x by summing up all user, item, and auxiliary features. The score function consists of three components.

- b is the global bias.
- w_i is the weight of the i-th variable. These parameters reflect linear (order-1) interactions among input variables.
- w_{ij} is the strength of a variable pair (variable i and variable j). These parameters indicate the pairwise (order-2) feature interactions.

However, the number of pairwise parameters w_{ij} would be very large. Given the n-dimensional input features, the number of w_{ij} approximates to $O(n^2)$. Furthermore, since the input feature vectors are extremely sparse, there is usually not enough data to effectively estimate interactions between variables separately.

To address this issue, the FM model factorizes the order-2 feature interaction weights as dot products of low-dimensional vectors.

$$\hat{y} = b + \sum_{i=1}^{n} w_i \cdot x_i + \sum_{i=1}^{n} \sum_{j=i+1}^{n} \langle V_i, V_j \rangle \cdot (x_i x_j). \tag{7.13}$$

Here, w_{ij} is replaced by the dot product of two vectors V_i and V_j, where $V \in \mathbb{R}^{n \times k}$. k is the size of the factorization vector and typically $k \ll n$. Hence, the number of parameters would be reduced from $O(n^2)$ to $O(n \cdot k)$. In practice, the FM only needs to train the non-zero elements when feature i and feature j occur at the same time. Therefore, the computation can be efficient in sparse settings.

7.6 Deep Learning Models

In recent years, deep learning techniques have been widely used in many applications, such as computer vision and natural language processing. It is natural to make full of the power of deep learning for recommender systems. Evidently, a variety of deep learning recommendation models have been developed [135] in both academia and industry.

Compared with the conventional recommendation approaches, deep learning models are able to effectively learn the non-linear relationships between users and items. They can capture complex relations from the massive user historical behavior data. Furthermore, deep learning models are able to integrate heterogeneous types of features, such as visual, textual, and contextual information, into recommending systems. Generally, deep learning models can achieve state-of-the-art recommendation results.

As an ongoing study, various deep learning recommendation methods are developed each year. In this section, we introduce one representative model, i.e., Neural Collaborative Filtering (NCF) [136].

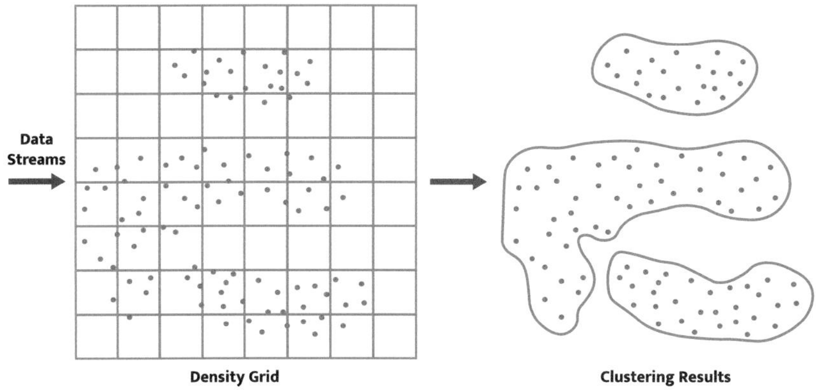

Fig. 7.2 Illustration of neural collaborative filtering (NCF) framework

Remember that latent factor recommendation models, such as matrix factorization, exploit inner products of user latent features and item latent features to capture the interactions between users and items. Different from latent factor models, NCF proposes a neural network architecture to extract the complex interactions between users and items. Specifically, it utilizes a multi-layer Perceptron to learn the user–item interaction function, rather than directly using the inner product of latent factors. In fact, the NCF framework is able to generalize the matrix factorization model to a non-linear setting, which makes the NCF more expressive.

Figure 7.2 [136] shows the NCF framework. The input layer consists of the feature vector of user u and feature vector of item i. Note that the input feature vectors can be extended to include various sources of information, such as user profiles and item attributes. To keep it simple, the input layer only considers the identities of users and items, which is a common setting in standard collaborative filtering models. NCF exploits the identity of user u and item i, converting them to binary one-hot encoding vectors. Next, the sparse one-hot representation is projected to a dense representation by an embedding layer. The obtained user embedding (item) can be viewed as the user (item) latent vector in the latent factor model.

The user latent vector and item latent vector are then fed into a multi-layer neural architecture, which is termed as neural collaborative filtering (NCF) layers, to map the latent vectors to predicted scores. After the last NCF layer, the output layer predicts a score $\hat{y}_{u,i}$, which is used to train with the target score $y_{u,i}$.

The NCF score function can be modeled as

$$\hat{y}_{u,i} = f(U^T x_u, V^T x_i | U, V, \Theta_f), \tag{7.14}$$

where x_u and x_i indicate input user features and item features, and U and V indicate user embedding matrix and item embedding matrix, respectively. Θ_f denotes the model parameters of the score function f, which is defined by a multi-layer neural network. The score function f is formulated as

$$f(U^T x_u, V^T x_i | U, V, \Theta_f) = \phi_{out}(\phi_X(...\phi_2(\phi_1(U^T x_u, V^T x_i))...)), \tag{7.15}$$

where ϕ_x denotes the projection function of the x-th NCF layer. ϕ_{out} is the mapping function of output layer, which estimates the final score.

Due to the limited space, we only cover one representative model, i.e., neural collaborative filtering. If the readers are interested in this domain, we recommend another two interesting models: wide and deep learning for recommender systems [137] and deep factorization machines [138].

Despite having been proven successful, deep learning recommender systems suffer from several drawbacks and limitations.

- **Data requirement**. Generally, deep learning methods require a large scale of data. However, in some practical scenarios, there may not have sufficient data to well train the deep neural networks.
- **Interpretability and explanation**. Deep learning techniques are well known as black-box. Hence, it is challenging to provide explanations or reasons for recommender systems.
- **Extensive computational cost**. There is no free lunch in data science. Deep learning can obtain satisfactory performance but need more sophisticated architectures and more trainable parameters.

Both deep learning and recommender systems are ongoing hot research topics in both academia and the industry community. It is worthy to keep an eye on the development of this direction.

7.7 Application and Resources

Recommendation techniques have been widely adopted in many applications. This section will introduce several typical scenarios including e-commerce platforms, news and articles, location-based services, and movie reviews.

7.7.1 Applications

E-commerce System Modern e-commerce platforms such as Amazon contain millions of products. Recommendation techniques are important tools for remedying the information overload in such platforms. The recommendation tasks in an e-commerce system can roughly be grouped into two types: (1) rating prediction and (2) click-through-rate (CTR) prediction.

On explicit rating data such as product reviews, the rating prediction problem is often studied. Product reviews often contain both a rating from 1–5 stars and a piece of text description of the user's opinions on different aspects of the product. Topic models [139–142] and attention networks [143–145] have been used to extract the aspects of products and to learn how a user likes different aspects. The extracted aspects and user interests will be used for accurate recommendations.

On implicit data, CTR prediction is performed. CTR implies how likely a user is interested in a product and clicks on the product. In order to improve the CTR, it is important to capture the diverse user interests from historical data. Guo et al. [138] introduce the DeepFM model which extends the Wide and Deep model [137] with factorization machine. As the prevalence of attention networks, deep interest networks [146, 147] are proposed to learn the importance of different historical transactions and model the diversity of user interests.

Article Recommendation In the early development of article recommendation models, both content-based approaches and collaborative filtering have been used [148–150]. Apart from separately applying the two techniques, some studies have proposed hybrid approaches that incorporate the content information in collaborative filtering [151–155]. Among them, a typical model is collective topic regression [152], which extracts topics from scientific articles and infers the latent factors of articles from the topics. As such, it solves the item cold-start problem in the sense that new articles can be mapped to the latent space given the topics extracted from its content. Recently, deep learning techniques such as autoencoders [156] and attention networks [157, 158] have been used to better extract the content information from the articles.

Location-Based Service Recent years have witnessed the rapid growth of location-based social networks (LBSNs), such as Foursquare and Facebook places. In LBSNs, the users share their locations by checking in at Points-of-Interest (POIs). With the increasing availability of check-in data, POI recommendation (e.g., [159–162]) has attracted extensive research interest, helping users to better explore their surroundings and find interesting locations based on their individual preferences.

Different from conventional recommendation tasks, POI recommendations need to incorporate geographical influence, i.e., the users tend to visit nearby POIs rather than far away locations. The main challenge of POI recommendation is to effectively model the personalized geographical preference from the sparse check-in data. To solve this task, various latent representation models have been proposed by exploiting factorized Markov chain [163], metric embedding [164, 165], or word2vec-based techniques [166]. Recently, several deep learning location recommendation models have been proposed, such as spatial-temporal recurrent neural networks (ST-RNNs) [167] and spatial-temporal LSTM model [168] that combines local temporal and spatial contexts into deep neural networks for POI recommendation.

Movie Recommendation In 2006, Netflix company released 100 million movie ratings for the Netflix prize (a grand prize of 1 million US dollars) [169]. The data and prize boosted extensive research interest for recommender systems. As one of the most popular applications for recommender systems, a large variety of methods have been developed for a movie recommendation, including the standard content-based filter and collaborative filter techniques. In addition to rating information, content features are commonly used to enhance the performance of movie recommendations, especially for the cold-start recommendations for newly released movies. For example, visual features (e.g., posters and still frames) [170] as well as audio features (e.g., conversations and background music) [171] have been incorporated into recommender systems. Recently, a social-aware movie recommendation system [172] has been developed to consider heterogeneous information, including textual description, poster images, user ratings, and social relationships.

7.7.2 Resources

Open Source Implementations:

- Surprise:[1] A Python scikit for recommender systems that includes both memory-based and model-based CF models
- LibFM:[2] Factorization machine library in C++
- pyFM:[3] Factorization machine library in Python
- NCF:[4] Python implementation of neural collaborative filtering

7.8 Putting It All Together

Recently, we are facing the situation that there is an enormous amount of irrelevant information on the Internet, which is commonly known as information overload. Recommender systems can well alleviate information overload by suggesting highly relevant information based on personal interests. A large variety of recommendation applications exist in our daily life, such as recommending products, news, locations, and movies. Recommender systems play a pivotal role in various large-scale online service providers, e.g., YouTube and Amazon. From the perspective of users, recommender systems enhance user experience and create user engagement. From the perspective of business owners, they can generate more revenue.

Recommender systems have been extensively studied in the past decade. In general, existing techniques can be roughly categorized into two groups: content-based approaches and collaborative filtering. The content-based methods generate recommendations by matching item content features to user preferences. The

[1] https://github.com/NicolasHug/Surprise.

[2] https://github.com/srendle/libfm.

[3] https://github.com/coreylynch/pyFM.

[4] https://github.com/hexiangnan/neural_collaborative_filtering.

collaborative filtering (CF) methods produce recommendations by utilizing the interactions between users and items. The user–item interaction can be explicit feedback or implicit feedback, depending on if explicit user preferences are expressed or not. Next, the recommendation methods introduced in this chapter will be briefly summarized and discussed.

The content-based technique can be regarded as a profile matching problem, which mainly includes three steps: (1) profiling items, which builds a feature vector for each item based on its attributes (e.g., category) or content information (e.g., TF-IDF of text content), (2) profiling users, which aggregates the profiles of the items that have been rated by the users, (3) matching item and user profiles, which directly calculates the similarity between the target user and each item. Content-based approaches are suitable for new items when item profiles can be obtained from descriptions and attributes. It can perform well even if no sufficient interaction data are available. However, it suffers from several disadvantages. It may not work for users with diverse or dynamic interests since it often suggests similar items to a user. Another issue would arise when new users lack a defined profile unless they are explicitly asked for information. In addition, it fails to detect the interdependency among different users or complex behaviors of the users. Therefore, its performance on real-world Recommender systems is not as good as other techniques, such as collaborative filtering.

As the most popular recommendation technique, collaborative filtering (CF) exploits the collaborative power of interactions provided by multiple users to make recommendations. CF approaches are generally divided into memory-based CF and model-based CF. User-based CF and item-based CF are two representative memory-based CF methods. In this chapter, we mainly present the user-based CF, which contains two steps: find the users with similar interests and then aggregate the ratings from similar users. A typical model-based CF method is the latent factor model, which learns low-dimensional vector representations for users and items given historical interactions. The latent factor model can be implemented by matrix factorization, such as the Singular Value Decomposition (SVD). Although CF approaches have been widely used, it suffers from several limitations. The first problem is cold-start since CF needs enough user–item interactions for the system to work. The second problem is that it is hard for CF approaches to incorporate heterogeneous information since user behaviors are very complex and can be influenced by many factors.

Factorization machine (FM) is another powerful framework for recommender systems. By considering the recommendation task as a regression task, it first represents user–item interactions as real-valued feature vectors, and then these constructed feature vectors are used as the input feature vectors to predict the target outputs. It is able to consider both order-1 and order-2 feature interactions. The FM model is also a flexible framework. For instance, the latent factor model can be viewed as a special case of FM. FM mainly has two benefits: handling sparse interaction data and scaling to large datasets.

Recently, deep learning (DL) methods have been developed for recommendation tasks, which have achieved state-of-the-art recommendation results. Compared with conventional techniques, these DL methods can effectively capture the non-linear relationships between users and items. Moreover, they can integrate multiple types of features, such as visual and textual features. One representative deep learning model is neural collaborative filtering (NCF), which combines collaborative filtering with deep learning techniques. Despite having been proven successful, DL methods suffer from several drawbacks: requiring a large scale of data, the difficulty of interpretability, and heavy computational cost.

If you are running an online business platform, you may be interested in exploiting Recommender systems to enhance user experiences and increase earnings. If only a small set of user–item interactions is available, but each user/item is associated with some descriptions or attributes, the basic content-based recommendation method could be a good starting point. When there are sufficient user–item interaction records, more powerful recommendation frameworks (CF, FM, and deep learning methods) can be considered. Specifically, when only user–item interaction (e.g., rating matrix) is available, CF would be an appealing choice. When the data is extremely sparse and auxiliary features exist, FM may be utilized. Deep learning-based recommender systems can be exploited to achieve excellent performance if there are multiple types of features or complex user behaviors. However, data learning methods require a large scale of data and expensive computational resources. Usually, the larger amount of data, the better the recommendation performance. Overall, data is the most important asset in recommender systems. We should properly select the most suitable recommendation framework based on the types and amounts of data.

7.9 Exercise Problems

Problem 7.1 Given the following product attribute features: $\mathbf{p}_1 = [1, 0, 1, 0]$, $\mathbf{p}_2 = [0, 1, 1, 0]$, and $\mathbf{p}_3 = [0, 0, 1, 1]$. In addition, several rating scores are observed: $(u_1, p_1, 5)$, $(u_1, p_2, 3)$, and $(u_1, p_3, 4)$. Please calculate the profile feature of user u_1.

Problem 7.2 Following the above question, given a new product $\mathbf{p}_4 = [1, 1, 0, 1]$, please compute the matching score of user u_1 and p_4.

Problem 7.3 Given the following transactions in the form of (user, item, rating) tuples in a recommender system:

$(u_1, p_1, 1.5)$, $(u_1, p_3, 4)$, $(u_1, p_5, 0.5)$, $(u_2, p_2, 4)$, $(u_2, p_4, 2)$, $(u_3, p_1, 4.5)$, $(u_3, p_4, 2.5)$, $(u_3, p_5, 5)$, $(u_4, p_2, 2)$, $(u_4, p_3, 3.5)$, $(u_4, p_4, 4)$, $(u_4, p_5, 2.5)$.

Let the set of users be $\{u_1, u_2, u_3, u_4\}$ and the set of items be $\{p_1, p_2, p_3, p_4, p_5\}$.

(1) Construct the rating matrix based on the above transactions.
(2) Apply the memory-based collaborative filtering with the Pearson correlation coefficient for user u_2 without considering bias. Give the top-1 recommended item to u_2 and the corresponding predicted rating.
(3) Give the predicted rating of user u_2 to item p_5 using the collaborative filtering that incorporates the bias mentioned in Sect. 7.4.

Problem 7.4 Given latent vectors of matrix factorization: $\mathbf{q}_u = [1.3, 0.9, 0.7]$ for user u and $\mathbf{p}_i = [0.6, 0.9, 1.2]$ for item i. The ground truth rating score is 5. Based on Eq. (7.9), calculate the loss for the predicted rating score.

Problem 7.5 Given latent vectors of matrix factorization: $\mathbf{q}_u = [1.1, 0.8, 0.5, 1.1]$ for user u and $\mathbf{p}_i = [0.5, 0.6, 0.9, 0.8]$ for item i. Consider the bias terms: global bias $b^{global} = 0.1$, user bias $b^{user} = -0.3$, and item bias $b^{item} = 0.5$. The ground truth rating score is 4. Based on Eq. (7.10), calculate the loss for the predicted rating score.

Problem 7.6 Implement the latent factor model in Eq. (7.11).

Problem 7.7 Prove that matrix factorization (MF) can be a special case of factorization machine (FM) model.

Problem 7.8 Prove that matrix factorization (MF) can be a special case of neural collaborative filtering (NCF) framework.

Problem 7.9 Prove that the overall time complexity of Factorization Machine is $O(kn)$. (k is the number of dimensionality of the factorization model, and n is the size of input features.)

Problem 7.10 What is the difference between Linear Regression (LR) and Factorization Machine (FM)?

Problem 7.11 Given three users $\{u_1, u_2, u_3\}$, four items $\{i_1, i_2, i_3, i_4\}$, and three interaction records $\{(u_1, i_2, 5), (u_2, i_1, 4), (u_3, i_3, 1)\}$, please construct the corresponding input features of FM model for these given interaction records.

Problem 7.12 Following the above problem, please calculate the total number of parameters to be learnt in FM model. (The dimension of latent factorization model is given as $k = 2$.)

Problem 7.13 Following the above 2 questions, assume that the learned parameters are global bias $b = 0.5$, weights of user variables $w_{u_1} = 0.1$, $w_{u_2} = -0.2$, $w_{u_3} = 0.5$, weights of item variables $w_{i_1} = 0.3$, $w_{i_2} = 1.2$, $w_{i_3} = 0.8$, $w_{i_4} = 1.4$, factorized latent vectors $V_{u_1} = [1.0, 0.5]$, $V_{u_2} = [0.7, 0.4]$, $V_{u_3} = [1.1, 0.7]$, $V_{i_1} = [0.5, 0.3]$, $V_{i_2} = [0.3, 1.2]$, and $V_{i_3} = [0.9, 0.3]$. Calculate the predicted scores for interactions (u_1, i_2) and (u_2, i_1).

Problem 7.14 Given four users $\{u_1, u_2, u_3, u_4\}$, three items $\{i_1, i_2, i_3\}$, and five implicit interaction records (e.g., click or visit, no explicit rating scores) $\{(u_1, i_2), (u_1, i_3), (u_2, i_1), (u_4, i_2)\}$, please construct the corresponding input features and target scores of FM model for this implicit feedback recommendation task.

Problem 7.15 Based on the introduced Neural Collaborative Filtering (NCF), design an approach to explicitly fuse NCF and matrix factorization (MF). (Hint: concatenate results of NCF and the inner product of MF latent factors.)

Chapter 8
Graph Learning

8.1 Introduction

Effective data processing necessitates an understanding of graph geometry. The non-Euclidean graph domain is used in a wide range of applications, such as transportation networks, power distribution networks, brain neural networks, and gene data on biological regulation networks.

A graph neural network is a sort of deep learning extension that is used to find patterns in the topology of a graph. Using topology and attribution data, we are currently characterizing node-level representation. The first several GNNs are built using a spectral graph theory notion borrowed from spectral analysis. With approximation theory, processing costs can be reduced further, making the algorithm useful for a wide range of applications. To make model generation even easier and more rapid, GNN switches to using spatial data. Mathematics shows that even though they appear to be opposites at first glance, the spectral and spatial domains have a lot in common. This makes it easier to compare different GNN approaches because the majority of them can be analyzed within a single framework.

8.2 Basics of Math

This section introduces fundamental concepts in mathematics, such as matrices and their manipulation, spectral decomposition, and approximation theory, all of which are essential to graph neural networks.

S. Rafatirad et al., *Machine Learning for Computer Scientists and Data Analysts*,
https://doi.org/10.1007/978-3-030-96756-7_8

8.2.1 Matrix Manipulation

Let A, B, C $\in \mathbb{R}^{N\times N}$; then they have basic properties including:

$$
\begin{aligned}
(AB)^{-1} &= B^{-1}A^{-1} \\
(ABC\ldots)^{-1} &= \ldots C^{-1}B^{-1}A^{-1} \\
\left(A^{\mathsf{T}}\right)^{-1} &= \left(A^{-1}\right)^{\mathsf{T}} \\
(A+B)^{\mathsf{T}} &= A^{\mathsf{T}} + B^{\mathsf{T}} \\
(AB)^{\mathsf{T}} &= B^{\mathsf{T}}A^{\mathsf{T}} \\
(ABC\ldots)^{\mathsf{T}} &= \ldots C^{\mathsf{T}}B^{\mathsf{T}}A^{\mathsf{T}}.
\end{aligned}
\tag{8.1}
$$

See more matrix properties in the matrix handbook [173].

8.2.2 Eigendecomposition on Matrix

A (non-zero) vector v in $\mathbb{R}^N$ is an eigenvector of A if it satisfies the linear equation:

$$\mathrm{Av} = \lambda \mathrm{v}. \tag{8.2}$$

Let A be decomposed by N linearly independent eigenvectors U_i (where $i = 1, \ldots, n$); A can be factorized as

$$\mathrm{A} = \mathrm{U}\,\Lambda\,\mathrm{U}^{-1}, \tag{8.3}$$

where U is the $N \times N$ matrix whose ith column is the eigenvector U_i of A, and Λ is the diagonal matrix whose diagonal elements are eigenvalues:

$$\Lambda = \begin{bmatrix} \lambda_1 & & & \\ & \lambda_2 & & \\ & & \ddots & \\ & & & \lambda_n \end{bmatrix}. \tag{8.4}$$

The eigenvectors are usually assumed to be normalized, but they need not be. In linear algebra, an orthogonal matrix is a square matrix whose columns and rows are orthogonal unit vectors, namely orthonormal vectors.

$$\mathrm{U}^{\mathsf{T}}\mathrm{U} = \mathrm{U}\mathrm{U}^{\mathsf{T}} = \mathrm{I}, \tag{8.5}$$

where $U^\intercal$ is the transpose of U, and I is the identity matrix. This leads to the equivalent characterization: a matrix U is orthogonal if its transpose is equal to its inverse:

$$U^\intercal = U^{-1}, \tag{8.6}$$

where U^{-1} is the inverse of U. The eigendecomposition allows for much easier computation of power series of matrices. If $f(x)$ is given by

$$f(x) = a_0 + a_1 x + a_2 x^2 + \cdots, \tag{8.7}$$

then we have

$$f(A) = U f(\Lambda) U^\intercal. \tag{8.8}$$

Similarly, we can have

$$\begin{aligned} A^2 &= \left(U\Lambda U^\intercal\right)\left(U\Lambda U^\intercal\right) = U\Lambda\left(U^\intercal U\right)\Lambda U^\intercal = U\Lambda^2 U^\intercal \\ A^n &= U\Lambda^n U^\intercal \\ \exp A &= U\exp(\Lambda)U^\intercal. \end{aligned} \tag{8.9}$$

If a matrix A can be eigendecomposed, then A is nonsingular, and its inverse is given by

$$A^{-1} = U\Lambda^{-1}U^\intercal. \tag{8.10}$$

Example 8.1 (Eigendecomposition)

Problem: Given a matrix $M = \begin{bmatrix} 1\ 2\ 3 \\ 4\ 5\ 6 \\ 7\ 8\ 9 \end{bmatrix}$, calculate its Eigendecomposition.

Solution: $M = U\,\Lambda\,U^{-1}$. We have

$$\Lambda = \begin{bmatrix} 16.1 & 0 & 0 \\ 0 & -1.1 & 0 \\ 0 & 0 & 0 \end{bmatrix}, U = \begin{bmatrix} -0.23197069 & -0.78583024 & 0.40824829 \\ -0.52532209 & -0.08675134 & -0.81649658 \\ -0.8186735 & 0.61232756 & 0.40824829 \end{bmatrix}.$$

```
# eigendecomposition
from numpy import array
from numpy.linalg import eig
# define matrix
A = array([[1, 2, 3], [4, 5, 6], [7, 8, 9]])
print(A)
```

```
# calculate eigendecomposition
values, vectors = eig(A)
print(values)
print(vectors)

> [[1 2 3]
  [4 5 6]
  [7 8 9]]

> [ 1.61168440e+01  -1.11684397e+00  -9.75918483e-16]

> [[-0.23197069 -0.78583024  0.40824829]
  [-0.52532209 -0.08675134 -0.81649658]
  [-0.8186735   0.61232756  0.40824829]]
```

8.2.3 Approximation Theory

Approximation theory focuses on approximating a function closely using simpler functions such as a polynomial of high degrees. One can calculate the optimal polynomials by expanding the given function in terms of Chebyshev polynomials. The types of Chebyshev polynomials are defined as

$$\begin{aligned} T_0(x) &= 1 \\ T_1(x) &= x \\ T_2(x) &= 2xT_1(x) - T_0(x) = 2x^2 - 1 \\ T_3(x) &= 2xT_2(x) - T_1(x) = 4x^3 - 3x \\ T_4(x) &= 2xT_3(x) - T_2(x) = 8x^4 - 8x^2 + 1 \\ &\dots \\ T_{n+1}(x) &= 2xT_n(x) - T_{n-1(x)} \quad n \geq 1. \end{aligned} \tag{8.11}$$

A Chebyshev expansion for a function is

$$f(x) \sim \sum_{i=0}^{\infty} c_i T_i(x) = \frac{1}{2} c_0 T_0(x) + c_1 T_1(x) + c_2 T_2(x) + \cdots, \tag{8.12}$$

where

$$c_i = \frac{2}{\pi} \int_{-1}^{1} \left(1 - x^2\right)^{-\frac{1}{2}} f(x) T_i(x) \mathrm{d}x. \tag{8.13}$$

In mathematics, the rational approximation is more powerful and accurate than polynomial, taking more computational overhead though. As the best approximation of a function by a rational function, the Padé approximant [174] of order $[m/n]$ is the rational function

$$R(x) = \frac{\sum_{j=0}^{m} a_j x^j}{1 + \sum_{k=1}^{n} b_k x^k} = \frac{a_0 + a_1 x + a_2 x^2 + \cdots + a_m x^m}{1 + b_1 x + b_2 x^2 + \cdots + b_n x^n}, \tag{8.14}$$

where $m \leq 0$ and $n \leq 1$. The parameters of a rational function can be obtained by Wynn's epsilon algorithm and sequence transformations. Both polynomial and rational approximations can be solved by the Remez algorithm [175]. See [176–178] for a complete theory of approximation techniques.

Example 8.2 (Polynomial Approximation)
Problem: Calculate a polynomial approximation with order 1 for $f(x) = e^x$.
Solution: $f(x) = e^x$ can be approximated by a polynomial:

$$f(x) \approx g(x) = \sum_{n=0}^{\infty} \frac{x^n}{n!} \approx 1 + x + \frac{x^2}{2!} + \frac{x^3}{3!} + \frac{x^4}{4!} + \cdots, x \in [0, 1].$$

Take $x = 1$ as example:

$$f(1) = e^1 \approx 2.718$$

$$g(1) = 1 + (1) + \frac{1}{2}(1)^2 + \frac{1}{6}(1)^3 + \frac{1}{24}(1)^4 + \frac{1}{120}(1)^5 = \frac{163}{60} \approx 2.717.$$

```
import math

x = 1
f_to_1 = math.exp(x)
print(f_to_1)

g_to_1 = x**0/math.factorial(0) + x**1/math.factorial(1) + x**2/math.factorial(2) + x**3/math.factorial(3) + x**4/math.factorial(4) + x**5/math.factorial(5)
print(g_to_1)

> 2.718281828459045
> 2.7166666666666663
```

8.2.4 *Graph Representations and Graph Signal*

A graph is defined as $\mathcal{G} = (\mathrm{V}, \mathrm{E}, \mathrm{A})$, where V is a set of n nodes and E represents edges. An entry $v_i \in \mathrm{V}$ denotes a node, and $e_{i,j} = \{v_i, v_j\} \in \mathrm{E}$ indicates an edge between nodes i and j. The adjacency matrix $\mathrm{A} \in \mathbb{R}^{N \times N}$ is defined by if $\mathrm{A}_{i,j} = 1$, there is a link between nodes i and j, and else 0. A multiple dimensional graph signal (or node features) $X \in \mathbb{R}^{N\times F}$ is a matrix with each entry $X_i \in \mathbb{R}^F$ representing the feature vector on node i. One-dimensional graph signal is denoted with $x \in \mathbb{R}^{N\times 1}$, and $x_i \in \mathbb{R}$ with $F = 1$. Another popular graph matrix is the graph Laplacian that is defined as $L = D - A \in \mathbb{R}^{N\times N}$, where D is the degree matrix. Due to its generalization ability [179, 180], the symmetric normalized Laplacian is often used, which is defined as $\tilde{L} = D^{-\frac{1}{2}} L D^{-\frac{1}{2}}$. Another option is random walk normalization: $\tilde{L} = D^{-1} L$. Note that normalization could also be applied to the adjacency matrix. Their relationship can be derived as $\tilde{L} = I - \tilde{A}$. L satisfies the following properties:

- L is symmetric and positive semi-definite.
- The smallest eigenvalue is 0 with the corresponding eigenvector of the constant one vector $\mathbb{1} := [1, 1, 1, ...1]$.
- L has n non-negative and real-valued eigenvalues.

The graph Laplacian can be treated as a difference operator:

$$L x_i = \sum_{j \in \mathcal{N}_i} a_{i,j}(x_i - x_j), \tag{8.15}$$

where $\mathcal{N}_i$ is the set of neighbors of the node i. The following also satisfies

$$x^\top L x = \frac{1}{2} \sum_{i,j=1}^{n} a_{ij} \left(x_i - x_j\right)^2, \tag{8.16}$$

which denotes the degree of smoothness of signal x over L. There are three types of normalized graph Laplacian operations denoted as $\tilde{L}_i$, $i = 1, 2, 3$, defined as

$$\begin{aligned} \tilde{L}_1 &:= D^{-\frac{1}{2}} L D^{-\frac{1}{2}} = I - D^{-\frac{1}{2}} A D^{-\frac{1}{2}} \\ \tilde{L}_2 &:= D^{-1} L = I - D^{-1} A \\ \tilde{L}_3 &:= L D^{-1} = I - A D^{-1}. \end{aligned} \tag{8.17}$$

The first type is also called *symmetric normalized Laplacian*. The second one is called *row-normalized Laplacian* or *random walk normalized Laplacian*. The third one is called *column-normalized Laplacian*. Similarly, normalized Laplacian satisfies the following properties:

- $\tilde{L}_{i\in\{1,2,3\}}$ are symmetric and positive semi-definite.
- 0 is an eigenvalue of $\tilde{L}_1$ with eigenvector $D^{-\frac{1}{2}}\mathbb{1}$. 0 is an eigenvalue of $\tilde{L}_2$, and the corresponding eigenvector is constant one vector.
- $\tilde{L}_{i\in\{1,2,3\}}$ have N non-negative real-valued eigenvalues.

8.2.5 Spectral Graph Theory

In mathematics, spectral graph theory is the study of the relationship between the properties of a graph and the eigenspace of the graph matrices such as its adjacency matrix and Laplacian matrix.

Given a graph Laplacian with eigendecomposition $\mathrm{L} = \mathrm{U}\,\Lambda\,\mathrm{U}^\intercal$ and graph signal x, the graph Fourier transform of x is defined as

$$\hat{x} = \mathrm{U}^\intercal x, \tag{8.18}$$

which transforms signal x into a new one $\hat{x}$ in the spectral domain, or expands x in terms of eigenvectors. Accordingly, the inverse graph Fourier transform is defined as

$$x = \mathrm{U}\hat{x}, \tag{8.19}$$

which transforms the spectral signal $\hat{x}$ back into the vertex domain.

For two continuous signals $x_1, x_2 \in \mathbb{R}^N$, the convolution is defined as

$$(x_1 * x_2)(i) \triangleq \int_{-\infty}^{\infty} x_1(\tau)x_2(i-\tau)d\tau. \tag{8.20}$$

Discrete signals are defined as

$$(x_1 * x_2)[i] = \sum_{k=-\infty}^{\infty} x_1[k]x_2[i-k] = \sum_{k=-\infty}^{\infty} x_1[i-k]x_2[k]. \tag{8.21}$$

Another definition of convolution is by convolution theorem in spectral domain:

$$x_1 * x_2 = \mathcal{F}^{-1}\{\mathcal{F}\{x_1\} \cdot \mathcal{F}\{x_2\}\}, \tag{8.22}$$

where $\mathcal{F}$ denotes Fourier transform, and $\mathcal{F}^{-1}$ denotes inverse Fourier transform. The convolution theorem can be expressed in two formats:

- The Fourier transform of a convolution is the product of the Fourier transforms.

$$\mathcal{F}\{x_1 * x_2\} = \{\mathcal{F}\{x_1\} \cdot \mathcal{F}\{x_2\}\}. \tag{8.23}$$

- The Fourier transform of a product is the convolution of the Fourier transforms.

$$\mathcal{F}\{x_1 \cdot x_2\} = \{\mathcal{F}\{x_1\} * \mathcal{F}\{x_2\}\}. \tag{8.24}$$

Convolution satisfies the following properties:

- $\alpha(x_1 * x_2) = (\alpha x_1) * x_2 = x_1 * (\alpha x_2)$.
- *Commutativity:* $x_1 * x_2 = x_2 * x_1$.
- *Distributivity:* $x_1 * (x_2 + x_3) = x_1 * x_2 + x_1 * x_3$.
- *Associativity:* $(x_1 * x_2) * x_3 = x_1 * (x_2 * x_3)$.

Accordingly, graph convolution is defined as

$$x * \mathrm{g} = \mathrm{U}\,\mathrm{g}(\Lambda)\mathrm{U}^{\mathsf{T}}x = \mathrm{U}\,\mathrm{g}(\Lambda)\mathcal{F}\{x\} = \mathcal{F}^{-1}\{\mathrm{g}(\Lambda)\mathcal{F}(x)\}, \tag{8.25}$$

where g is a function of eigenvalues Λ of L, and it is in the spectral domain. This spectral domain is constructed by graph topology via its basis U, which is the eigenvectors of graph Laplacian L. Therefore, we have

$$\mathrm{g(L)\,X = U\,g}(\Lambda)\,\mathrm{U^{\mathsf{T}}\,X, \text{ or } g(L) = U\,g}(\Lambda)\,\mathrm{U}^{\mathsf{T}},$$

where $\mathrm{X} \in \mathbb{R}^{N \times F}$, and g is polynomial or rational function.

Example 8.3 (Graph Convolution)
Problem: Given a graph that contains four nodes, v_1, v_2, v_3, v_4 and three links $(v_1, v_2), (v_1, v_3), (v_1, v_4)$, verify $\mathrm{g(L) = U\,g}(\Lambda)\mathrm{U}^{\mathsf{T}}$.
Solution: The adjacency and Laplacian matrices are

$$\mathrm{A} = \begin{bmatrix} 0\,1\,1\,1 \\ 1\,0\,0\,0 \\ 1\,0\,0\,0 \\ 1\,0\,0\,0 \end{bmatrix}, \mathrm{L} = \begin{bmatrix} 3 & -1 & -1 & -1 \\ -1 & 1 & 0 & 0 \\ -1 & 0 & 1 & 0 \\ -1 & 0 & 0 & 1 \end{bmatrix},$$

and it is easy to obtain the eigenvalues matrix of Laplacian:

$$diag(\Lambda) = \begin{bmatrix} 4\,0\,0\,0 \\ 0\,0\,0\,0 \\ 0\,0\,1\,0 \\ 0\,0\,0\,1 \end{bmatrix}.$$

In the coding, we will show that:

$$\mathrm{g(L) = U\,g}(\Lambda)\mathrm{U}^{\mathsf{T}},$$

which shows that learning a function on eigenvalues is equivalent to learning the same function on the Laplacian matrix.

```
import numpy as np
from numpy.linalg import eig

# g = x^2 + 1

adj = np.array([
      [0,1,1,1],
      [1,0,0,0],
      [1,0,0,0],
      [1,0,0,0]
])

# calculate degree matrix
degrees = np.sum(adj, axis=0)
degree_matrix = np.diag(degrees)

# calculate Laplacian matrix
Laplacian_matrix = np.matrix(degree_matrix - adj)
values, vectors = eig(Laplacian_matrix)

g_L = Laplacian_matrix ** 2 + np.eye(4)
print(g_L)

# @ denotes matrix multiplication
U_g_Ut = vectors @ np.diag(values ** 2 + 1) @ vectors.T
print(U_g_Ut)


>[[13. -4. -4. -4.]
 [-4.  3.  1.  1.]
 [-4.  1.  3.  1.]
 [-4.  1.  1.  3.]]

>[[13. -4. -4. -4.]
 [-4.  3.  1.  1.]
 [-4.  1.  3.  1.]
 [-4.  1.  1.  3.]]
```

8.3 Graph Neural Network Models

GNNs are fundamental techniques for graph representation learning, which tries to learn node-level representations by combining non-Euclidean graph topological and Euclidean node features. Graph convolution, the most common type of GNN, is a learning method for generating node-level embeddings:

$$\mathcal{G}, \mathrm{X} \rightarrow \mathrm{Z}, \tag{8.26}$$

where Z denotes the fused feature of nodes, $\mathcal{G}$ is the graph representation (e.g., adjacency or Laplacian matrix), f indicates graph convolution, and $\mathrm{X} \in \mathbb{R}^{N \times F}$ is an F-dimensional graph signal.

Take graph convolutional networks (GCNs) [181] as a representative GNN, and node i has three neighbors, nodes j, k, and l. GCNs perform neighbor aggregation on node i with mean function:

$$Z(i) = \frac{X_i + X_j + X_k + X_l}{4},$$

which means the updated representation of i equals to the average of itself and its neighbors. Writing in a general case, GCN executes mean neighbor aggregation on node i that has M neighbors:

$$Z(i) = \frac{X_i + \overbrace{X_j + X_k + X_l + \cdots}^{M}}{1 + \underbrace{(1 + 1 + 1 + \cdots)}_{M}}. \tag{8.27}$$

Rewriting Eq. (8.27) in matrix form, we have

$$Z = \tilde{A} X, \tag{8.28}$$

where $\tilde{A} = D^{-1}(A + I)$ is normalized adjacency matrix, and $D_{ii} = \sum_j \tilde{A}_{ij}$. $\tilde{A}$ conducts normalization as in the denominator of Eq. (8.27).

Example 8.4 (Mean Graph Aggregation and Matrix Implementation)
Problem: Given a graph that contains four nodes, v_1, v_2, v_3, v_4 and, v_1, has three neighbors (v_1, v_2), (v_1, v_3), (v_1, v_4), calculate mean aggregation by one-by-one entry and matrix form.
Solution: Therefore, we have its adjacency matrix, and features of all nodes are initialized as X:

$$A = \begin{bmatrix} 0 & 1 & 1 & 1 \\ 1 & 0 & 0 & 0 \\ 1 & 0 & 0 & 0 \\ 1 & 0 & 0 & 0 \end{bmatrix}, D = \begin{bmatrix} 3 & 0 & 0 & 0 \\ 0 & 1 & 0 & 0 \\ 0 & 0 & 1 & 0 \\ 0 & 0 & 0 & 1 \end{bmatrix}, A + I = \begin{bmatrix} 1 & 1 & 1 & 1 \\ 1 & 1 & 0 & 0 \\ 1 & 0 & 1 & 0 \\ 1 & 0 & 0 & 1 \end{bmatrix}, \hat{D} = \begin{bmatrix} 4 & 0 & 0 & 0 \\ 0 & 2 & 0 & 0 \\ 0 & 0 & 2 & 0 \\ 0 & 0 & 0 & 2 \end{bmatrix}, X = \begin{bmatrix} 2 \\ 4 \\ 6 \\ 8. \end{bmatrix}$$

Take note that $\hat{D}$ is distinct from the formal degree matrix D that does not include self-counting elements. Following Eq. (8.27), the updated representation of node v_1 is

$$Z(v_1) = \frac{2 + 4 + 6 + 8}{1 + 1 + 1 + 1} = 5.$$

Similarly, we can calculate the others one by one. The remaining three nodes have only one neighbor (i.e., v_1), so we have

(continued)

Example 8.4 (continued)

$$Z(v_2) = \frac{4+2}{1+1} = 3, Z(v_3) = \frac{6+2}{1+1} = 4, Z(v_4) = \frac{8+2}{1+1} = 5.$$

Alternatively, the same result can be obtained by Eq. (8.28):

$$Z = D^{-1}(A+I)\,X = \begin{bmatrix} 4 & 0 & 0 & 0 \\ 0 & 2 & 0 & 0 \\ 0 & 0 & 2 & 0 \\ 0 & 0 & 0 & 2 \end{bmatrix}^{-1} \begin{bmatrix} 1 & 1 & 1 & 1 \\ 1 & 1 & 0 & 0 \\ 1 & 0 & 1 & 0 \\ 1 & 0 & 0 & 1 \end{bmatrix} \begin{bmatrix} 2 \\ 4 \\ 6 \\ 8 \end{bmatrix} = \begin{bmatrix} 5 \\ 3 \\ 4 \\ 5 \end{bmatrix}.$$

Therefore, matrix implementation gets the same result as one-by-one calculation.

```
import numpy as np
from numpy.linalg import inv


adj = np.array([
        [0,1,1,1],
        [1,0,0,0],
        [1,0,0,0],
        [1,0,0,0]
])
X = np.array([
        [2],
        [4],
        [6],
        [8]
])

adj_with_self_loop = adj + np.eye(4)

degrees = np.sum(adj_with_self_loop, axis=0)
degree_matrix = np.diag(degrees)
degree_matrix_inv = inv(degree_matrix)
print(degree_matrix)

Z = degree_matrix_inv @ adj_with_self_loop @ X

print(Z)


>[[4. 0. 0. 0.]
 [0. 2. 0. 0.]
 [0. 0. 2. 0.]
 [0. 0. 0. 2.]]

>[[5.]
 [3.]
 [4.]
 [5.]]
```

In the remaining of this section, we will show different aggregation functions except *mean* in the example above.

8.3.1 Spatial-Based Graph Convolution Networks

Spatial-based methods characterize the behaviors of models in the spatial or vertex domain, and most of them focus on neighbors aggregation.

Definition 8.1 (Spatial Method) By integrating graph connectivity $\mathcal{G}$ and node features X, the updated node representations (Z) are defined as

$$\mathrm{Z} = f(\mathcal{G})\,\mathrm{X}, \tag{8.29}$$

where $\mathcal{G}$ is often implemented with A or L in the existing works. Therefore, spatial methods focus on finding a **node aggregation function** $f(\cdot)$ that learns how to aggregate node features to obtain an updated node embedding Z.

According to the type of node aggregation function, GCNNs can be classified into three categories: linear, polynomial, and rational:

- When the function f is **linear**, GCNNs update representations with the aggregation of first-order neighbors (i.e., direct neighbors). Examples include GCN [181] and GraphSAGE [182] and GIN [183].
- When the function f is **polynomial**, GCNNs involve higher order of neighbors with customized weights [128, 184–189].
- When the function f is **rational**, GCNNs add reverse propagation, which means that the propagation could teleport back to itself with a certain probability [190, 191].

Linear Aggregation Function

A number of works [128, 182, 183, 192, 193] can be treated as learning the aggregation scheme among first-order neighbors (i.e., direct neighbors). This aspect focuses on adjusting the weights for each node and its neighbors to reveal the patterns regarding the supervision signal. Formally, the updated node embeddings $\mathrm{Z}(v)$ can be written as

$$\mathrm{Z}(v_i) = \Phi(v_i)\,\mathrm{h}(v_i) + \sum_{u_j \in \mathcal{N}(v_i)} \Psi(u_j)\,\mathrm{h}(u_j), \tag{8.30}$$

where u_j denotes a neighbor of node v_i, and $\mathrm{h}(\cdot)$ is the representation of a node. Φ and Ψ denote the weight functions. The first item on the right-hand side denotes the representation of node v_i, while the second represents the update from its neighbors.

Applying random walk normalization (i.e., dividing neighbors by the degree of the current node), Eq. (8.30) can be written as

$$Z(v_i) = \Phi(v_i)\, h(v_i) + \sum_{u_j \in \mathcal{N}(v_i)} \Psi(u_j) \frac{h(u_j)}{d_i}, \tag{8.31}$$

or symmetric normalization:

$$Z(v_i) = \Phi(v_i)\, h(v_i) + \sum_{u_j \in \mathcal{N}(v_i)} \Psi(u_j) \frac{h(u_j)}{\sqrt{d_i d_j}}, \tag{8.32}$$

where d_i represents the degree of node v_i. Normalization is deemed to have better generalization capacity [194]. In a simplified configuration, the weights for the neighbors (Ψ) are the same. Therefore, they can be rewritten in a matrix form as follows:

$$Z = \phi\, X + \psi\, D^{-1}\, A\, X = (\phi\, I + \psi\, D^{-1^{-1}}\, A)\, X \tag{8.33}$$

or

$$Z = \phi\, X + \psi D^{-\frac{1}{2}}\, A\, D^{-\frac{1}{2}}\, X = (\phi\, I + \psi D^{-\frac{1}{2}}\, A\, D^{-\frac{1}{2}})\, X, \tag{8.34}$$

where ϕ and ψ are the weights. Equations (8.33) and (8.34) can be generalized as the same form:

$$Z = (\phi\, I + \psi \tilde{A})\, X, \tag{8.35}$$

where $\tilde{A}$ denotes the normalized A, which could be implemented by random walk or symmetric normalization. The new representation of the current node is updated as the sum of the previous representations of itself and its neighbors, and it may adjust the weights of the neighbors. Several state-of-the-art methods are selected to illustrate this schema:

Graph Convolutional Network (GCN)

Graph convolutional network (GCN) [181] adds a self-loop to nodes and applies a *renormalization* trick that changes degree matrix from $D_{ii} = \sum_j A_{ij}$ to $\hat{D}_{ii} = \sum_j (A + I)_{ij}$. Specifically, GCN can be written as

$$Z = \hat{D}^{-\frac{1}{2}} \hat{A} \hat{D}^{-\frac{1}{2}} X = \hat{D}^{-\frac{1}{2}} (I + A) \hat{D}^{-\frac{1}{2}} X = (I + \tilde{A})\, X, \tag{8.36}$$

where $\hat{\mathrm{A}} = \mathrm{A}+\mathrm{I}$, and $\tilde{\mathrm{A}}$ is the normalized adjacency matrix with self-loop. Therefore, Eq. (8.36) is equivalent to Eq. (8.35) when setting $\phi = 1$ and $\psi = 1$ with the *renormalization* trick, and GCN takes the sum of each node and average of its neighbors as new node embeddings.

GraphSAGE

GraphSAGE [182] applies an aggregation among its immediate neighbors; then we have

$$\mathrm{Z} = \mathrm{D}^{-\frac{1}{2}}(\mathrm{I}+\mathrm{A})\mathrm{D}^{-\frac{1}{2}}\,\mathrm{X} = (\mathrm{I}+\tilde{\mathrm{A}})\,\mathrm{X}, \tag{8.37}$$

which is equivalent to Eq. (8.35) with $\phi = 1$ and $\psi = 1$. Note that the key difference between GCN and GraphSAGE is the normalization: the former is symmetric normalization and the latter is random walk normalization.

Example 8.5 (Graph Representation with GCN and GraphSAGE)
Problem: Given a graph that contains four nodes, v_1, v_2, v_3, v_4 and, v_1, has three neighbors (v_1, v_2), (v_1, v_3), (v_1, v_4), calculate one iteration of Z with GCN and GraphSAGE.
Solution: Therefore, we have its adjacency matrix, and features of all nodes are initialized as X:

$$\mathrm{A} = \begin{bmatrix} 0\,1\,1\,1 \\ 1\,0\,0\,0 \\ 1\,0\,0\,0 \\ 1\,0\,0\,0 \end{bmatrix}, \mathrm{D} = \begin{bmatrix} 3\,0\,0\,0 \\ 0\,1\,0\,0 \\ 0\,0\,1\,0 \\ 0\,0\,0\,1 \end{bmatrix}, \mathrm{A}+\mathrm{I} = \begin{bmatrix} 1\,1\,1\,1 \\ 1\,1\,0\,0 \\ 1\,0\,1\,0 \\ 1\,0\,0\,1 \end{bmatrix}, \hat{\mathrm{D}} = \begin{bmatrix} 4\,0\,0\,0 \\ 0\,2\,0\,0 \\ 0\,0\,2\,0 \\ 0\,0\,0\,2 \end{bmatrix}, \mathrm{X} = \begin{bmatrix} 2 \\ 4 \\ 6 \\ 8. \end{bmatrix}$$

For GCN, the adjacency matrix is normalized by renormalized degree matrix that is

$$\mathrm{Z} = \tilde{\mathrm{D}}^{-\frac{1}{2}}(\mathrm{A}+\mathrm{I})\tilde{\mathrm{D}}^{-\frac{1}{2}}\,\mathrm{X} = \begin{bmatrix} 6.86396103 \\ 2.70710678 \\ 3.70710678 \\ 4.70710678 \end{bmatrix}.$$

For GraphSAGE:

(continued)

Example 8.5 (continued)

$$Z = D^{-\frac{1}{2}}(A+I)D^{-\frac{1}{2}}X = \begin{bmatrix} 11.05897151 \\ 5.15470054 \\ 7.15470054 \\ 9.15470054 \end{bmatrix}.$$

As we can see, renormalization increases the weight of current nodes, resulting in increased values for current nodes.

```
import numpy as np
from scipy.linalg import fractional_matrix_power
import scipy

adj = np.array([
    [0, 1, 1, 1],
    [1, 0, 0, 0],
    [1, 0, 0, 0],
    [1, 0, 0, 0]
])
X = np.array([
    [2],
    [4],
    [6],
    [8]
])

adj_with_self_loop = adj + np.eye(4)

# GCN
degrees = np.sum(adj_with_self_loop, axis=0)
degree_matrix = np.diag(degrees)

degree_matrix_half_inv = scipy.linalg.fractional_matrix_power(degree_matrix,
    -0.5)
Z_GCN = degree_matrix_half_inv @ adj_with_self_loop @ degree_matrix_half_inv
    @ X

# GraphSAGE
degrees = np.sum(adj, axis=0)
degree_matrix = np.diag(degrees)

degree_matrix_half_inv = scipy.linalg.fractional_matrix_power(degree_matrix,
    -0.5)
Z_SAGE = degree_matrix_half_inv @ adj_with_self_loop @ degree_matrix_half_inv
    @ X

print(Z_GCN, Z_SAGE)

>> [[6.86396103]
 [2.70710678]
 [3.70710678]
 [4.70710678]]

>> [[11.05897151]
 [ 5.15470054]
 [ 7.15470054]
 [ 9.15470054]]
```

Graph Isomorphism Network (GIN)

Graph isomorphism network (GIN) [183] is inspired by Weisfeiler–Lehman (WL) test [195]. GIN develops conditions to maximize the power of GNN, proposing a simple architecture, graph isomorphism network (GIN). With strong theoretical support, GIN updates node representations as

$$\mathrm{Z} = (1+\epsilon)\cdot \mathrm{h}(v) + \sum_{u_j \in \mathcal{N}(v_i)} \mathrm{h}_(u_j) = [(1+\epsilon)\,\mathrm{I} + \mathrm{A}]\,\mathrm{X}, \tag{8.38}$$

which is equivalent to Eq. (8.35) with $\phi = 1+\epsilon$ and $\psi = 1$. Note that GIN does not perform normalization on graph adjacency.

Polynomial Aggregation Function

To collect richer local structure, several studies [185, 186, 189, 196, 197] involve higher orders of neighbors, since immediate first-order neighbors are not always sufficient for representing the node. On the other hand, large order usually averages much more node representations, causing an over-smoothing issue and losing its focus on the local neighborhood [198]. This motivates many models to tune the aggregation scheme on different orders of neighbors. Therefore, proper constraint and flexibility of orders are critical for node representation. High order of neighbors has been proved to characterize challenging signals such as Gabor-like filters [199]. Formally, this type of works can be written as

$$\mathrm{Z} = \left(\phi\,\mathrm{I} + \sum_{j=1}^{k} \psi_j\,\mathrm{A}^j\right)\mathrm{X} = \mathrm{P}(\mathrm{A})\,\mathrm{X}, \tag{8.39}$$

where $\mathrm{P}(\cdot)$ is a polynomial function. Several existing works are analyzed below, showing that they are variants of Eq. (8.39):

ChebNet

ChebNet [196] applies truncated Chebyshev polynomial for estimating the filtering function; then we have

$$\sum_{k=0}^{K-1} \theta_k T_k(\tilde{\mathrm{L}})\,\mathrm{X} = [\tilde{\theta_0}\,\mathrm{I} + \tilde{\theta_1}(\mathrm{I}-\tilde{\mathrm{A}}) + \tilde{\theta_2}(\mathrm{I}-\tilde{\mathrm{A}})^2 + \cdots]\,\mathrm{X} = \mathrm{P}(\tilde{\mathrm{A}})\,\mathrm{X}\,. \tag{8.40}$$

Node2Vec

Node2Vec [186] defines a second-order random walk to control the balance between *breath first search* and *depth first search*. As analyzed before,

$$Z = \left(\frac{1}{p} \cdot \overbrace{\mathrm{I}}^{\text{source}} + \overbrace{\tilde{\mathrm{A}}}^{\text{BFS}} + \frac{1}{q} \overbrace{(\tilde{\mathrm{A}}^2 - \tilde{\mathrm{A}})}^{\text{DFS}} \right) \mathrm{X}, \tag{8.41}$$

which can be transformed and reorganized into

$$Z = \left[\frac{1}{p} \mathrm{I} + \left(1 - \frac{1}{q}\right) \tilde{\mathrm{A}} + \frac{1}{q} \tilde{\mathrm{A}}^2 \right] \mathrm{X} = \mathrm{P}(\tilde{\mathrm{A}})\, \mathrm{X}, \tag{8.42}$$

where transition probability $\tilde{\mathrm{A}} = \mathrm{D}^{-1}\mathrm{A}$ is a random walk normalized adjacency matrix.

Simple Graph Convolution (SGC)

Simple Graph Convolution (SGC) [189] removes non-linear function between neighboring graph convolution layers and combines graph propagation in one single layer:

$$Z = \tilde{\mathrm{A}}^K \mathrm{X}, \tag{8.43}$$

where $\tilde{\mathrm{A}}$ is the renormalized adjacency matrix, i.e., $\tilde{\mathrm{A}} = \tilde{\mathrm{D}}^{-\frac{1}{2}} \mathrm{A} \tilde{\mathrm{D}}^{-\frac{1}{2}}$, and $\tilde{\mathrm{D}}^{-\frac{1}{2}}$ is the degree matrix with self-loop (same as in GCN). Therefore, it can be rewritten as

$$Z = (0 \cdot \mathrm{i} + 0 \cdot \tilde{\mathrm{A}} + 0 \cdot \tilde{\mathrm{A}}^2 + \cdots + 1 \cdot \tilde{\mathrm{A}}^K)\, \mathrm{X} = \mathrm{P}(\tilde{\mathrm{A}})\, \mathrm{X}\,. \tag{8.44}$$

Rational Propagation Function

Most spatial-based GCNNs merely consider label propagation from the node to its neighbors, namely gathering information from its neighbors. However, they usually ignore propagation in the reverse direction. Reverse propagation means that labels or attributes can be propagated back to themselves with some probability, or restart propagating with a certain probability. This reverse behavior can avoid the over-smoothing issues [190]. Note that polynomial propagation can also alleviate the over-smoothing issue by manually adjusting the order number, while rational propagation automatically fits the proper order number. Several works explicitly or implicitly implement reverse propagation by applying a rational function on the adjacency matrix [190, 191, 200–204]. Since general label propagation is implemented by multiplying graph Laplacian, reverse propagation could be implemented by multiplying inverse graph Laplacian as

$$Z = P(\tilde{A})\, Q(\tilde{A})^{-1}\, X = \frac{P(\tilde{A})}{Q(\tilde{A})} X, \tag{8.45}$$

where P and Q are two different polynomial functions, and the bias of Q is often set to 1.

Autoregressive

Autoregression is a type of label propagation (LP) and is written as

$$Z = (I + \alpha\tilde{L})^{-1}\, X = \frac{I}{I + \alpha(I - \tilde{A})} X = \frac{I}{(1+\alpha)\, I - \alpha\tilde{A}} X, \tag{8.46}$$

which is equivalent to the form of Eq. (8.45), i.e., $P = I$ and $Q = (1+\alpha)\, I - \alpha\tilde{A}$.

Personalized PageRank (PPNP)

Personalized PageRanking (PPNP) [190] is inspired by page rank algorithm. Then we have

$$Z = \alpha \left(I - (1-\alpha)\tilde{A}\right)^{-1} X = \frac{\alpha}{I - (1-\alpha)\tilde{A}} X, \tag{8.47}$$

where $\tilde{A} = D^{-1} A$ is the random walk normalized adjacency matrix with self-loop. Equation (8.47) is with a rational function whose numerator is a constant.

Example 8.6 (Graph Representation with PPNP)
Problem: Given a graph that contains four nodes, v_1, v_2, v_3, v_4 and, v_1, has three neighbors (v_1, v_2), (v_1, v_3), (v_1, v_4), calculate one iteration of Z with PPNP.
Solution: Therefore, we have its adjacency matrix, and features of all nodes are initialized as X:

$$A = \begin{bmatrix} 0\,1\,1\,1 \\ 1\,0\,0\,0 \\ 1\,0\,0\,0 \\ 1\,0\,0\,0 \end{bmatrix}, D = \begin{bmatrix} 3\,0\,0\,0 \\ 0\,1\,0\,0 \\ 0\,0\,1\,0 \\ 0\,0\,0\,1 \end{bmatrix}, A + I = \begin{bmatrix} 1\,1\,1\,1 \\ 1\,1\,0\,0 \\ 1\,0\,1\,0 \\ 1\,0\,0\,1 \end{bmatrix}, \hat{D} = \begin{bmatrix} 4\,0\,0\,0 \\ 0\,2\,0\,0 \\ 0\,0\,2\,0 \\ 0\,0\,0\,2 \end{bmatrix}, X = \begin{bmatrix} 2 \\ 4 \\ 6 \\ 8. \end{bmatrix}$$

(continued)

Example 8.6 (continued)
Setting $\alpha = 0.1, 0.5$, and 0.9, we have

$$Z = \alpha\left(\mathrm{I} - (1-\alpha)\tilde{\mathrm{A}}\right)^{-1}\mathrm{X} = \begin{bmatrix} 5.9753023 \\ 3.50485815 \\ 3.70485815 \\ 3.90485815 \end{bmatrix}, \begin{bmatrix} 4.79743495 \\ 5.38490018 \\ 3.38490018 \\ 4.38490018 \end{bmatrix}, \begin{bmatrix} 2.7629368 \\ 3.75951823 \\ 5.55951823 \\ 7.35951823 \end{bmatrix}.$$

For big α, the propagation is much reduced, resulting in very close to the original values. X

```
import numpy as np
from numpy.linalg import inv
from scipy.linalg import fractional_matrix_power

adj = np.array([
    [0, 1, 1, 1],
    [1, 0, 0, 0],
    [1, 0, 0, 0],
    [1, 0, 0, 0]
])
X = np.array([
    [2],
    [4],
    [6],
    [8]
])

alpha = [0.1, 0.5, 0.9]
degrees = np.sum(adj, axis=0)
degree_matrix = np.diag(degrees)
degree_matrix_half_inv = fractional_matrix_power(degree_matrix, -0.5)
adj_norm = degree_matrix_half_inv @ adj @ degree_matrix_half_inv

for a in alpha:
    Z = a * inv(np.identity(4) - (1 - a) * adj_norm) @ X
    print(Z)

>> [[5.9753023 ]
 [3.50485815]
 [3.70485815]
 [3.90485815]]

>> [[4.79743495]
 [3.38490018]
 [4.38490018]
 [5.38490018]]

>> [[2.7629368 ]
 [3.75951823]
 [5.55951823]
 [7.35951823]]
```

ARMA Filter [191]

Utilize ARMA filter for approximating any desired filter response function that can be written in the spatial domain as

$$Z = \frac{b}{I - a\tilde{A}} X. \tag{8.48}$$

Note that ARMA filter is an unnormalized version of PPNP. When $a + b = 1$, ARMA becomes PPNP.

8.3.2 Spectral-Based Graph Convolution Networks

Spectral methods are based on graph Fourier transform [205–207], and they characterize GCNNs in terms of eigenvalue function. The graph Laplacian L can be diagonalized by the Fourier basis $U^\top$ (i.e., graph Fourier transform) [205, 206]: $\tilde{L} = U \Lambda U^\top$, where Λ is the diagonal matrix whose diagonal elements are the corresponding eigenvalues (i.e., $\Lambda_{ii} = \lambda_i$), and U is also called eigenvectors. The graph Fourier transform of a signal X is defined as $\hat{X} = U^\top X \in \mathbb{R}^{N\times N}$ and its inverse as $X = U\hat{X}$.

Definition 8.2 (Spectral Method) A graph convolution operation is defined in the Fourier domain such that

$$f_1 * f_2 = U\left[\left(U^\top f_1\right) \odot \left(U^\top f_2\right)\right],$$

where $\odot$ is the element-wise product, and f_1/f_2 are two signals defined on nodes. It follows that a node signal $f_2 = X$ is filtered by spectral signal $\hat{f}_1 = U^\top f_1 = g$ as

$$Z = g(\tilde{L})\, X = U\left[g(\Lambda) \odot \left(U^\top X\right)\right] = U\, g(\Lambda) U^\top X, \tag{8.49}$$

where g is known as **frequency response function**. Therefore, the objective of spectral methods is to learn a function $g(\cdot)$.

According to the function type of g, GCNNs can be classified into three categories: linear, polynomial, and rational:

- When the filter function g is **linear**, most GCNNs employ negative slope for g, which makes it low-pass filtering and assigns larger value for smaller eigenvalues. Since smaller eigenvalues correspond to low-frequency components, this category implements a smoothing process among neighbors. Examples include GCN [181] and GraphSAGE [182] and GIN [183].
- When the filter function g is **polynomial**, it has higher accuracy with reasonable computational cost since the polynomial function is the most popular approxi-

mation [205, 208]. Many GNNs fall into this category with the predefined order numbers [128, 184–189].
- When the filter function g is **rational**, it achieves the highest accuracy with significant computational overhead. Compared with polynomial, the rational function has exponentially less error in approximating signal discontinuities [208]. Relatively few GNNs [190, 191, 200] belong to this category due to its efficiency, and it is affordable when applying an iterative algorithm [190, 191].

Linear Filter Function

There exist numerous works that can be boiled down to adjusting weights of frequency components in the spectral domain. The goal of the filter function is to adjust eigenvalues (i.e., the weights of eigenvectors) to fit the target output. Many of them are deemed low-pass filters [201], meaning that only low-frequency components are emphasized, namely, the first few eigenvalues are enlarged, and the others are reduced. There exist a large number of works that can be understood as adjusting weights of frequency component during aggregation. All the examples in Sect. 8.3.1 will be rewritten in this section:

Graph Convolutional Network (GCN)

$$Z = D^{-\frac{1}{2}}(A + I)D^{-\frac{1}{2}} X = D^{-\frac{1}{2}}(D - L + I)D^{-\frac{1}{2}} X = (I - L + I) X = U(2 - \Lambda)U^{\mathsf{T}} X.$$

Therefore, the frequency response function is $g(\Lambda) = 2 - \Lambda$, which is a low-pass filter, i.e., a smaller eigenvalue will be adjusted to a large value, in which a small eigenvalue corresponds to a low-frequency component.

GraphSAGE

$$Z = D^{-\frac{1}{2}}(I + A)D^{-\frac{1}{2}} X = (I + \tilde{A}) X = (2I - \tilde{L}) X = U(2 - \Lambda)U^{\mathsf{T}} X, \tag{8.50}$$

so the frequency response function is $g(\Lambda) = 2 - \Lambda$. Note that GraphSAGE's normalization is different from GCN, since GCN has renormalization trick.

Graph Isomorphism Network (GIN)

$$Z = D^{-\frac{1}{2}}[(1 + \epsilon) I + A]D^{-\frac{1}{2}} X = D^{-\frac{1}{2}}[(2 + \epsilon) I - \tilde{L}]D^{-\frac{1}{2}} X = U(2 + \epsilon - \Lambda)U^{\mathsf{T}} X.$$

GIN can be seen as a generalization of GCN and GraphSAGE without normalized adjacency matrix A. The frequency response function is $g(\Lambda) = 2 + \epsilon - \Lambda$.

Table 8.1 Frequency response functions are grouped by an approximation theory

	GNN model	Frequency response function
Linear	GCN	$1 - \Lambda$
	GraphSAGE	$2 - \Lambda$
	GIN	$1 + \epsilon + \Lambda$
Polynomial	ChebNet	$\tilde{\theta_0} \cdot 1 + \tilde{\theta_1}\ \Lambda + \tilde{\theta_2}\Lambda^2 + \cdots$
	Node2Vec	$\frac{1}{p} + (1 - \frac{1}{q})\ \Lambda + \frac{1}{q}\ \Lambda^2$
	SGC	$\sum_i^n \begin{pmatrix} K \\ i \end{pmatrix} \Lambda^i$
Rational	AR	$\frac{1}{1+\alpha(1-\Lambda)}$
	PPNP	$\frac{\alpha}{\alpha+(1-\alpha)\ \Lambda}$
	ARMA	$\frac{b}{1-a+a\ \Lambda}$

The above-mentioned methods apply linear low-pass filtering, and the only difference among them is that the bias is different (i.e., 2 for GCN, 2 for GraphSAGE, and $2 + \epsilon$ for GIN). All the examples are rewritten in the spectral domain, and they are listed in Table 8.1.

Polynomial Filter Function

Considering higher order of frequency, filter function g can approximate any smooth filter function, as it is equivalent to applying the polynomial approximation. Therefore, introducing higher order of frequencies boosts the representation power of filter function in simulating spectral signals. Formally, this type of work can be written as

$$\mathrm{Z} = \left(\sum_{i=0}^{l} \sum_{j=0}^{k} \theta_j \lambda_i^j\ \mathrm{u}_i\ \mathrm{u^T}_i \right) \mathrm{X} = \mathrm{U}\,\mathrm{P}_\theta(\Lambda)\mathrm{U^T}\,\mathrm{X}, \tag{8.51}$$

where $\mathrm{g}(\cdot) = \mathrm{P}_\theta(\cdot)$ is a polynomial function.

In theory, the polynomial approximation becomes more accurate as the order increases [176–178, 208, 209]. Note that linear approximation can be treated as a polynomial approximation with an order of 1. We study polynomial approximation on $sign(x)$ function, showing the difference among all the examples listed. As shown in Fig. 8.1, the linear function cannot well approximate $sign(x)$, since it is difficult for any straight line to fit a jump signal. When applying the polynomial approximation, the situation becomes much better. If the order of the polynomial function increases, the variance will significantly be reduced. In sum, the higher order of polynomial approximation is more accurate and yet incurs higher computational complexity. Therefore, Node2Vec with an order of 2 has relatively low approximation power than the others (i.e., ChebNet, DeepWalk, Diffusion CNN,

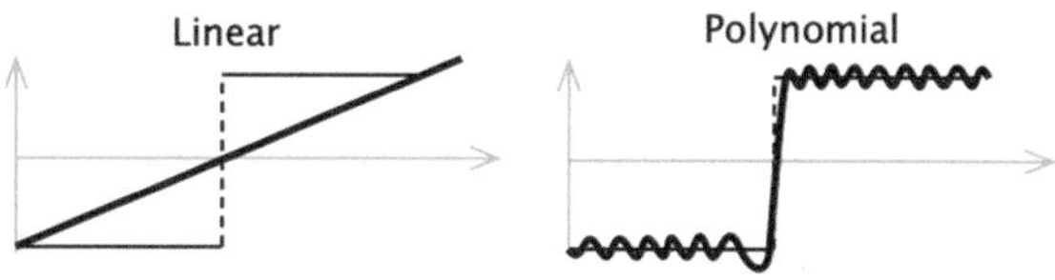

Fig. 8.1 Approximation for $sign(x)$: linear and polynomial approximation

Simple Graph Convolution), since the order of the latter is predefined and could be as large as possible.

Rational Filter Function

Although polynomial approximation is widely used and empirically effective, it only works when applying to a smooth signal in the spectral domain. However, there is no guarantee that any real-world signal is smooth. Therefore, the rational approximation is introduced to improve the accuracy of non-smooth signal modeling. The rational kernel-based method can be written as

$$\mathrm{Z} = \left(\sum_{i}^{l} \frac{\sum_{j=0}^{k} \theta_j \lambda_i^j}{\sum_{m=1}^{n} \phi_m \lambda_i^m + 1} \mathrm{u}_i \, \mathrm{u}^{\mathsf{T}}{}_i \right) \mathrm{X} = \mathrm{U} \frac{\mathrm{P}_\theta(\Lambda)}{\mathrm{Q}_\phi(\Lambda)} \mathrm{U}^{\mathsf{T}} \, \mathrm{X}, \tag{8.52}$$

where $\mathrm{g}(\cdot) = \frac{\mathrm{P}_\theta(\cdot)}{\mathrm{Q}_\phi(\cdot)}$ is the rational function, and P, Q are independent polynomial functions. Spectral methods process graph as a signal in the frequency domain. Existing works include generalized rational filter [200] as shown in Eq. (8.52).

Time Complexity and Expressive Power

Linear filter function has a time complexity of $\mathcal{O}(N^2 F)$ due to the matrix multiplication of A X. Accordingly, polynomial and rational methods are analyzed in Table 8.2, where K is the order number. To compare their expressive powers, the convergence rate on challenging jump signal is employed as a benchmark [200] (the simple signal cannot distinguish them). As shown in Table 8.2, rational filters converge exponentially faster than linear filters, and polynomial filters converge linearly faster than linear filters. Therefore, there is a trade-off between expressive power and computational efficiency. Linear filters have the best efficiency but only capture the linear relationship. Rational filters consume the most considerable overhead but could tackle more challenging signals. Generally, linear or polynomial filters with small order numbers should be first considered, since the time complexity of rational filters is dramatically high.

Table 8.2 Comparison of time complexity and expressive power

	Linear	Polynomial	Rational
Time	$\mathcal{O}(N^2F)$	$\mathcal{O}(N^{K+1}F)$	$\mathcal{O}(N^{K+1}F + N^3)$
Expressivity	$\mathcal{O}(1)$	$\mathcal{O}(1/K)$	$\mathcal{O}(\exp^{-\sqrt{K}})$

8.3.3 Other Graph Neural Networks

Sequential GNN
Combining LSTM [66] and GRU [210], GNN [211–213] improves the robustness of information propagation across the graph topology. As a type of graph, tree structure can be model with graph LSTM [212, 214]. GNN employs reinforcement learning to obtain sequential modeling [215, 216].

Generative Graph Model
By integrating generative models such as generative adversarial net and variational autoencoder, GNN is used to generate new topology given a number of observed structures [217–221]. A comprehensive review about graph generative models has summarized all state-of-the-art works [222].

8.4 Application and Resources

Graph neural networks have been applied in various domains, as listed in several comprehensive surveys on GNNs [223–228]. In this section, we categorize them into several groups:

Physics Graph is a powerful tool to model the physics of objects. DeepMind [229] provides a toolkit, the graph network, to generalize approaches that operate on graphs, including manipulating structured knowledge and producing structured behaviors. Sanchez-Gonzalez et al. [230] simulate complex physics including fluids, rigid solids, and deformable materials with graph neural networks. Ju et al. [231] reconstruct particles in high-energy physics detectors with graph neural networks. Seo et al. [232] propose physics-aware difference graph networks (PA-DGN), utilizing neighboring information to learn finite differences inspired by physics equations. Alet et al. [233] exploit graph neural networks to characterize spatial processes when no prior graphical structure exists.

Chemistry The chemical structure is a natural graph data, and a graph neural network is an appropriate tool to represent this complex connectivity. Duvenaud et al. [234] and Kearnes et al. [235] represent the molecular structure and [236, 237] model protein interfaces. Do et al. [238] and Dai et al. [239] predict the chemical reaction and retrosynthesis.

Computer Vision A point cloud is a group of 3D points scanned by LiDAR, and it can be modeled by graph neural networks [240–242]. Question-specific interactions can be model as graph connectivity in visual question answering [243, 244]. Object and/or human interaction could be modeled by their connection [245, 246, 246–248].

Natural Language Processing Relations among documents and words could be modeled as a graph to infer their properties [249]. A syntactic dependency tree can characterize the syntactic relations inside documents [250, 251]. Modeling semantic structure, GNN enriches the relationship among words and improves the generalization of machine translation [252, 253]. As a typical structural problem, the recommendation system can also model with GNN [254–256].

Traffic Network Predicting traffic flow is a fundamental problem in transportation intelligence. This problem can be modeled as a spatiotemporal graph and integrate the sequential model with GNN to solve the problem [257–260].

Knowledge Graph Schlichtkrull et al. [261], Shang et al. [262], and Nathani et al. [263] model the relationship among entities. Wang et al. [264] and Xu et al. [265] utilize multiple languages to conduct knowledge graph alignment. Balazevic et al. [266] allow multiple relations in one link.

Major benchmarks include citation network: Cora [267], Citeseer [267], PubMed [267], DBLP [267]; Social Networks: Reddit [182], BlogCatalog [268]; Biology/-Chemistry [269–274]; MNIST [275, 276]. Two popular implementations of GNNs are PyTorch Geometric (PyG) [277], which is a geometric deep learning extension library for PyTorch, and Deep Graph Library (DGL) [278], which builds models on PyTorch, TensorFlow, or MXNet. Most commonly used libraries for operations on graph include Networkx[1] and PyGSP.[2]

8.5 Put It All Together

In total, we summarize the advantages and disadvantages of each combination as illustrated in Fig. 8.2. The category selection is based on the data complexity and the efficiency required, as there is a trade-off between computational efficiency and generalization capability.

This subsection illustrates the hierarchical relationship between the spatial and spectral domains. Three categories of spatial-based approaches exist, each with its own specialization and generalization relationship:

LINEAR AGGREGATION ⇄ POLYNOMIAL AGGREGATION ⇄ RATIONAL AGGREGATION,

[1] https://networkx.github.io.

[2] https://github.com/epfl-lts2/pygsp.

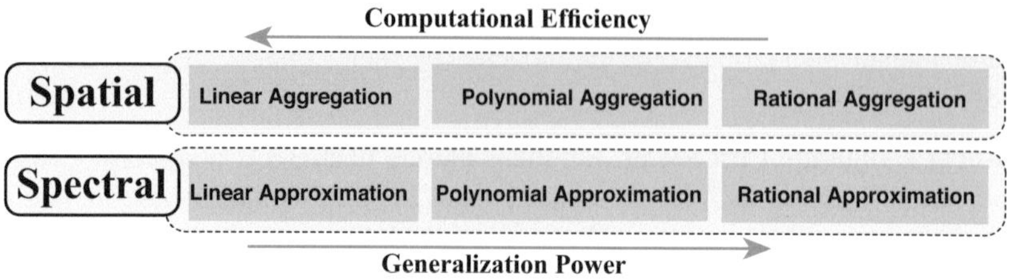

Fig. 8.2 Comparison of different categories of GNNs

where it is a generalization from left to right and specialization from right to left. More precisely, linear aggregation refers to all strategies for learning a linear function among first-order neighbors. Polynomial aggregation includes higher-order neighbors, and the order number is defined by the polynomials. Furthermore, rational aggregation makes use of self-aggregation. Due to the fact that the inclusion of higher-order neighbors transforms linear aggregation into polynomial aggregation, and polynomial aggregation into rational aggregation when self-aggregation is included. The methodologies that fall under the broad heading of spectral analysis can be classified into three basic categories:

LINEAR APPROXIMATION ⇄ POLYNOMIAL APPROXIMATION ⇄ RATIONAL APPROXIMATION,

which, similarly, includes left-to-right generalization and right-to-left specialization.

8.6 Exercise Problems

Problem 8.1 Calculate the inverse, transpose, and trace of the matrix

$$A = \begin{bmatrix} 1 & 2 & 4 \\ 5 & 6 & 4 \\ 4 & 3 & 3 \end{bmatrix}.$$

Problem 8.2 Given

$$A = \begin{bmatrix} 1 & 2 & 4 \\ 5 & 6 & 4 \\ 4 & 3 & 3 \end{bmatrix}, \text{ and } B = \begin{bmatrix} 5 & 2 & 2 \\ 6 & 9 & 4 \\ 1 & 3 & 8 \end{bmatrix}.$$

Calculate $A + B$, AB, $(AB)^{-1}$, $B^{-1}A^{-1}$.

Problem 8.3 Write down the degree matrix D, adjacency matrix A, and Laplacian matrix L of the graph in Fig. 8.3.

Fig. 8.3 Graph structure

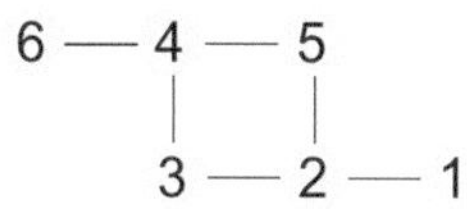

Problem 8.4 Given the graph in Fig. 8.3, calculate three different normalized Laplacian matrices: $\mathrm{D}^{-\frac{1}{2}}\,\mathrm{L}\,\mathrm{D}^{-\frac{1}{2}}$, $\mathrm{D}^{-1}\,\mathrm{L}$, $\mathrm{L}\,\mathrm{D}^{-1}$.

Problem 8.5 Given the adjacency matrix A of the graph in Fig. 8.3, calculate A^2, $\mathrm{A}^\intercal$, $Tr(\mathrm{L})$, $Tr(\mathrm{L}^\intercal)$ and compare A^2 and $\mathrm{A}^{\intercal 2}$, $Tr(\mathrm{L})$, and $Tr(\mathrm{L}^\intercal)$.

Problem 8.6 Given the graph in Fig. 8.3 and its unnormalized Laplacian matrix Ł, perform eigendecomposition on $\mathrm{L} = \mathrm{U}\,\Lambda\,\mathrm{U}^\intercal$ and show the eigenvalues.

Problem 8.7 Given the graph in Fig. 8.3 and its normalized Laplacian matrix $\tilde{\mathrm{L}} = \mathrm{D}^{-\frac{1}{2}}\,\mathrm{L}\,\mathrm{D}^{-\frac{1}{2}}$, perform eigendecomposition on $\mathrm{L} = \mathrm{U}\,\Lambda\,\mathrm{U}^\intercal$ and show the eigenvalues.

Problem 8.8 Given the graph in Fig. 8.3 and its normalized Laplacian matrix $\tilde{\mathrm{L}} = \mathrm{D}^{-1}\,\mathrm{L}$, perform eigendecomposition on $\mathrm{L} = \mathrm{U}\,\Lambda\,\mathrm{U}^\intercal$ and show the eigenvalues.

Problem 8.9 Given the graph in Fig. 8.3 and its normalized adjacency matrix $\tilde{\mathrm{A}} = \mathrm{D}^{-\frac{1}{2}}\,\mathrm{A}\,\mathrm{D}^{-\frac{1}{2}}$, perform eigendecomposition on $\mathrm{L} = \mathrm{U}\,\Lambda\,\mathrm{U}^\intercal$ and show the eigenvalues.

Problem 8.10 Write a program to calculate unnormalized graph Laplacian and normalized graph Laplacian of the graph in Fig. 8.3 and compare their eigendecomposition.

Problem 8.11 (1) Calculate the eigenvalue of the Laplacian matrix for the graph in Fig. 8.4 and find the median of its eigenvalues. (2) Randomly create another bipartite graph and calculate its median eigenvalues. (3) Compare those two median eigenvalues.

Problem 8.12 Assign X_1 and X_2 to the graph Fig. 8.4, respectively:

$$X_1 = \left[1\ 2\ 3\ 2\ 3\ 3\ 5\ 6\ 1\ 3\right]^\intercal,\ X_2 = \left[1\ 2\ 3\ 2\ 3\ 100\ 101\ 99\ 98\ 97\right]^\intercal.$$

In X_1, the values of nodes are close, which can be called a smooth graph signal. In X_2, neighbors of every node are very different from themselves, which is a typical non-smooth graph signal. Calculate the energy of two graph signals with Eq. (8.16).

Problem 8.13 Create graphs that contain two, three, four disconnected subgraphs, respectively. Calculate the multiplicities of their zero eigenvalues.

Problem 8.14 (1) Implement a program to calculate the polynomial and rational approximation with order of 6 for $sign(x)$, $abs(x)$, and x^2 where $x \in [0, 1]$. (2)

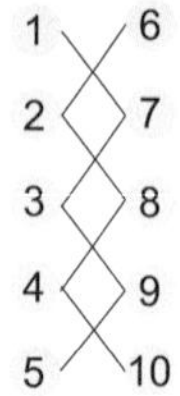

Fig. 8.4 Bipartite graph

Implement a neural network to approximate $sign(x)$, $abs(x)$, and x^2 where $x \in [0, 1]$.

Problem 8.15 Implement ChebNet (orders of 3, 5, and 10) [196] and GCN [181] and compare their efficiency in Cora and Citeseer dataset [267, 279].

Problem 8.16 Label propagation (LP) [201, 280–282] is the most popular method for semi-supervised learning on graphs. Plot the filtering function of LP and tune its parameters. Plot the filtering function of graph convolutional network [181] and compare it with LP.

Chapter 9
Adversarial Machine Learning

9.1 Introduction

Machine learning (ML), especially the deep neural networks (DNNs) and the convolutional neural networks (CNNs) have transformed the processing capabilities of the present-day computing systems. These techniques are widely deployed in different domains ranging from computer vision to hardware security. For instance, autonomous driving is envisaged due to the advancements in the field of ML and computer vision [283–285]. Similarly, ML has made its impact on malware and side-channel attack detection toward securing the computing systems [286–290]. Despite the benefits offered by the advancements in the ML, it has also been exploited for the vulnerabilities in the existing ML techniques.

Though the ML techniques are shown to be robust to the noise, the exposed vulnerabilities have shown that the outcome of the ML can be manipulated by adding crafted perturbations to the input data [291–294], often referred as *Adversarial samples*. These adversarial samples are constructed by perturbing the inputs in one or multiple cycles iteratively under certain constraints in order to amplify the classification error rate.

A simple adversarial sample generated from the MNIST digit dataset [295] for digit "9" is shown in Fig. 9.1. Figure 9.1a is the normal sample that is classified as 9 by the DNN classifier, presented in Sect. 9.3. The images in Fig. 9.1b, c are generated by the fast gradient sign method (FGSM) and Carlini Wagner (CW) attack techniques, respectively. One can observe from Fig. 9.1a, b, and c that the normal and adversarial samples look similar for human observation. It needs to be noted that the noise in Fig. 9.1b and c can be increased or reduced by tuning the parameters of the attack. With the change in attack parameters, the classifier output and its confidence will be modified. More details on generating the adversarial attacks are presented in Sect. 9.2.1, and the details regarding the classifier architecture and the dataset are presented in Sect. 9.3.

S. Rafatirad et al., *Machine Learning for Computer Scientists and Data Analysts*,
https://doi.org/10.1007/978-3-030-96756-7_9

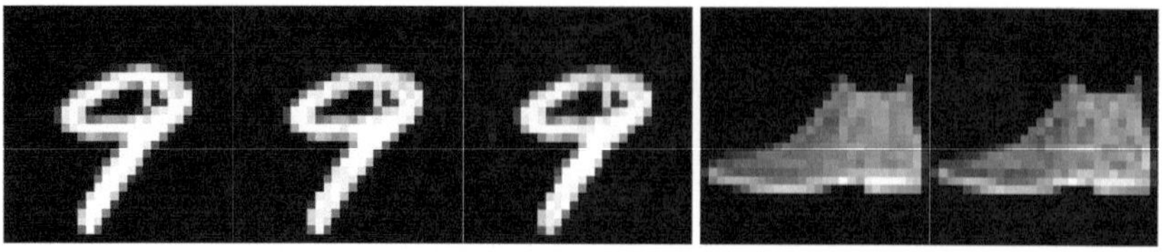

(a) (b) (c) (d) (e)

Fig. 9.1 (**a**) A normal MNIST Digit image classified correctly classified as Digit "9"; (**b**) FGSM generated adversarial sample image in (**a**) classified as Digit "4"; (**c**) CW generated adversarial sample image in (**a**) classified as Digit "4"; (**d**) Normal MNIST Fashion image classified as shoe; (**e**) FGSM generated image of (**e**) classified as sneakers

Though the term *adversarial samples* in the context of ML is introduced in the recent few years, similar concepts date back to 2004 [296] in the context of spam filtering. The work in [297] has shown that the linear classifiers can be fooled by crafted modifications in the content of spam emails to classify them as normal emails. Similar work on biometric recognition fooling is proposed in [298]. The adversarial attacks can be broadly classified into two categories: (a) poisoning attacks and (b) evasion attacks. Poisoning attacks are attacks on the ML classifier during the training phase [299–303], and the evasive attacks are targeted for the inference stage of ML techniques. As the poisoning attacks focus on attacking the classifiers during the training phase it is more suitable for online environments. Thus, this work focuses on the evasive attacks, as many of the existing ML works are primarily offline learning-based and are constrained by resources and the processing time requirements.

In this work, we first provide an overview of evasive attacks on the ML classifiers. Further, we present different existing defense techniques for adversarial attacks. As FGSM is one of the fastest evasive attacks, an in-depth discussion regarding the FGSM adversarial attack is provided. In this work, we look at initially introduced defense against adversarial samples. Adversarial training is one of the defense techniques introduced for adversarial attacks. Adversarial training [304] is similar to a brute force solution, where one generates an ample number of adversaries and trains the classifier to alleviate the impact of adversarial attacks. Though, adversarial training is shown to be confined to be efficient for one or few attacks, it is not always the case. Further, in this work, we show how to efficiently utilize the adversarial training in order to enhance the robustness of the ML classifier even against the recent and powerful adversarial attacks such as CW. Contrarily, we also provide the information regarding under which circumstances this robust adversarial training fails. We show that adversarial training with FGSM can show high robustness to even CW attacks, under certain conditions by having a classification accuracy of up to 97% against adversarial attacks. Having said this, it fails when the number of binary steps as well as the number of iterations is increased, in a nutshell if the attacker has more computational capability.

9.2 Adversarial Attacks and Defenses

Adversarial samples are the samples that are generated by introducing crafted perturbations into the normal input data generated by introducing optimum yet worst-case perturbations in order to make the adversarial data look similar to the normal input data, but still the ML model mispredicts the class with a high probability. These adversarial samples can be considered as an optical illusion for the ML classifiers. In this section, we present different techniques widely used for generating the adversarial samples and review some of the popular defense techniques deployed.

9.2.1 Adversarial Attacks

The adversarial sample is introduced as an optimization problem, mathematically defined as follows:

$$\underset{\epsilon}{\operatorname{argmin}} f(x+\epsilon) = t \quad s.t. \begin{cases} (x+\epsilon) \in D, \\ f(x+\epsilon) \neq f(x). \end{cases} \tag{9.1}$$

In this optimization problem, f is a classifier that maps image pixel vectors x to a discrete k-label set t, i.e., $f : \mathbb{R}^m \rightarrow \{1...k\}$. The goal of this optimization formula is to find the minimum perturbation ϵ, such that by applying it to the original data sample x, the under-attack machine learning model f misclassifies the perturbed sample $x + \epsilon$ as the target class t, $f(x + \epsilon) = t$. The obtained perturbed sample $x + \epsilon$ also needs to remain in the acceptable input domain, i.e., $D \in [0, 1]^m$. In Szegedy et al. [291], this problem was solved using the LBFGS algorithm, which is a second-order gradient method for solving an optimization problem. Although their offered solution is effective, it is a time-extensive process to achieve the adversarial perturbation. We present an overview of different adversarial attacks that are effective against the ML classifiers here.

Fast Gradient Sign Method (FGSM)

The most common adversarial attack technique is to perturb the image with gradient of the loss with respect to the image or input, gradually increasing the magnitude of the perturbation until the image is misclassified.

The Fast Gradient Sign method (FGSM) [292] is one of the first known adversarial attacks. The complexity to generate an FGSM attack is lower compared to other adversarial attacks, even against deep learning models. Some of the advantages of this technique are its low complexity and fast implementation.

Consider an ML classifier model with θ as the parameter, x being the input to the model, and y is the output for a given input x, and $L(\theta, x, y)$ be the cost function used to train the neural networks. Then the perturbation with FGSM is computed as the sign of the model's cost function gradient. The adversarial perturbation generated with FGSM [292] is mathematically given as

$$x^{adv} = x + \epsilon sign(\nabla_x L(\theta, x, y)), \tag{9.2}$$

where ϵ is a scaling constant ranging between 0.0–1.0 is set to be very small such that the variation in x (δx) is undetectable. One can observe that in FGSM the input x is perturbed along each dimension in the direction of gradient by a perturbation magnitude of ϵ. Considering a small ϵ leads to well-disguised adversarial samples. Also, a large ϵ is likely to introduce large perturbations.

Example 9.1 (FGSM Attack)
Problem: What is the effect of changing ϵ of FGSM attack on the accuracy of the under-attack model, on the code snippet 9.2.1?
Solution: By changing the parameter *eps* in the code snippet 9.2.1, we can investigate the effect of changing the ϵ of FGSM attack. We selected 7 different ϵs in the range of [0.0,0.30]. The image on the left shows that by increasing ϵ the accuracy drops dramatically. The images on the right show some samples of MNIST dataset which FGSM attack has applied on them with different parameters ϵ. On top of each image, there is two number in the format "n -> m" that mean the actual image has the label n and after the FGSM attack, the model classifies that sample as a class with label m.

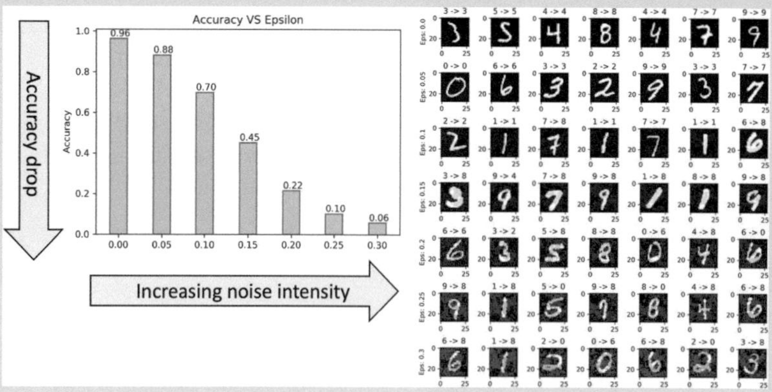

The code snippet for the targeted and non-targeted FGSM attack has been shown at the Listing below. At lines 11 to 15 of this piece of code, the input image is prepared in order that the gradients of the model are reflected in the input image.

After this part, based on the goal of the attack, i.e., targeted or not-targeted, in lines 17 to 28, the obtained gradient is applied to the input image.

```
1
2 def attack(model, criterion, img, label, eps, targeted=False):
3     '''
4     model: the model under attack
5     criterion: the optimization function
6     img: the input image to the model
7     label: the label corresponding to img
8     eps: intensity of the applied perturbations
9     targeted: type of FGSM attack
10     '''
11     adv = img.detach()
12     adv.requires_grad = True
13
14     iterations = 1
15     step = eps
16
17     for j in range(iterations):
18         out_dev = model(adv.clone())
19         loss = criterion(out_adv, label)
20         loss.backward()
21
22         noise = adv.grad
23         if targeted == False:
24             adv.data = adv.data + step * noise.sign()
25         else:
26             adv.data = adv.data - step * noise.sign()
27         adv.data.clamp_(0.0, 1.0)
28         adv.grad.data.zero_()
29
30     return adv.detach()
```

Note that, unlike the LBFGS, FGSM is very fast and effective. The downside of FGSM is that it perturbs all the input pixels (features) for obtaining the adversarial example. But, it has been shown that a subset of input pixels can be found that has a similar effect and at the same time lead to a more imperceptible adversarial perturbation.

Basic Iterative Method (BIM)

As seen previously, FGSM adds perturbation in each of the dimensions; however, no optimization on perturbations is performed. Kurakin proposed an iterative version of FGSM, called as Basic iterative method (BIM) in [305]. BIM is an extension of FGSM technique, where instead of applying the adversarial perturbation once with ϵ, the perturbation is applied multiple times iteratively with small ϵ. This produces a recursive noise on the input and optimized application of noise, given mathematically as follows:

$$\begin{aligned} x_0^{adv} &= x \\ x_{N+1}^{adv} &= Clip_{x,\epsilon}(x_N^{adv} + \epsilon sign(\nabla_x L(\theta, x_N^{adv}, y)). \end{aligned} \tag{9.3}$$

In the above expression, $Clip_{x,\epsilon}$ represents the clipping of the adversarial input magnitudes such that they are within the neighborhood of the original sample x. This technique allows more freedom for the attack compared to the FSGM method, as the perturbation can be controlled and the distance of the adversarial sample from the classification boundary can be carefully fine-tuned. The simulations in [305] have shown that BIM can cause higher misclassifications compared to the FGSM attack on the Imagenet samples.

Example 9.2 (BIM Attack)
Problem: What is the effect of changing ϵ of BIM attack on the accuracy of the under-attack model, on the code snippet 9.2.1?
Solution: By changing the parameter *eps* in the code snippet 9.2.1, we can investigate the effect of changing the ϵ of BIM attack. We selected 7 different ϵs in the range of [0.0,0.30]. The adversarial perturbation remains imperceptible to the human eyes but causes the neural network to misclassify the input image.

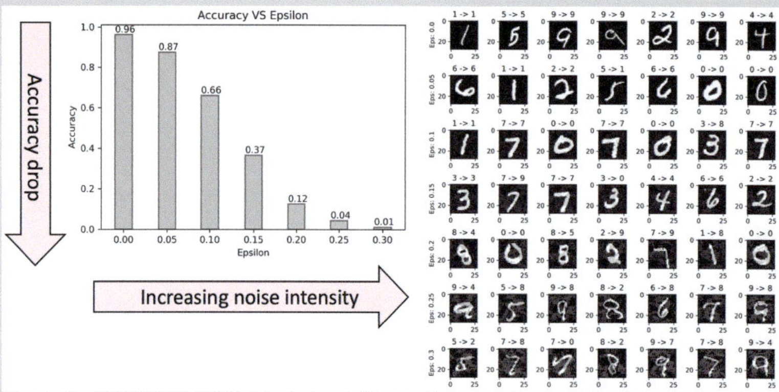

The code snippet for the targeted and non-targeted BIM attack has been shown at Listing 9.2.1. At lines 11–15 of this piece of code, the input image is prepared in order that the gradients of the model are reflected in the input image. After this part, based on the goal of the attack, i.e., targeted or not-targeted, in lines 17–28, the obtained gradient is applied to the input image.

```
1
2 def attack(model, criterion, img, label, eps, targeted=False):
3     '''
4     model: the model under attack
5     criterion: the optimization function
6     img: the input image to the model
7     label: the label corresponding to img
8     eps: intensity of the applied perturbations
9     targeted: type of BIM attack
10    '''
```

```
    adv = img.detach()
    adv.requires_grad = True

    iterations = 256
    step = eps / iterations

    for j in range(iterations):
        out_dev = model(adv.clone())
        loss = criterion(out_adv, label)
        loss.backward()

        noise = adv.grad
        if (targeted == False):
            adv.data = adv.data + step * noise.sign()
        else:
            adv.data = adv.data - step * noise.sign()
        adv.data.clamp_(0.0, 1.0)
        adv.grad.data.zero_()

        return adv.detach()
```

Momentum Iterative Method (MIM)

The momentum method is an accelerated gradient descent technique that accumulates the velocity vector in the direction of the gradient of the loss function across multiple iterations. In this technique, the previous gradients are stored, which aids in navigating through narrow valleys of the model, and alleviate problems of getting stuck at local minima or maxima. This momentum method also shows its effectiveness in stochastic gradient descent (SGD) to stabilize the updates. This MIM principle is deployed in [306] to generate adversarial samples. MIM has shown a better transferability and shown to be effective compared to FGSM attack.

In the Momentum Iterative Method (MIM) [307], the momentum is also considered when calculating the adversary perturbation and is mathematically expressed as

$$\begin{aligned} g_0 = 0,\ g_n = \mu g_{n-1} + \frac{\nabla_x J(\theta, x_{n-1}, y)}{||\nabla_x J(\theta, x_{n-1}, y)||_1} \\ x'_n = x'_{n-1} + \epsilon sign(g_n) \end{aligned} \tag{9.4}$$

in which μ is the momentum, and $||\nabla_x J(\theta, x_{n-1}, y)||_1$ is the L_1 norm of the gradient, and g_n is the momentum gradient. Similar to BIM, the Momentum Iterative Method changes all the input pixels based on the sign of the gradient at each iteration. However, at MIM, the momentum term prevents the abrupt change of the gradient sign, and consequently, an adversarial perturbation can be obtained using fewer iterations compared to the BIM.

Example 9.3 (MIM Attack)
Problem: What is the effect of changing ϵ of MIM attack on the accuracy of the under-attack model, on the code snippet 9.2.1?
Solution: By changing the parameter *eps* in the code snippet 9.2.1, we can investigate the effect of changing the ϵ of BIM attack. We selected 7 different ϵs in the range of [0.0,0.30]. By increasing the intensity of attack the performance of attack drops dramatically. The adversarial perturbation remains imperceptible to the human eyes but causes the neural network to misclassify the input image.

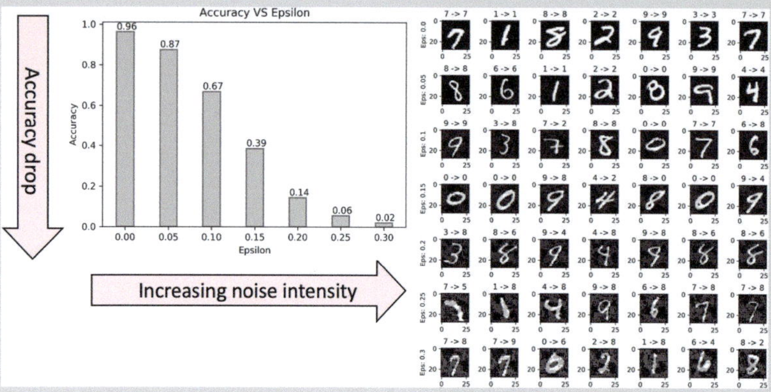

The code snippet for the targeted and non-targeted MIM attack has been shown below. At lines 12–17 of this piece of code, the input image is prepared in order that the gradients of the model are reflected in the input image. After this part, based on the goal of the attack, i.e., targeted or not-targeted, in lines 19–35, the obtained gradient is applied to the input image.

```
1
2  def attack(model, criterion, img, label, eps, u, targeted=False):
3      '''
4      model: the model under attack
5      criterion: the optimization function
6      img: the input image to the model
7      label: the label corresponding to img
8      eps: intensity of the applied perturbations
9      u: momentum parameter
10     targeted: type of MIM attack
11     '''
12     adv = img.detach()
13     adv.requires_grad = True
14
15     iterations = 256
16     step = eps / iterations
17     noise = 0
18
19     for j in range(iterations):
20         out_dev = model(adv.clone())
21         loss = criterion(out_adv, label)
```

```
        loss.backward()

        adv_mean= torch.mean(torch.abs(adv.grad), dim=1)
        adv_mean= torch.mean(torch.abs(adv_mean), dim=2)
        adv_mean= torch.mean(torch.abs(adv_mean), dim=3)
        adv.grad = adv.grad / adv_mean
        noise = u * noise + adv.grad

        if (targeted == False):
            adv.data = adv.data + step * noise.sign()
        else:
            adv.data = adv.data - step * noise.sign()
        adv.data.clamp_(0.0, 1.0)
        adv.grad.data.zero_()

    return adv.detach()
```

Projected Gradient Descent Attack

Projected Gradient Descent (PGD) [308] is one of the strongest first-order iterative attacks, which is similar to BIM, but with the difference that it forces the obtained adversarial example to stay within a γ-neighborhood of input sample x at each iteration. The objective function of PGD is defined as

$$\begin{aligned} x'_0 = x,\; x'_n &= x'_{n-1} + \epsilon sign(\nabla_x J(\theta, x_{n-1}, y)) \\ x'_n &= clip(x'_n, x'_n - \gamma, x'_n + \gamma). \end{aligned} \tag{9.5}$$

Example 9.4 (PGD Attack)
Problem: What is the effect of changing ϵ of PGD attack on the accuracy of the under-attack model, on the code snippet 9.2.1?
Solution: By changing the parameter *eps* in the code snippet 9.2.1, we can investigate the effect of changing the ϵ of BIM attack. We selected 7 different ϵs in the range of [0.0,0.30]. By increasing the intensity of attack the performance of attack drops dramatically. The adversarial perturbation remains imperceptible to the human eyes but causes the neural network to misclassify the input image.

(continued)

Example 9.4 (continued)

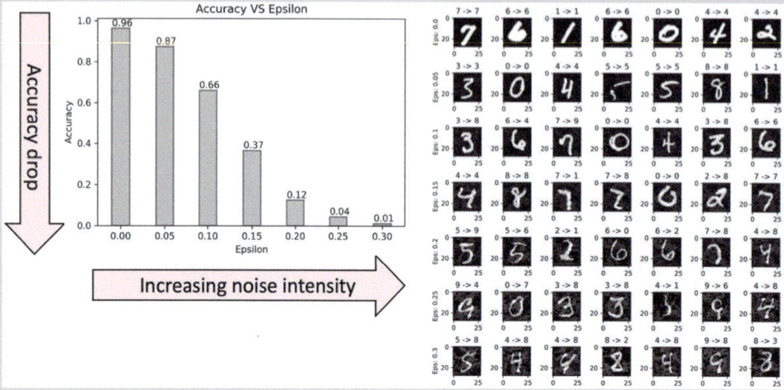

The code snippet for the targeted and non-targeted PGD attack has been shown in the following listing. At lines 11–15 of this piece of code, the input image is prepared in order that the gradients of the model are reflected in the input image. After this part, based on the goal of the attack, i.e., targeted or not-targeted, in lines 17–36, the obtained gradient is applied to the input image.

```
1
2  def attack(model, criterion, img, label, eps, targeted=False):
3      '''
4      model: the model under attack
5      criterion: the optimization function
6      img: the input image to the model
7      label: the label corresponding to img
8      eps: vicinity diameter
9      targeted: type of PGD attack
10     '''
11     adv = img.detach()
12     adv.requires_grad = True
13
14     iterations = 256
15     step = 0.01
16
17     for j in range(iterations):
18         out_dev = model(adv.clone())
19         loss = criterion(out_adv, label)
20         loss.backward()
21
22         noise = adv.grad
23
24         if (targeted == False):
25             adv.data = adv.data + step * noise.sign()
26         else:
27             adv.data = adv.data - step * noise.sign()
28
29         adv.data = torch.where(adv.data > img.data + eps,
30                                img.data + eps,
31                                adv.data)
32         adv.data = torch.where(adv.data < img.data - eps,
33                                img.data - eps,
34                                adv.data)
```

```
            adv.data.clamp_(0.0, 1.0)
            adv.grad.data.zero_()

        return adv.detach()
```

Jacobian-Based Saliency Map Attack (JSMA)

In contrast to application of noise in each of the directions, Papernot in [309] proposed a simple iterative method where the forward derivative of DNN is exploited for adding the perturbations. Consider a neural network F with input x, and the output of class j, denoted by $F_j(x)$. The main principle of this work is: In order to achieve a target class t, the probability for $F_t(X)$ must be increased, on the other hand the probabilities of $F_j(X)$ for all the other classes $j \neq t$ have to be decreased, until $t = arg\,max_j F_j(X)$ is achieved. This is a targeted attack; however, it can be used as an untargeted attack as well. This is accomplished by exploiting the saliency map, as defined below

$$S(X,t)[i] = \begin{cases} 0, \text{ if} \frac{\partial F_t(X)}{\partial X_i} < 0 \text{ or} \sum_{j \neq t} \frac{\partial F_j(X)}{\partial X_i} > 0 \\ \left(\frac{\partial F_t(X)}{\partial X_i}\right) | \sum_{j \neq t} \frac{\partial F_j(X)}{\partial X_i} |, \text{ otherwise.} \end{cases} \tag{9.6}$$

For an input feature i starting with the normal input x, we determine the pair of features $\{i, j\}$ that maximizes $S(X,t)[i] + S(X,t)[j]$ and perturb each of the features by a constant offset ϵ. This process is repeated iteratively until the target misclassification is achieved.

DeepFool Attack

DeepFool (DF) is an untargeted adversarial attack optimized for L_2 norm, introduced in [310]. DF is efficient and produces adversarial samples that are more similar to the original inputs as compared to the discussed adversarial samples generated by FGSM and BIM attacks. The principle of the Deepfool attack is to assume neural networks as completely linear with a hyper-plane separating each class from another. Based on this assumption, an optimal solution to this simplified problem is derived to construct adversarial samples. As the neural networks are non-linear in reality, the same process is repeated considering the non-linearity into the model. This process is repeated multiple times for creating the adversaries. This process is terminated when an adversarial sample is found, i.e., misclassification happens. Considering the brevity and focus of the current work, we limit the details in this draft. However, the interested readers can refer to the work in [310] for exact formulation of DF.

Carlini and Wagner Attack (CW)

One of the most recent adversarial attacks is introduced by Carlini and Wagner in [311], popularly called as Carlini and Wagner (CW) attack. The CW attack is shown to outperform adversarial defense techniques such as defensive distillation. It is an iterative attack that finds adversarial samples against multiple defenses as compared to other attacks. At a high level, this attack is iterative using Adam optimizer and a specially chosen loss function to find adversarial examples with lower distortions than the other attack. This comes at the cost of speed as this attack is much slower than the other attacks. It encompasses a range of attacks based on the norms, all cast through the same optimization framework, thus resulting in 3 powerful attacks, that are designed employing L_0, L_2, and L_∞ norms. For the L_2 attack, which is considered in this work, the perturbation in the input, i.e., δ is defined in terms of an auxiliary variable ω. The objective of the CW attack with L_2 norm can be mathematically defined as

$$\delta_i^* = \frac{1}{2}(\tanh(\omega_i + 1)) - x_i. \tag{9.7}$$

Then, the δ_i^* that is an unrestricted perturbation is optimized over ω as follows:

$$\min_{\omega} ||\frac{1}{2}(\tanh(\omega) + 1) - x||_2^2 + cf\Big(\frac{1}{2}\tanh(\omega) + 1\Big). \tag{9.8}$$

Similarly, if the L_2 is considered, the optimization becomes

$$\min_{\delta} ||\delta||_2 + c \cdot f(x + \delta) \tag{9.9}$$

$$S.T. x + \delta \in [0, 1]^n, \tag{9.10}$$

where f (objective function) is defined as

$$f(x') = \max(\max\{Z(x') : i \neq t\} - Z(x') - k). \tag{9.11}$$

Here, $Z(x')$ is the pre-softmax output for class i, t is the target class, and k is the parameter that controls the confidence with which the misclassification occurs. The parameter k encourages the solver to find an adversarial instance x' that will be classified as class t with high confidence. The three variants of this attack were shown to be quite effective in comparison to other attacks on a network trained with defensive distillation [311].

One-Pixel Attack

Unlike the previous attacks where the gradient of the underlying model was needed, in the one-pixel attack [312] they used an evolutionary algorithm so called differential evolution (DE) in order to search in the input domain of the under-attack model for finding one pixel that can turn the input image into an adversarial example. The attack is carried out as follows:

First, generate several adversarial samples by modifying a random pixel and run the images through the neural network. Next, combine the previous pixels' positions and colors together, generate several more adversarial samples from them, and run the new images through the neural network. If there were pixels that lowered the confidence of the network from the last step, replace them with the current best-known solutions. Repeat these steps for a few iterations, then on the last step return the adversarial image that reduced the network's confidence the most. If successful, the confidence would be reduced so much that a new (incorrect) category now has the highest classification confidence.

Universal Perturbation

Till now, all the previously described adversarial attacks find an adversarial perturbation that is tailored for a specific input sample. Meaning, adversarial perturbation on the input sample x_1 may not be also an adversarial perturbation for the input sample x_2. So, for each input sample x_i, the adversarial attack needs to be run from scratch. However, in 2017, Dezfooli et al. in [313] debuted an attack for building an universal adversarial perturbation. In the proposed solution, the optimization process synthesizes a perturbation that is universal, meaning it could be added to any image of any class and significantly increase the chances of the model's misclassification.

9.2.2 *Adversarial Defenses*

Several works have investigated defense mechanisms against adversarial attacks. One of the preliminary works on adversarial defenses is proposed in [292], termed as adversarial training proposed to enhance the robustness of the model based on the training. In [314, 315], autoencoders are employed to remove the adversarial perturbations and reconstruct a perturbation with less input. In [309] distillation is used to hide the gradients of the network from the attacker. Other approaches such as [316–318] are introduced to defend adversarial samples. Here, we review the existing defenses against adversarial examples.

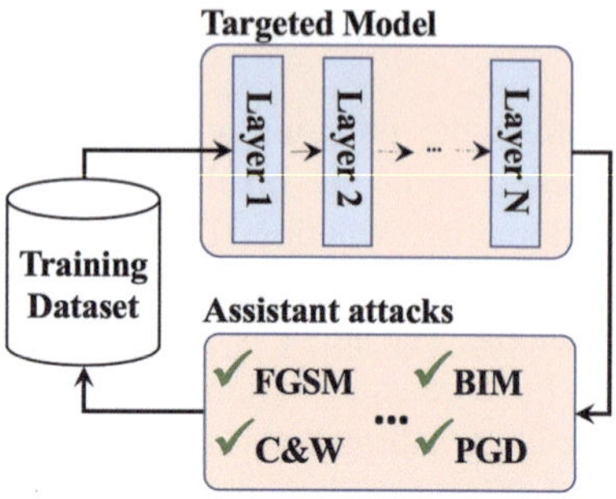

Fig. 9.2 Adversarial training in a nutshell

Adversarial Training

Adversarial training is one of the preliminary solutions for making the ML classifiers robust against the adversarial examples, proposed in [304]. The preliminary idea is to train the ML classifier with the adversarial examples so that the ML classifier can have adversarial information [291, 292, 310] and adapt its model based on the learned adversarial data. One of the major drawbacks of this technique is to determine what kind of attack is going to happen and train the classifier based on those attacks and determine the criticality of the adversarial component.

This idea of adversarial training is shown in Fig. 9.2, in which a targeted model is being hardened through an iterative procedure. At each iteration, the target model is being trained based on the training dataset, and then different attacks are applied to the model and the extracted adversarial examples are added to the training dataset. This procedure continues till reaching an acceptable level of robustness.

The process of adversarial training using two attacks FGSM and PGD has shown in the code snippet below. In adversarial training, the size of each batch expands to an extent that is defined by the number of helper attacks. For example, in this code snippet, that two attacks FGSM and PGD have been utilized, for each batch we have added the corresponding adversarial samples to the fetched batch using FGSM and PGD. Subsequently, the model proceeds with the training procedure using the dataset that has a combination of clean data, adversarial samples using FGSM, and adversarial samples using PGD.

```

def adv_training(model, loader, optimizer, criterion, device):
    '''
    model: the model under attack
    loader: training/test dataset
    optimizer: the algorithm for updating the trainable params
    criterion: the used loss function
    device: CPU or GPU
    '''
    for (x, y) in loader:
        x = x.to(device)
        y = y.to(device)

        FGSM_adv = FGSM_attack(model, criterion, x, y, eps=0.1)
        x = torch.cat((x, FGSM_adv), 0)
        y = torch.cat((y, y))
```

```
        PGD_adv = PGD_attack(model, criterion, x, y, eps=0.1)
        x = torch.cat((x, PGD_adv), 0)
        y = torch.cat((y, y))

        optimizer.zero_grad()
        fy = model(x)
        loss = criterion(fy, y)
        loss.backward()
        optimizer.step()
```

Although this method is simple and effective, it has two drawbacks: (1) It can only make the model robust against the assistant attacks (see Fig. 9.2); (2) it also increases the training time significantly.

Defensive Distillation

In [309], distilling was originally proposed to train a smaller student model from a larger teacher model with the objective that the smaller network predicts the probability similar to that of the bigger network. The distillation technique takes advantage of the fact that a probability vector contains more information than class labels, hence, it is a more effective means for training a smaller network. For defensive distillation, the second network is the same size as the first network [309]. An abstract view of this method is shown in Fig. 9.3, in which the main idea is to hide the gradients between the pre-softmax and softmax layers to make the attacker's job more difficult. However, it was illustrated in [311], that this defense can be broken by using the pre-softmax layer outputs in the attack algorithm and/or choosing a different loss function.

In [319], intelligent JPEG compression is suggested for feature distillation. JPEG is a form of lossy compression based on the fact that human eyes are more sensitive to low-frequency components than high-frequency ones. Liu et al. [319] uses JPEG compression in a way that removes the noise and also adversarial perturbation. Comdefend [320] suggests a reconstruction module based on two neural networks.

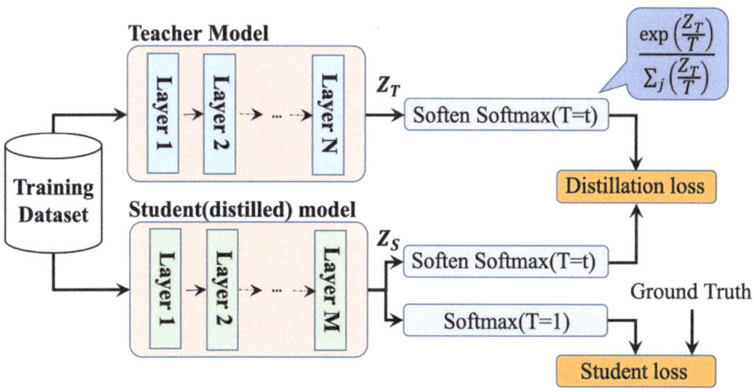

Fig. 9.3 Defensive distillation in a nutshell

The first neural network compresses the image, and the second one reconstructs the image. However, the image compression does not solve the robustness issue of the neural network.

Random Ensemble

In [318] an approach is presented to switch between random modules during inference to protect the neural network against adversarial attacks. Hierarchical Random Switching divides the base model to several random blocks of parallel channels. The architectures of all blocks are similar, but they have been trained to have different weights. In the test time, the network randomly switches between blocks to make the network more robust to adversarial attacks [318].

Gradient Regularization

Input gradient regularization was first introduced in [321] to improve the generalization of training in neural networks by a double backpropagation method. The work in [309] mentions the double backpropagation as a defense and [316] evaluates the effectiveness of this idea to train a more robust neural network. This approach intends to make sure that if there is a small change in the input, the change in Kullback–Leibler (KL) divergence between the predictions and the labels also will be small. However, this approach is sub-optimal because of the blindness of the gradient regulation.

MagNet

MagNet is proposed in [315], where a two-level strategy with detector and reformer is proposed. In the detector phase(s), the system learns to differentiate between normal and adversarial examples by approximating the manifold of the normal examples. This is performed with the aid of autoencoders. Further, in the reformer, the adversarial samples are moved close the manifold of normal samples with small perturbations. Further using the diversity metric, the MagNet can differentiate the normal and adversarial samples. MagNet is evaluated against different adversarial attacks presented previously and has shown to be robust in [315].

Detecting Adversaries

Another idea of defense proposed in the existing works is to detect adversarial examples with the aid of statistical features [322] or separate classification networks [323]. In [323], for each adversarial technique, a DNN classifier is built to classify whether the input is a normal sample or an adversary. The detector was directly

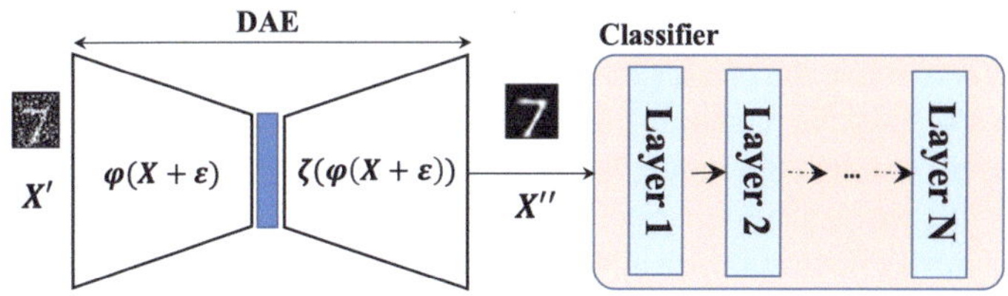

Fig. 9.4 Autoencoders for defense

Table 9.1 Comparison of defense methods

Defense	Disadvantage
Adversarial training [292, 307, 308, 310]	Prone to unseen adversarial examples
Defensive distillation [309]	Beaten by attacks that use the pre-softmax layer
Random ensemble [318]	Computational complexity
Gradient regularization [316]	Sub-optimal due to blindness of gradient regulation
Image compression [319, 320]	Does not solve the robustness issue of the classifier
Adversarial detection [317, 325]	Not effective for large size inputs
Autoencoders [314, 315]	Computational overhead

trained on both normal and adversarial examples. The detector showed good performance when the training and testing attack examples were generated from the same process and the perturbation is large enough. However, it does not generalize well across different attack parameters and attack generation processes.

Autoencoders

Authors in [314] analyze the use of normal and denoising autoencoders as a defense method against adversarial samples. Autoencoders are neural networks that code the input and then try to reconstruct the original image as their output. Meng and Chen [315], as illustrated in Fig. 9.4, use a two-level module and uses autoencoders to detect and reform adversarial images before feeding them to the target classifier. However, this method may change the clean images and also add a computational overhead to the whole defense-classifier module. To improve the method introduced in [315, 324] presents an efficient autoencoder with a new loss function, which was learned to preserve the local neighborhood structure on the data manifold.

In essence, the existence of adversarial examples is due to the lack of adequate generality of the underlying model. Among all of the mentioned defenses, the adversarial training increases the generality of the model, while other existing defenses consider adversarial examples as abnormal data samples that can be detected or removed. Thus, these defenses either are not effective for all of the attack scenarios, like white box or black box, or are not effective for all the attack types such as FGSM and PGD. Despite being effective, adversarial training requires large training dataset and is confined to trained attacks. In contrast, the proposed technique, Gravity, addresses these issues by increasing the generality of the model. This idea is to move the classes to be far apart, i.e., decreasing the overlap between the latent spaces of the model. In this manner, not only the accuracy of the model can be improved, but also the model learns a simplified decision boundary that in turn enhances the generalization capacity. And at the same time, unlike the adversarial training, Gravity does not need a large dataset of all available adversarial examples in order to increase the generality of the model. A comparison of all the adversarial defenses is outlined in Table 9.1.

9.3 Experimental Results

The impact of adversarial training on different attacks is analyzed here. We evaluated the accuracy on the MNIST Digit [295] and MNIST Fashion [326] datasets. The adversarial attacks are generated using Cleverhans library [327]. The source code to reproduce the experiments presented in this work can be found on github at the URL found at bottom of this page.[1]

9.3.1 Network Architecture

We used the ML classifier, i.e., DNN architecture similar to the [328] for classifying the MNIST Digits dataset. The MNIST dataset comprises 60,000 examples for training and 10,000 examples for testing. On a normal classifier, we achieve an accuracy of 98.15% on MNIST Digits dataset and 89.36% on MNIST fashion dataset with the employed classifier architecture, which are close to the state-of-the-art results. More details on network architecture and configuration are presented in Tables 9.2 and 9.3, respectively. For generating the adversarial attacks, we have employed the L_2 norms, and the most non-trivial parameters influencing the accuracy are reported in Table 9.4.

9.3.2 Performance with Adversarial Attacks

Table 9.4 reports the performance of the employed neural network on MNIST Digits dataset.

Normal Classification Accuracy In the absence of adversarial samples, the classifier achieves an accuracy of 98.15%, loss of 0.088, precision of 0.98, and recall of

[1] https://github.com/saimanojpd/ICCAD-18_Adversarial_Training.git.

Table 9.2 Architectural details of employed ML classifier (DNN)

Parameter	MNIST Digits	MNIST Fashion
Input	28×28	28×28
# hidden layers	2	3
Input layer	784 neurons	784 neurons
Hidden layer 1 (ReLu)	512 neurons	512 neurons
Dropout	0.2	0.2
Hidden layer 2 (ReLu)	512 neurons	512 neurons
Dropout	0.2	0.2
Hidden layer 3 (ReLu)		512 neurons
Dropout		0.2
Output layer (Softmax)	10 neurons	

Table 9.3 Training parameters of the employed classifiers

Parameter	MNIST Digits	MNIST Fashion
Optimization method	ADAM	ADAM
Batch size	128	128
Epochs	20	20
Learning rate	0.001	0.001
Loss	Cross-entropy	Cross-entropy

Table 9.4 Accuracy of the classifier after and before adversarial attacks

Attack	Parameter	No attack	With attack
FGSM	$\epsilon = 0.3$	98.15%	6.59%
	$\epsilon = 0.5$	98.15%	3.09%
BIM	$\epsilon = 0.3$	98.15%	1.41%
	$\epsilon = 0.5$	98.15%	1.33%
MIM	$\epsilon = 0.3$	98.15%	1.46%
	$\epsilon = 0.6$	98.15%	1.29%
JSMA	$\theta = 0.1, \gamma = 1$	98.15%	3.60%
	$\theta = 1, \gamma = 1$	98.15%	2.26%
DF	MI[a] = 10	98.15%	1.36%
	MI[a] = 100	98.15%	1.29%
CW	BS[b] = 10, MI[a] = 300	98.15%	4.32%
	BS[b] = 5, MI[a] = 1000	98.15%	1.41%

[a]Maximum iterations
[b]Binary step

0.98. Similarly, for MNIST Fashion, the classifier achieves an accuracy of 89.36%, loss of 0.3144, precision of 0.89, and recall of 0.89.

Effect of Adversaries The adversarial samples generated from the discussed adversarial attacks are shown in Fig. 9.5. As one can observe the adversarial samples generated with FGSM, MIM, and BIM look alike and the adversarial samples from JSMA, DF, and CW look more? alike. It needs to be noted that the digit "4" is classified as "9" in all the cases. The noise in each of them can be altered, which leads to differences in the confidence of output.

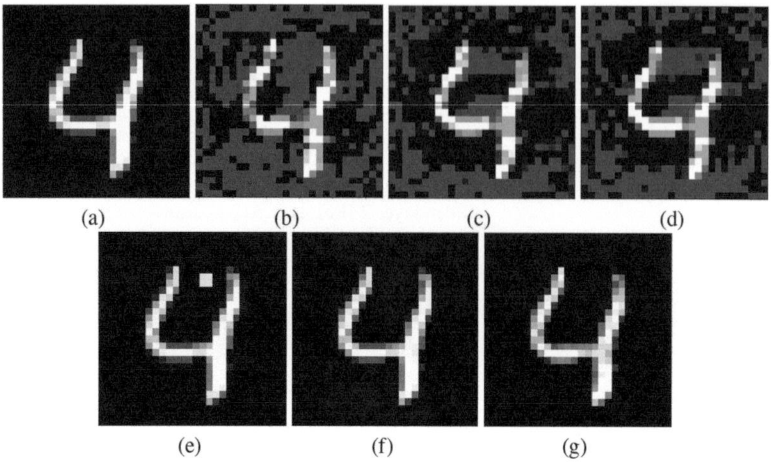

Fig. 9.5 One of the normal MNIST Digit image is represented in (**a**), followed by generated adversarial images generated from (**b**) FGSM; (**c**) BIM; (**d**) MIM; (**e**) JSMA; (**f**) DF; and (**g**) CW attacks

Table 9.4 shows the accuracy of the classifier in the presence of difference attacks. The number of adversarial samples is 10,000 in each case, and one can observe that in the presence of adversaries the classification accuracy falls to as low as 1.3%. With the increase in ϵ, the accuracy decreases in the case of FGSM, MIM, and BIM. With the number of iterations, the accuracy decreases for DF and CW attacks. The step size for each attack iteration ϵ_{iter} is set to 0.06 in the simulations. For the FGSM, with the increase in the θ, γ the attack can hamper the classification accuracy of a neural network classifier.

9.3.3 Effective Adversarial Training

As seen from Fig. 9.1 and Sect. 9.2.1, the FGSM samples are generated by perturbing almost all the pixels in the original input. As such, the other attacks can be seen as selective tweaking of the FGSM. Thus, the adversarial training with FGSM can enhance the robustness of the classifier. However, the perturbations based on correlations and optimization might not be fully captured in FGSM samples, as there is no specific optimization involved. Here, we analyze the effect of adversarial training when the classifier is trained with the samples generated by different attacks. We consider six different attacks presented in Sect. 9.2.1 for adversarial training.

Table 9.5 Accuracy (%) for MNIST digit classification under different adversarial attacks on different adversarial trained networks

	BIM			MIM			FGSM		
	$\epsilon = 0.3$	$\epsilon = 0.5$	$\epsilon = 0.9$	$\epsilon = 0.3$	$\epsilon = 0.5$	$\epsilon = 0.9$	$\epsilon = 0.3$	$\epsilon = 0.5$	$\epsilon = 0.9$
Adv. training with									
FGSM	96.38	89.63	48.69	97.35	94.84	60.13			
CW	62.85	44.34	34.12	63.61	45.29	30.80	51.80	32.60	26.59
JSMA	9.86	1.28	0.59	8.06	1.62	0.64	23.83	13.50	10.18
DF	53.40	31.68	25.68	54.25	32.13	22.92	41.04	24.95	18.94
MIM	99.86	97.22	71.52				87.60	61.69	40.31
BIM				99.17	91.99	76.00	84.45	53.46	34.33

	DF			JSMA (θ, γ)			CW (BS, MI)		
	MI = 50	MI = 100	MI = 10	(1, 1)	(0.9, 0.9)	(1, 0.1)	(5, 1000)	(9, 200)	(10, 300)
FGSM	97.33	90.66	90.66	81.47	86.15	92.25	97.65	88.34	86.75
CW	99.70	44.74	44.74	72.36	80.83	92.27			
JSMA	92.40	6.22	6.22				93.20	88.70	85.40
DF				73.43	82.29	89.40	99.75	92.15	90.12
MIM	98.07	44.33	44.33	78.55	85.91	92.68	98.28	88.39	85.32
BIM	97.58	47.65	47.65	78.67	83.98	91.16	97.86	87.79	84.36

Performance Evaluation and Comparison

Table 9.5 presents the performance (accuracy) of the employed classifier (DNN) when trained with adversarial samples generated from different attacks and tested with all the six attacks for the MNIST Digits dataset. For instance, the row with `FGSM` indicates that the classifier is trained with adversarial samples generated by FGSM attack. The classifier is provided with adversarial samples generated with the attacks mentioned in the top row of Table 9.5. As the training and testing with same kind of attacks have shown accuracies of nearly 99%, we have not reported them in the Table to avoid confusion and wrong analysis. The following observations can be made from the reported results in Table 9.5:

- FGSM, MIM, and BIM based adversarial training achieves good classification accuracy even when tested with attacks such as CW and DF.
- However, the FGSM based adversarial training outperforms MIM and BIM. For instance, with DF attack, only FGSM based adversarial training achieves higher accuracy compared to MIM and BIM.
- The classifier trained with CW/JSMA/DF performs better compared to normal classifier when attacked with any of the CW/JSM/DF attacks. However, the samples generated by the FGSM, MIM, and BIM still keeps the misclassification rate high.

Based on these observations and Fig. 9.5, one can notice that the FGSM, MIM, and BIM have similar characteristics. Also, FGSM based adversarial training outperforms the others and can aid in achieving robustness (to some extent) even against the most advanced (unseen) attacks such as CW and DF. On the other hand, CW, JSMA, and DF have shown similar performances and trends. As such, when trained with one of them, it can aid to achieve robustness against the other two, compared to no defense classifiers.

In addition to MNIST Digits, we have also tested with MNIST Fashion dataset. The deployed DNN achieves an accuracy of 88%, whereas state-of-the-art network [329] achieves an accuracy of 96%, at max. However, the major intention of the work is not on performance improvement, rather on adversarial analysis. We have performed adversarial training on the MNIST Fashion dataset as well. It has shown to follow similar trend as what is observed with MNIST Digits test case.

A glance at the results is presented below:

- With the FGSM based adversarial training, the adversarial training achieves accuracies of 80%, 84%, and 81% when the number of iterations of DF is kept 50, 100, and 10 respectively. Similar trends are obtained when tested with CW and JSMA.
- On the other hand, when the adversarial training is performed with DF and tested with FGSM, the accuracies are 29%, 17%, and 11% with ϵ of 0.3, 0.5, and 0.9, respectively. Similar trends are obtained when tested with BIM and MIM.

In this work, we performed the adversarial training and testing on the same kind of network, as the adversaries are generated for the same or similar network architecture as the testing network architecture. From the above analysis, it needs to be noted that the FGSM based adversarial training enhances the robustness even against unseen attacks such as CW and DF.

9.4 Putting It All Together

Adversarial machine learning is an emerging topic with numerous and defenses evolving over time-varying in terms of their impact, and complexity. Among the techniques discussed in this chapter, FGSM is a low complex adversarial attack compared to other techniques, but, requires large perturbation to achieve the adversarial capability. On the other hand, CW attack is considered to be one of the sophisticated adversarial attack with smaller perturbations.

In terms of defenses, adversarial training, though involves retraining of the ML, is capable of defending against adversarial attacks to some extent. Though the adversarial training, especially FGSM retrained classifier has shown robustness to the adversarial attacks, it has some of the shortcoming in addition to what is exposed in literature. Table 9.6 reports the classification accuracies for MNIST Digits dataset showing the pitfalls of adversarial training based approach. Though, FGSM retrained classifier is robust to adversarial attacks caused by MIM and BIM based adversarial samples, it fails when the ϵ, that is if the perturbation is increased rapidly, i.e., the magnitude of perturbations increase drastically. This can

Table 9.6 Accuracies of adversarial training

	BIM		MIM	
	$\epsilon = 0.6$,	$\epsilon = 0.9$	$\epsilon = 0.7$,	$\epsilon = 0.7$
	$\epsilon_{iter} = 0.6$	$\epsilon_{iter} = 0.6$	$\epsilon_{iter} = 0.5$	$\epsilon_{iter} = 0.7$
FGSM	0.7345	0.4869	0.6013	0.7352
DF	0.3044	0.2568	0.2292	0.2564
JSMA	0.0063	0.0059	0.0064	0.0066
CW	0.3971	0.3412	0.308	0.3385

also be observed from Table 9.5. Similarly, the classifier trained with DF/JSM/CW fails fatally when the number of iterations is increased with additional processing capabilities. However, under certain scenarios such as maximum perturbations and if attacker has more computational power, the FGSM based retraining still has to be enhanced. As such the main pitfall of the adversarial training is its ineffectiveness to the large perturbations and increased iteration based (optimized) advanced attacks.

9.5 Exercise Problems

Problem 9.1 Are the adversarial examples unique?

Problem 9.2 Is it possible to entirely harden a model against adversarial examples?

Problem 9.3 How can the intensity of adversarial attack be regulated at each one of the attacks FGSM, BIM, MIM, C&W, PGD, and DeepFool?

Problem 9.4 For each one of the attacks FGSM, BIM, MIM, C&W, PGD, and DeepFool two versions targeted and untargeted can be implemented. For a fixed amount of perturbation, which one of these versions needs less amount of perturbation?

Problem 9.5 What are the differences between FGSM and BIM from the magnitude of the used ϵ?

Problem 9.6 Explain how the used momentum in MIM prevents the abrupt change in the obtained perturbation?

Problem 9.7 Explain the role of parameters C and k in the C&W attack.

Problem 9.8 What is the role of the parameter temperature (showed with T in Fig. 9.3) in Defensive Distillation method?

Problem 9.9 Can we use autoencoder structure for generating adversarial examples? (hint: autoencoders cannot construct the input images perfectly and always there are some degree of reconstruction error in the output of the autoencoders)

Problem 9.10 There are plenty of frameworks written in Pytorch that have implemented different adversarial attacks. From such frameworks FoolBox [330] and AdverTorch [331] are the most popular ones. Having said that,

- Investigate the available functions for FGSM, BIM, MIM, and PGD in these two frameworks. Apply each one of these attacks on MNIST and CIFAR10 dataset. Compare the fooling rate of each one of the attacks between these two frameworks.
- In some cases, the implementations of frameworks are slightly different from each other. In the first part, the fooling rate of each attack has been compared between different frameworks. In this part, the quality of resulted adversarial examples needed to be compared. Different metrics are available for measuring the quality of an image, from such PSNR and SSIM are the most popular ones. Repeat part one for comparing PSNR and SSIM of the resulted adversarial image.

Problem 9.11 In the mentioned attacks like FGSM, MIM, BIM, PGD, the scalar parameter ϵ is used as a knob for adjusting the severity of adversarial perturbation, i.e., the more the ϵ is the more severe perturbation obtains. How can we implement multi-dimensional ϵ instead of a scalar? (hint: second-order adversarial attacks)

Problem 9.12 Perform the adversarial training of a 5-layer DNN using FGSM, MIM, BIM, and CW attack data for MNIST dataset and evaluate the impact on the samples generated through the JSMA. (Note: Parameters of the attacks can be same as in Table 9.4)

Part III
Machine Learning in the Field

Chapter 10
SensorNet: An Educational Neural Network Framework for Low-Power Multimodal Data Classification

10.1 Introduction

Time-series data is a generalized form of data that is gathered in different kinds of domains from healthcare where one can track a patient's vital signs (heart rate, blood pressure), to fitness and wellness, where one can monitor a person's activity, to engines in cars and power plants using sensors. Modeling and classifying time series thus has a wide range of applications. All these datasets are represented by a time series that is univariate or multivariate depending on the number of sensor modalities being measured. Multivariate (multimodal) signals are generated by different sensors, usually with different sampling frequencies such as accelerometers, magnetometers, gyroscopes, and heart rate monitors.

Traditionally, time-series classification problems have been solved with approaches such as Dynamic Time Warping (DTW) and k-nearest neighbor (k-NN). These methods or a combination of them provide a benchmark for current time-series classification research. Different signal processing techniques such as feature extraction and classification are employed to process the data generated by each sensor modality that: (1) can lead to a long design time, (2) requires expert knowledge in designing the features, (3) requires new algorithm development and implementation if new sensors are employed, which is tedious, (4) needs extensive signal preprocessing, and (5) is unscalable for different real-time applications.

Deep convolutional neural networks (DCNN) have become extremely popular over the last few years after their success during the Imagenet challenge. Supervised CNNs are used to perform a large number of tasks such as object detection, image segmentation and are combined with recurrent neural networks (RNN) to generate captions for images as well as to recognize speech. Inspired by these developments, deep networks are applied to classify time-series data, perform event detection, and engineer features from the input data. However, these solutions encounter various challenges such as low detection accuracy, high latency, large and power-hungry architectures when deployed at Internet of Things (IoT) and wearable devices.

S. Rafatirad et al., *Machine Learning for Computer Scientists and Data Analysts*,
https://doi.org/10.1007/978-3-030-96756-7_10

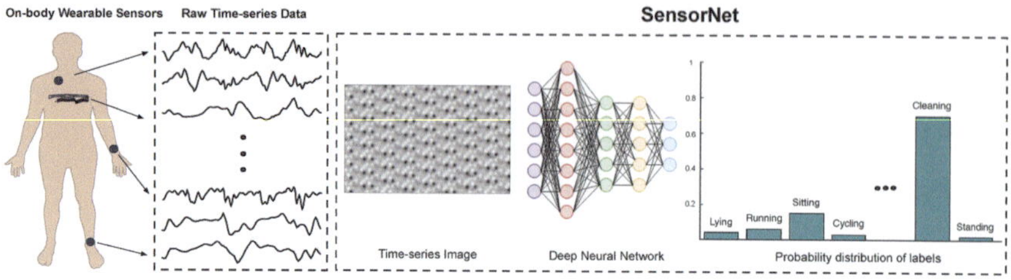

Fig. 10.1 High-level diagram of the proposed SensorNet

In this chapter, SensorNet shown in Fig. 10.1 is proposed, which is a scalable deep convolutional neural network (DCNN) designed to classify multimodal time-series signals in embedded, resource-bounded settings that have strict power and area budgets. SensorNet: (1) is scalable as it can process different types of time-series data with a variety of input channels and sampling rates, (2) does not need to employ separate signal processing techniques for processing the data generated by each sensor modality, (3) does not require expert knowledge for extracting features for each sensor data, (4) achieves very high detection accuracy for different case studies, and (5) has a very efficient architecture that makes it suitable to be deployed at low-power and resource-bounded embedded devices.

10.2 SensorNet Architecture

10.2.1 Deep Neural Networks Overview

Most deep neural networks consist of various layers including Convolutional, Fully connected, Pooling, and Batch normalization layers, etc. Also, there are activation functions such as Sigmoid, Tanh, and ReLU, which can be considered separate layers. Among the neural network layers, fully connected and convolutional layers are often the most highly utilized and contain the majority of the complexity in the form of computation and memory requirements. A brief explanation about the most commonly used layers, including their mathematical formulation and complexity requirements in terms of computation and memory, is provided in the following section.

Convolutional Layers

Convolutional layers are the core building block of a convolutional neural network. The layers consist of learnable filters banks (sets), which have a small receptive field that extends through the full depth of the input. During the forward pass, each filter is convolved across the width and height of the input, computing the dot product

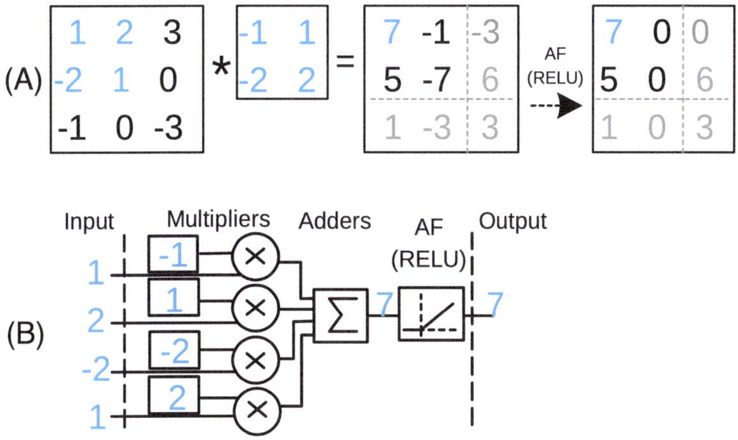

Fig. 10.2 (**a**) An example of convolving a 3 × 3 image by a 2 × 2 filter, (**b**) A hardware schematic that demonstrates one single convolution operation

between the entries of the filter and the input and producing a feature map of that filter. Feature maps for all filters along the depth dimension of the input data form the full output of the convolution layer. Figure 10.2 shows convolution operation for a 3 × 3 image by a 2 × 2 filter followed by an activation function. A hardware schematic that demonstrates one single operation is also depicted. The convolutional layers use a non-linear activation function that will be discussed later.

For a 1-D input $X_{M,C_{in}}$ of length M and with input channels C_{in}, a 1-D convolutional layer with stride S, filter length F, weight $W_{C_{out},C_{in},F}$, feature maps C_{out}, an output signal $Y_{N,C_{out}}$ with length $N = 1 + \lfloor (M - F)/S \rfloor$, and output channels C_{out}, the output of a single element of a feature channel is computed by

$$Y_{i,j} = \sum_{c=1}^{C_{in}} \left(\sum_{f=1}^{F} (X_{f+iS,c} W_{j,c,f}) \right) \quad for\ i = 0..N-1,\ j = 1..C_{out}. \tag{10.1}$$

The total amount of memory requirements by the layer corresponds to the number of weights for all of the filters, which is $C_{out}C_{in}F$. The total computation required for the layer is $2FC_{in}C_{out}N$.

Pooling Layers

Pooling layers are usually used immediately after convolutional layers and perform dimensionality reduction. These layers are also referred to as downsampling layers. What the pooling layers do is simplify the information in the output from the convolutional layer. There are different pooling layers such as max-pooling and

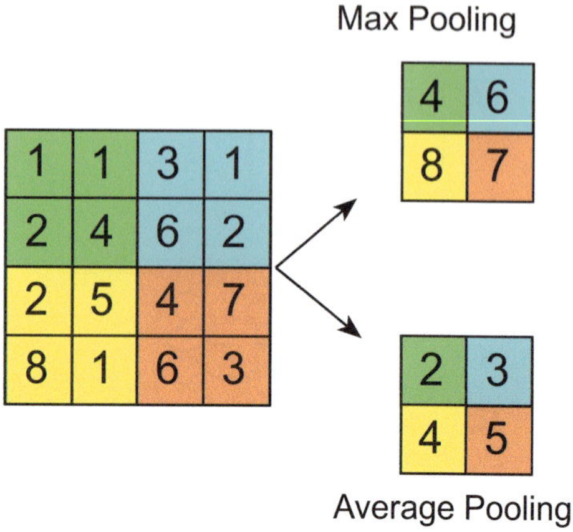

Fig. 10.3 Max-pooling and average-pooling examples with a 2 × 2 window and stride = 2

average-pooling. Max-pooling reduces the size of the image and and also helps the network learn abstract features in the signal by maximizing the value across the pooling window. The pooling layers are usually applied independently to each input channel. Given a 1-D input $X_{M,C_{in}}$ of length M and with C_{in} input channels, a 1-D pooling layer with stride S and pooling length P will produce an output signal $Y_{N,C_{in}}$ with length $N = 1 + \lfloor (M - P)/S \rfloor$. This layer does not require any memory and significantly less computation compared to convolution layers because it is applied independently to each input channel (Fig. 10.3).

Fully Connected Layers

The fully connected layer is a traditional Multi-Layer Perceptron (MLP) that connects every neuron in the previous layer to every neuron on the next layer (Fig. 10.4). Their activations can thus be computed with a matrix multiplication followed by a bias offset.

The main issue with fully connected layers is that the layer requires a significant amount of memory and computation complexity. Given a 1-D input X_M of length M, a fully connected layer with N neurons, weight $W_{N,M}$, and a 1-D output Y_N with length N, the output for a single neuron is computed by

$$Y_j = \sum_{m=1}^{M} \left(X_m W_{j,m} \right) \; for \; j = 1 ... N. \tag{10.2}$$

The total amount of memory required for the layer corresponds to the total number of weights, NM, and the total computation is approximately $2MN$. Therefore, the memory and computation contribute equally in terms of complexity.

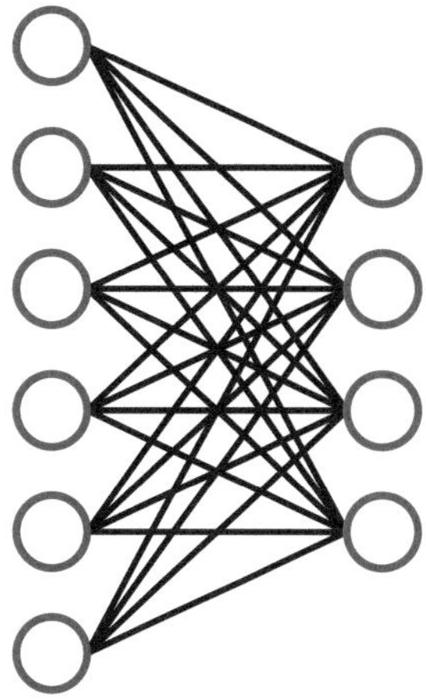

Fig. 10.4 Fully connected layer

Usually, after several convolutional and max-pooling layers, the high-level reasoning in the neural network is performed through fully connected layers. Also, a fully connected layer with softmax activation function is used in the output layer for the final classification.

Activation Functions

In biologically inspired neural networks, the activation function is usually an abstraction representing the rate of action potential firing of the cell. Activation functions play an important role in the Artificial Neural Network to learn and make sense of non-linear complex functional mappings between the inputs and response variables and the ability to satisfy the profound universal approximation theorem. Figure 10.5 shows some common activation functions used in the neural networks including Rectified Linear Unit (ReLU), hyperbolic tangent (Tanh), and Sigmoid. Convolutional and fully connected layers use non-linear activation functions. Recently, the most common activation functions are ReLUs that have been shown to provide better performance compared to others. A ReLU is represented with the following function:

$$f(x) = \begin{cases} x & x > 0 \\ 0 & x \le 0. \end{cases}$$

In the ReLUs, the activation is linear when the output is positive and hence does not suffer from a vanishing gradient problem. Also, ReLUs are very efficient for hardware implementation because they require few logics and operations to perform.

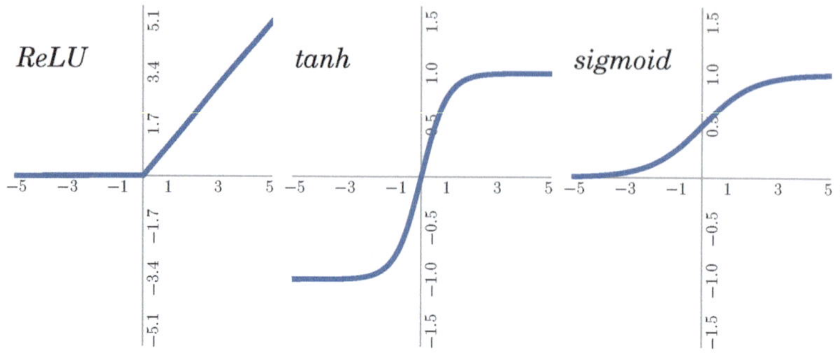

Fig. 10.5 Some common activation functions used in neural networks

10.2.2 Signal Preprocessing

Consider a given time series that consists of M modalities/variables with the same or different sampling frequency. Prior to training, each variable is independently normalized using the $l2$ norm. To generate an image from the normalized variables, a sliding window of size W and step size S is passed through all variables, creating a set of images of shape $1 \times W \times M$ (single channel image). The label associated with this image depends on the dataset. The datasets used to test the network in this chapter contain a label for every time step. Since a single label is assigned to each image, the label of the current time step is taken as the label of the image (and the label that needs to be predicted subsequently while testing). A given image generated at time step I_t has the prior states of each variable from $(t - W + 1)...t$. Thus, the network can look back W prior states of each variable and, given the current state of each variable, predicts the label.

10.2.3 Neural Network Architecture

Figure 10.6 shows SensorNet architecture. It consists of 5 convolutional layers, 1 fully connected, and a softmax layer that is equivalent in size to the number of class labels (depending on the case study). In the preprocessing stage, SensorNet takes the input time-series data and fuses them into images. Then, the images are passed into the convolutional layers, and some features that are shared across multiple modalities are generated using a set of local filters. Then, these features are fed into the fully connected and the softmax layers. SensorNet architecture including a number of layers, a number of filters, and filter shapes for each layer is chosen based on an extensive hyperparameter optimization process that will be discussed in detail in Sect. 10.4.

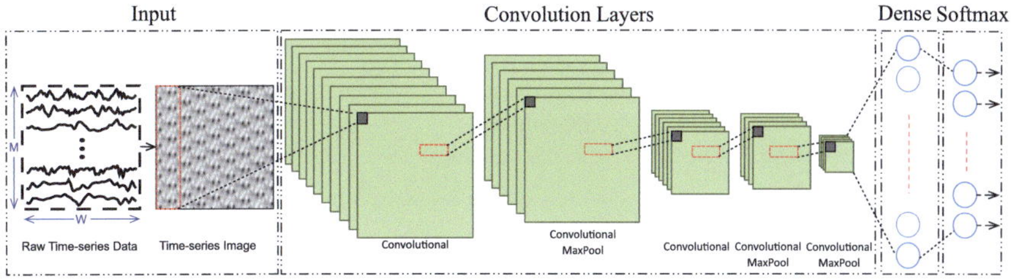

Fig. 10.6 The proposed SensorNet architecture that consists of convolutional, fully connected (dense), and softmax layers

The first, second, third, fourth, and fifth convolutional layers contain 32, 16, 16, 8, and 8 filter sets, respectively. The convolution filters have a height of either M or 1 because it is assumed that there are no spatial correlations between the variables. Also, the ordering of variables prior to generating images does not affect the ability of the network to perform classification. A filter of height M or 1 remains unaffected by the ordering of the variables. Therefore, the filter size for the first convolutional layer is $M \times 5$, where M is the number of input modalities. For other layers, a filter shape of 1×5 is chosen.

Max-pooling is applied thrice, once after the second convolutional layer, then after the fourth convolutional layer, and the last one after the fifth convolutional layer. A max-norm regularization of 1 is used to constrain the final activation output. The pooling size for all max-pooling layers is 1×2. Once the convolution operations have been performed, the image is flattened into a single vector so that a fully connected layer can be added.

Two fully connected layers are employed in SensorNet, in which the first one has a size of 64 nodes and the second one has a size equivalent to the number of class labels with softmax activation. All the layers of the network have their weights initialized from a normal distribution. A learning rate of 0.0001 is used to train the network. Rectified Linear Unit (ReLU) is used as activation function for all the layers. The network is trained using backpropagation and optimized using RMSprop. Categorical cross-entropy is used as the loss function. Following is the loss function:

$$L(y_p, y_a) = \frac{-1}{N} * \sum_{i=1}^{N}[y_a^i \log y_p^i - (1 - y_a^i) \log(1 - y_p^i)], \tag{10.3}$$

where y_p is the predicted label and y_a is the expected label.

As shown in Algorithm 5, there are three main functions defined:

- **Reshape:** This function reshapes a given tensor into another form. We transform a 2D matrix into a 3D tensor with the first axis as 1 representing a single channel.
- **Forwardpass:** It is a single complete processing of the input image to predict the label (for the given dataset).

Algorithm 5 Train *SensorNet* to predict labels (actions)

Input: The Network N as defined in Fig. 10.6. An input dataset D of size k $(d_1..d_k)$ sampled from various sensors with each point having M attributes.
Output: Predict the class label l_i for a datapoint d_i
 # Consider the training batch size to be b, the learning rate LR and reshape() changes the shape of the tensor.
 # W is the size of the sliding window.
 # epochs is the number of epochs for which the model is trained.
 # For categorical crossentropy refer to Eq.10.3
x_{train} is a list of images and y_{train} has the expected labels.

for $i \leftarrow W$ **to** D **do** # After reshape the tensor is of shape (1, W, M)
x^i_{train} = reshape$(d_{i-W+1}..d_i)$
$y^i_{train} = l_i$
Train the model.

for $e \leftarrow 1$ **to** *epochs* **do**
for $j \leftarrow 1$ **to** $\lfloor \frac{D}{b} \rfloor$ **do** batch = $x_{train}[j * b : (j+1) * b]$
y_{pred} =*forwardpass(N, batch)*
loss = L(y_{pred}, $y_{train}[j * b : (j+1) * b]$)
g = backwardpass(loss)
gradientupdate(g) **return** N

- **Backwardpass:** It computes the gradient of weights with respect to the loss function (required to perform gradient descent).
- **GradientUpdate:** This function updates the weights using the gradient and learning rate that is defined.

10.3 SensorNet Evaluation using Three Case Studies

SensorNet is evaluated using three real-world case studies including Physical Activity Monitoring [332], stand-alone dual-mode Tongue Drive System (sdTDS) [333], and Stress Detection [334] and in-depth analysis, and experimental results are provided.

The information for all the case studies is shown in Table 10.1. As it can be seen from the table, the sampling rates of the sensors for each case study are different in the range of 1–100 Hz. Also, the sensors are placed in a variety of spots on human body including Chest, Arm, Ankle, Head, and Hand Fingers. The number of channels for each case study refers to the number of input time-series signals that are received simultaneously by SensorNet. Physical Activity Monitoring, sdTDS, and Stress Detection case studies can be considered to generate large, medium, and small-size datasets.

Table 10.1 Information for three different case studies including physical activity monitoring, sdTDS, and stress detection

Application	# of activity labels	Sensors position	Sampling rate (Hz)	# of subjects	# of channels
Physical activity	12	Chest and arm and ankle	100 and 9	8	40
sdTDS	12	Headset	50	2	24
Stress detection	4	Wrist and finger	8 and 1	20	7

For all the case studies, SensorNet is trained using Keras with the TensorFlow as backend on an NVIDIA 1070 GPU with 1664 cores, 1050 MHz clock speed, and 8 GB RAM. Models are trained in a fully supervised way, backpropagating the gradients from the softmax layer through to the convolutional layers.

10.3.1 *Case Study 1: Physical Activity Monitoring*

Dataset

Physical Activity Monitoring dataset (PAMAP2) [332] records 12 physical activities performed by 9 subjects. The physical activities are, for instance: "standing," "walking," "lying," and "sitting." Three IMUs (inertial measurement units) and one heart rate monitor are placed on chest, arm, and ankle to record the data. The sampling frequency of the IMU sensors is 100 Hz, and the heart rate monitor sensor has a sampling frequency of 9 Hz. In total, the dataset includes 52 channels of data, but 40 channels are valid according to [332]. Also, out of 9 subjects, the data of 8 subjects are used, as subject 9 has a very small number of samples.

Experiment Setup and Results

SensorNet utilizes 5 convolutional layers, followed by 2 fully connected layers. The First convolutional layer has 32 filter sets, and each filter size is 40×5. Other convolutional layers have 16, 16, 8, and 8 filter sets with a size of 1×5. For this experiment, 80%, 10%, and 10% of the entire data for each subject are chosen randomly as the training, validation, and testing set, respectively. To determine the number of required epochs for the training, we train SensorNet for 150 epochs and plot validation and training loss and accuracy results. As is shown in Fig. 10.7, after 100 epochs, the validation loss and accuracy are stable and satisfactory. Therefore, for all the experiments for this dataset, we train SensorNet with 100 epochs.

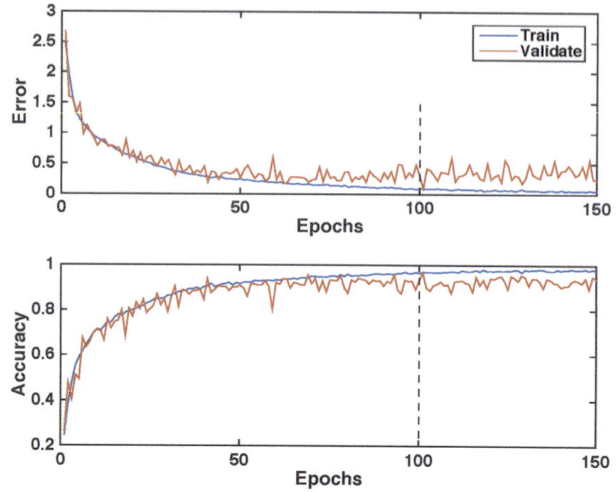

Fig. 10.7 Error and accuracy of the training and validation sets for Physical Activity Monitoring case study over 150 epochs. The vertical dashed line indicates the determined epoch

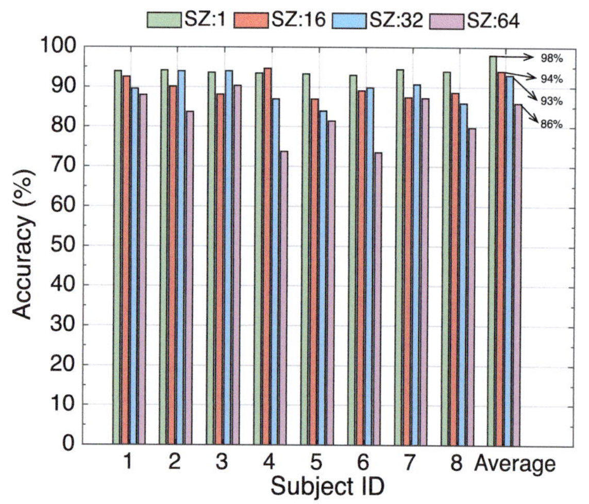

Fig. 10.8 Comparison of SensorNet classification accuracy for Physical Activity Monitoring case study. The results are for different subjects with a sliding window of size 64 samples and step size (SZ) of 1-16-32-64

After training SensorNet, we evaluate the trained model to determine the detection accuracy. Figure 10.8 shows the classification accuracy of SensorNet for the Physical Activity Monitoring case study for different subjects with a sliding window of size 64 samples and step size of 1-16-32-64. As can be seen from the figure, all subjects with step size 1 achieve a high detection accuracy. However, as the step size increases from 1 to 64, the detection accuracy decreases. The average accuracy of all subjects with step sizes of 1, 16, 32, and 64 are 98%, 94%, 93%, and 86%, respectively.

10.3.2 Case Study 2: Stand-Alone Dual-Mode Tongue Drive System (sdTDS)

sdTDS Overview and Experimental Setup

In [333], a stand-alone Tongue Drive System (sTDS) was proposed and developed, which is a wireless wearable headset, and individuals with severe disabilities can use it to potentially control their environments such as computer, smartphone, and wheelchair using their voluntary tongue movements. In this chapter, in order to expand the functionality of sTDS, a stand-alone dual-mode Tongue Drive System (sdTDS) is introduced by adding head movements detection. Figure 10.9 shows sdTDS prototype that includes a local processor, four magnetic and acceleration sensors, a BLE transceiver, a battery, and a magnetic tracer that is glued to the user's tongue. Two magnetic and acceleration sensors are placed on each side of the headset, and the processor is placed in a box at backside of the headset. The box is also used for placing a battery, and its weight is around 0.14 lb. The box is designed using 3D printing technology. In order to generate user-defined commands, the user should move his/her tongue to 7 specific teeth or move his/her head to 5 different directions. The raw data generated by 4 magnetometers and accelerometers are transferred into an FPGA processor where the entire signal processing including feature extraction and classification is performed by SensorNet and 12 different user-defined commands can be generated.

Experiment Results

Several different datasets are captured using sdTDS for training and testing purposes. sdTDS generates 24 channels of time-series data that corresponds to tongue and head movements. As it was mentioned in Sect. 10.2.1, SensorNet utilizes 5

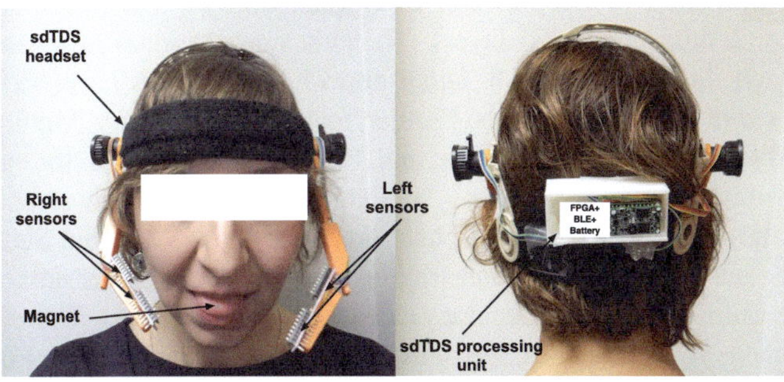

Fig. 10.9 sdTDS prototype placed on a headset that includes a FPGA, four acceleration and magnetic sensors, a Bluetooth low-energy transceiver, a battery, and a magnetic tracer that is glued to the user's tongue

convolutional layers, followed by 2 fully connected layers. For the sdTDS, the first convolutional layer has 32 filter banks, and each filter size is 24 × 5. Other convolutional layers have 16, 16, 8, and 8 filter banks with a size of 1 × 5. For this experiment, 80%, 10%, and 10% of the entire data for each trivial are chosen randomly as the training, validation, and testing sets, respectively. We train SensorNet for 100 epochs.

After training SensorNet using sdTDS dataset, we evaluate the trained model to determine the detection accuracy. Based on previous experiments, we train and test the sdTDS with a sliding window of size 64 samples and a step size of 1, as the step size of 1 gives better detection accuracy consistently. For sdTDS case study, SensorNet detection accuracy for tongue and head movements detection is approximately 96.2%.

10.3.3 Case Study 3: Stress Detection

Dataset

This database contains non-EEG physiological signals used to infer the neurological status including physical stress, cognitive stress, emotional stress, and relaxation of 20 subjects. The dataset was collected using non-invasive wrist-worn biosensors. A wrist-worn Affectiva collects electrodermal activity (EDA), temperature, and acceleration (3D); and a Nonin 3150 wireless wristOx2 collects heart rate (HR) and arterial oxygen level (SpO2) data [334]. Therefore, in total, the dataset includes 7 channels of data. The sampling frequency of wrist-worn Affectiva is 8 Hz, and wristOx2 has a sampling frequency of 1 Hz.

Experiment Setup and Results

As it was discussed in Sect. 10.2.1, SensorNet utilizes 5 convolutional layers, followed by 2 fully connected layers. The first convolutional layer has 32 filter sets, and each filter size is 7 × 5. Other convolutional layers have 16, 16, 8, and 8 filter sets with a size of 1 × 5. Similar to Physical Activity Monitoring case study, for this experiment, 80%, 10%, and 10% of the entire data for each subject are chosen randomly as the training, validation, and testing set, respectively. To determine the number of required epochs for the training, we train SensorNet for 150 epochs and plot validation and training loss and accuracy results. After 100 epochs, the validation loss and accuracy are stable and satisfactory. Therefore, for all the experiments for this dataset, we train SensorNet with 100 epochs. Figure 10.10 shows the classification accuracy of SensorNet for Stress Detection case study for 20 different subjects. As is shown in the figure, most of the subjects have a high detection accuracy of more than 90%. The average accuracy of all 20 subjects is approximately 94%.

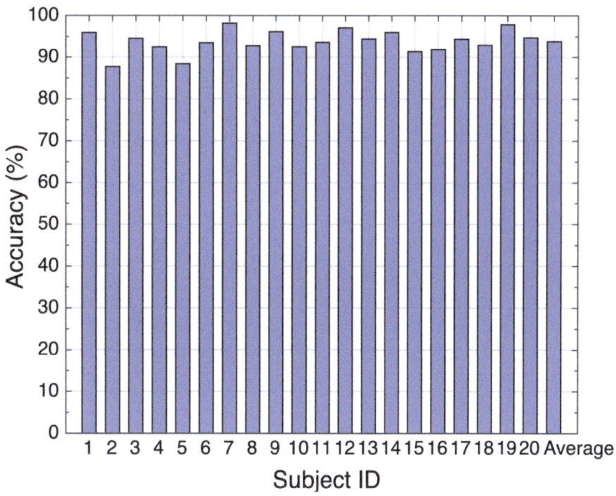

Fig. 10.10 Comparison of SensorNet classification accuracy for Stress Detection case study. The results are for different subjects with a sliding window of size 64 samples and a step size (SZ) of 1-16-32-64

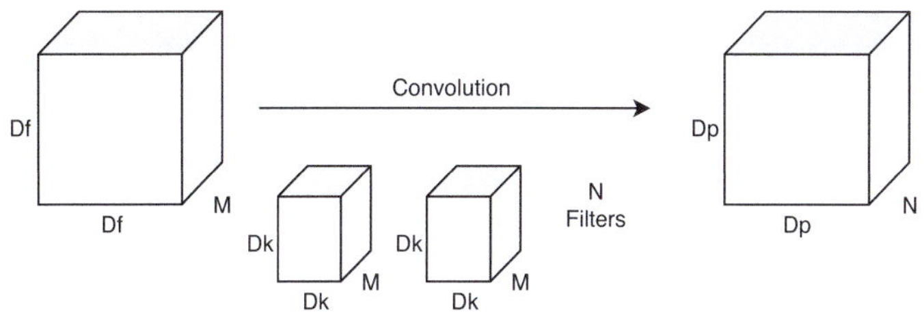

Fig. 10.11 Traditional convolution layer with input shape of $D_f \times D_f \times M$ and output shape of $D_p \times D_p \times N$

10.4 SensorNet Optimization and Complexity Reduction

In the traditional convolution layer, if the input is of size $D_f \times D_f \times M$ and N is the number of filters applied to this input of size $D_k \times D_k \times M$, then the output of this layer without zero-padding applied is of size $D_p \times D_p \times M$. If the stride for the convolution is S, then D_p is determined by the following equation:

$$D_p = \frac{D_f - D_k}{S} + 1. \tag{10.4}$$

In this layer, the filter convolves over the input by performing element-wise multiplication and summing all the values. A very important note is that the depth of the filter is always the same as the depth of the input given to this layer. The computational cost for traditional convolution layer is $M \times D_k^2 \times D_p^2 \times N$ (Fig. 10.11).

The specified equations in Table 10.2 provide the number of parameters and computations for one forward pass in terms of traditional convolution and fully connected layers. For convolutional layers, the number of parameters and computations is dependent upon the stride, which is determined by Eq. 10.4. The input to the fully

Table 10.2 Parameter and operations calculation for convolution and fully connected layers. Here, D_f, D_k, D_p, M, and N represent input height/width, filter height/width, output height/width, the number of input channels, and the number of filters, respectively

Layers	Convolution	Fully connected
Input height	D_f	D_f
Input width	D_f	*1*
# input channels	M	*1*
# filters	N	N
Filter height	D_k	
Filter width	D_k	
Output height	D_p	N
Output width	D_p	*1*
# parameters	$(M \times D_k^2 \times N) + N$	$(D_f \times N) + N$
# computations	$M \times D_k^2 \times D_p^2 \times N$	$D_f \times N$

connected layer is flattened. Hence, it becomes a one-dimensional vector where the computations depend on the number of filters or neurons.

In this section, the impact of changing the following parameters or configurations on SensorNet performance is specifically explored: *(1) the number of convolutional layers, (2) the number of filters, (3) filter shapes, (4) input zero-padding, and (5) activation functions.*

10.4.1 The Number of Convolutional Layers

In this experiment, six SensorNet configurations with an increasing number of convolutional layers for the three different case studies were compared. These 6 configurations are depicted in Fig. 10.12. The comparison has been made in terms of detection accuracy, the number of convolutional operations, the number of parameters (model weights), and memory requirements. Figure 10.13a, b, c and d depict the impact of increasing the number of convolutional layers on the number of model parameters and memory requirements. As is shown in the figures, by increasing the number of convolutional layers, the number of model parameters and memory requirements decrease, which is desired. The reason that three max-pooling layers after the convolutional layers are used comes from the intuition that by adding more convolutional layers the size of the time-series images shrinks and the fully connected layer needs to process a less number of data and thus requires less memory. Figure 10.13e shows the impact of increasing the number of convolutional layers on detection accuracy. As seen from the figure, if the neural network is too shallow, high-level features cannot be learned. Therefore, the detection accuracy is low. However, the results show that, by increasing the number of convolutional layers, detection accuracy increases but up to 5 convolutional layers. After that, for Activity Monitoring and sdTDS case studies, the accuracy improves slightly but for the Stress Detection reduces because the useful features may be filtered out during the convolutional and max-pooling processes. Also,

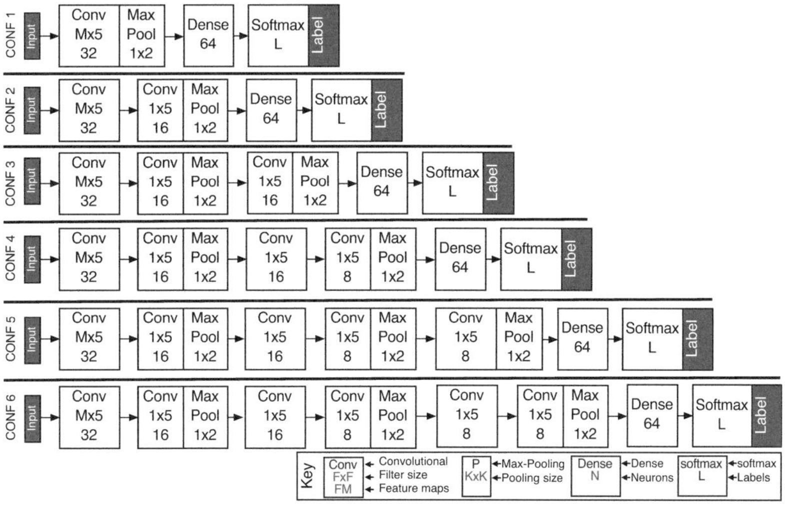

Fig. 10.12 Comparison of six different SensorNet configurations. M and L are the number of input data channels and labels for different case studies, respectively

by adding additional convolutional layers, the number of operations to finish a classification task increases slightly, which is shown in Fig. 10.13f. These analysis results show that SensorNet with 5 convolutional layers is the best candidate with regard to detection accuracy, the number of convolutional operations, and memory requirements.

10.4.2 The Number of Filters

The number of filters (weights) is another important hyperparameter for implementing SensorNet on low-power and resource-limited embedded devices because the number of model weights affects the memory requirements and also the number of required computations to finish a classification task. The number of required computations has a direct effect on energy consumption. In this experiment, the number of convolutional layers was fixed (5 layers), and the number of filters for each layer is increased as shown in Fig. 10.14. The goal of this experiment is to find the impact of the number of filters on the detection accuracy, the number of convolutional operations, the number of parameters (model weights), and memory requirements for Physical Activity Monitoring, sdTDS, and Stress Detection case studies. Therefore, SensorNet is trained and tested using four different configurations with a different number of filter sizes. Figure 10.15 shows a comparison of the number of required parameters (model weights) for different trained models. Model weights include the parameters for convolution, fully connected, and softmax

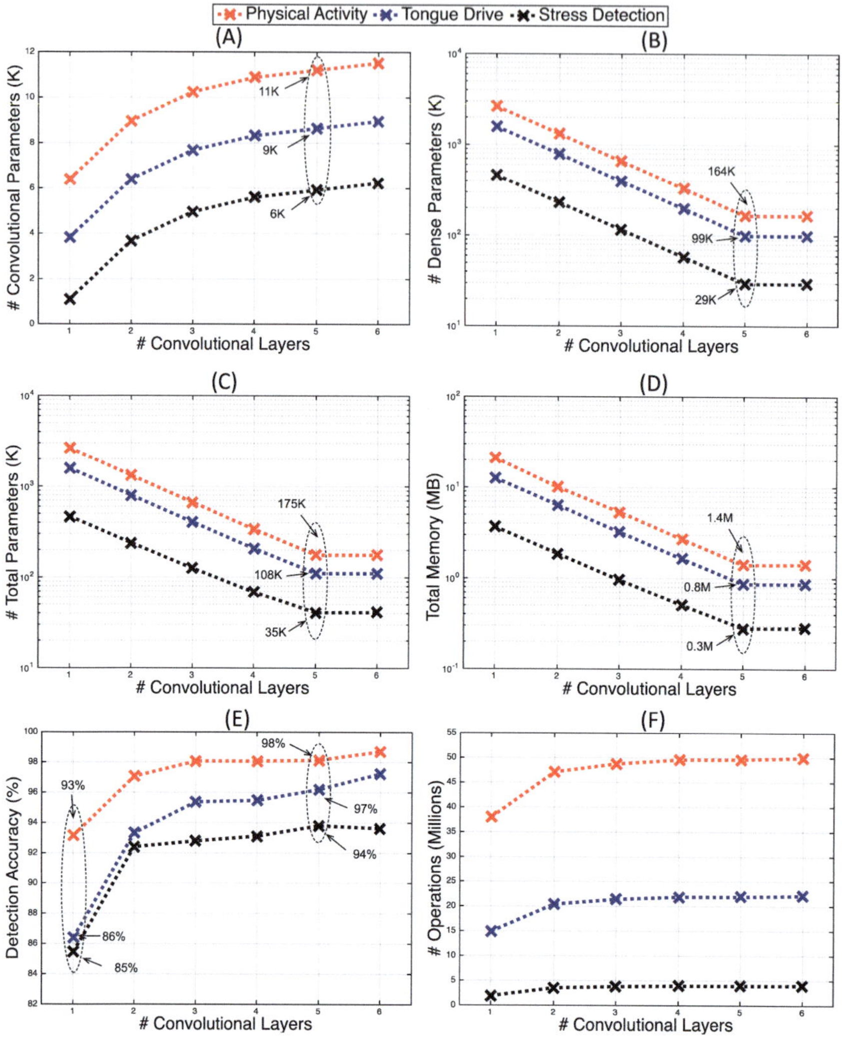

Fig. 10.13 Impact of increasing the number of convolutional layers on memory requirements, detection accuracy, and the number of operations of SensorNet, for three different applications

layers. As is shown in the figure, as the number of filters for each layer is increased, the detection accuracy improves. However, the number of operations, memory requirements, and the number of model parameters increase, which is not desired for hardware implementation in a resource-limited embedded platform.

Based on the results, SensorNet with different filter sets achieves similar detection accuracies, but Set 4 needs a lower number of parameters and requires smaller memory compared to other filter sets and therefore is chosen to be implemented on hardware.

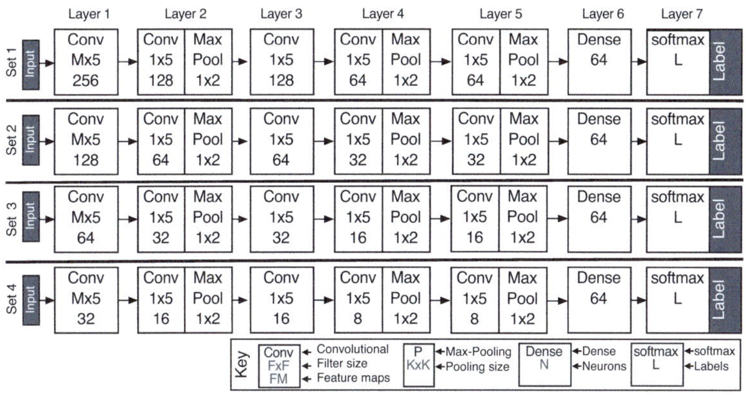

Fig. 10.14 SensorNet configurations with four different filter sets. The number of filters for convolutional layers is doubled for each filter set. M and L are the number of data channels and labels for different case studies, respectively

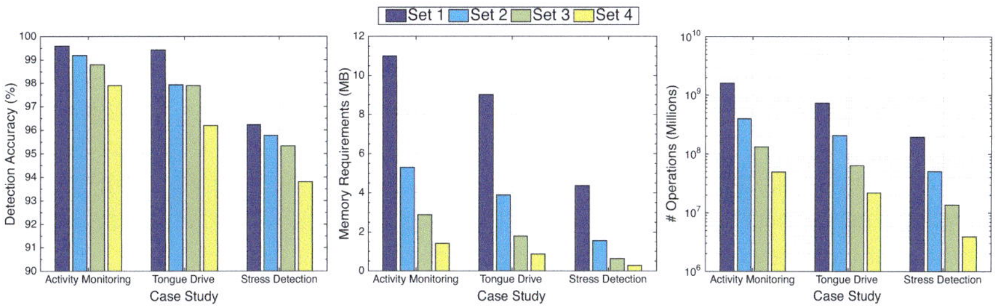

Fig. 10.15 Comparison of four different SensorNet configurations in terms of detection accuracy, memory requirements, and the number of operations. By adding additional model weights to each layer, the computation and memory grow dramatically with only modest improvement in detection accuracy

10.4.3 Filter Shapes

Another important parameter for implementing SensorNet on low-power embedded platforms is the filter shape. As it was explained in Sect. 10.2.1, the idea is to generate some shared features across different input modalities. Therefore, the filters with size $M \times 5$, where M is the number of input modalities, are chosen, for the first convolutional layer. For other convolutional layers, the filters are 1×5. By employing this size of the filter without zero-padding, the outputs of the first layer are 1-D vectors, and the following layers will also be 1-D vectors. This improves the memory requirements on an embedded platform drastically because the feature maps are 1-D signals that compared to an image are much smaller. Also, a smaller number of model weights are needed as the dense layer takes 1-D vectors rather than images. Furthermore, it reduces the number of operations, which directly

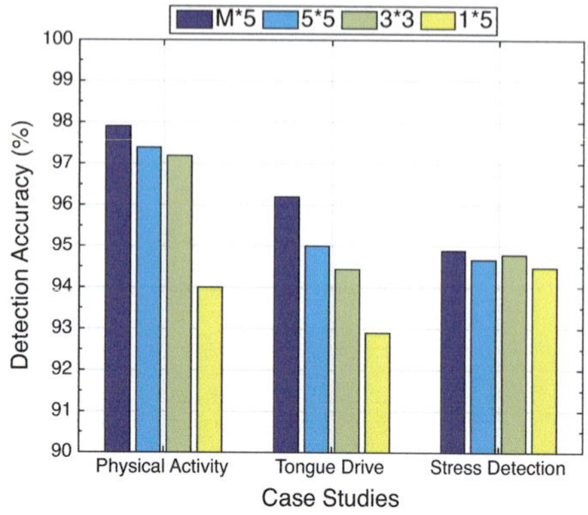

Fig. 10.16 Comparison of SensorNet detection accuracy for four different filter shapes

affects energy consumption of the framework when is implemented on an embedded platform.

In this experiment, the filter shapes are changed for the first convolutional layer to $M \times 5$, 5×5, 3×3, and 1×5 for Physical Activity Monitoring, sdTDS, and Stress Detection case studies. Based on the results, filter size of $M \times 5$ gives better detection accuracy compared to 5×5, 3×3, and 1×5 filter sizes, as is shown in Fig. 10.16. Also, another interesting finding is that, for the dataset with more number of input channels, choosing $M \times 5$ filter size gives better accuracy compared to smaller datasets because the small-size filters can cover most of the input channels in the smaller dataset but not in the dataset with many input channels.

10.4.4 Zero-Padding

In this experiment, the impact of input data zero-padding in the first convolutional layer on detection accuracy is explored, for Physical Activity Monitoring, dTDS, and Stress Detection case studies. Input zero-padding makes the output of the convolutional layer to be similar or the same as the input to the layer. Based on the results shown in Fig. 10.17, zero-padding the input data helps with accuracy, although it increases the number of parameters and memory requirements. As it can be seen from the figure, by applying the zero-padding, the detection Accuracy increases by 4.6%, 3.4%, and 3.8% for Physical Activity Monitoring, sdTDS, and Stress Detection case studies, respectively. However, the total memory requirements and the number of operations are increased 9×, 40×, on average, which will affect the power consumption negatively as well. Therefore, SensorNet without zero-padding was implemented on the hardware.

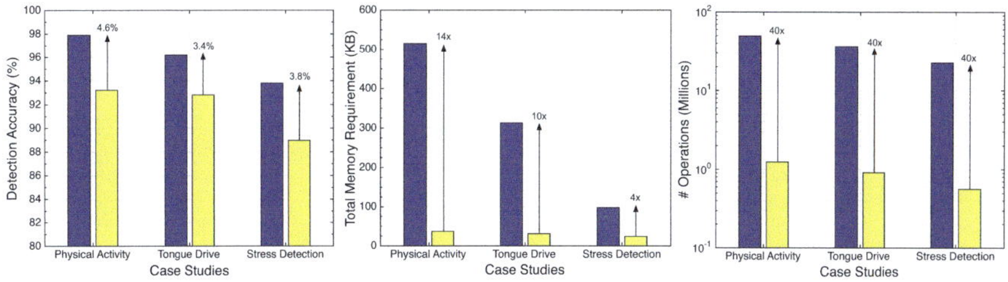

Fig. 10.17 Impact of zero-padding on SensorNet detection accuracy, memory requirements, and the number of computations, for Physical Activity Monitoring, sdTDS, and Stress Detection case studies

Fig. 10.18 Impact of different activation functions including Sigmoid, Tanh, and ReLU in the fully connected layer on SensorNet accuracy

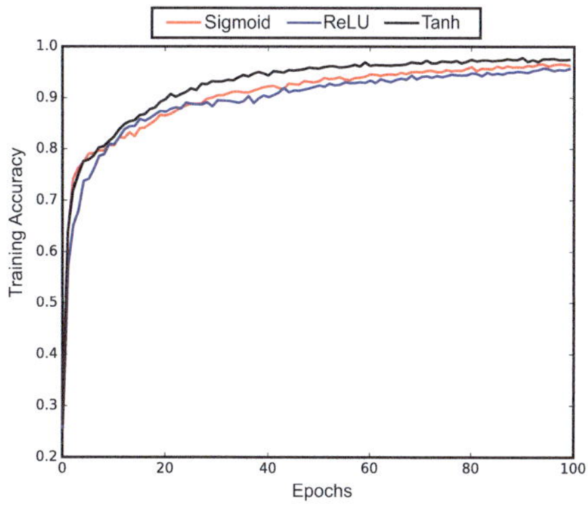

10.4.5 Activation Functions

In Sect. 10.2.1, it was mentioned that Rectified Linear Unit (ReLU) activation functions are efficient because it requires few operations to perform. Therefore, it reduces the hardware complexity on a hardware embedded setting. In all the convolutional layers used in this chapter, ReLU is employed as the activation function. Typically, Sigmoid is used as the activation function for the fully connected layer. However, Sigmoid introduces hardware complexity to the design that is not desired. Thus, in this section, SensorNet detection accuracy by employing different activation functions in the fully connected layer is explored. In this experiment, the SensorNet for stand-alone dual-mode Tongue Drive System case study using ReLU as the activation function for all the convolutional layers and using three activation functions including Sigmoid, Tanh, and ReLU for the fully connected layer was trained. The performance results in terms of training accuracy during 100 epochs are shown in Fig. 10.18. As it can be seen from the figure, SensorNet using any of Sigmoid, Tanh, and ReLU activation functions achieves similar accuracies eventually and using different activations functions does not affect what SensorNet can learn. Therefore, ReLU is chosen as the activation function for all the layers

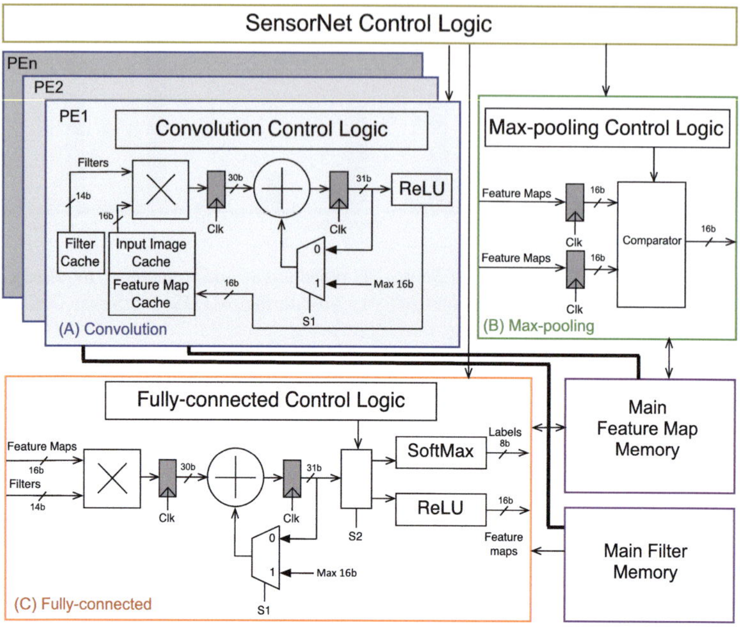

Fig. 10.19 Block diagram of SensorNet hardware architecture that includes convolution, max-pooling, and fully connected blocks and also a top-level state machine that controls all the blocks. PE refers to convolution processing engine (PE)

because it has less hardware complexity compared to other activation functions and achieves comparable training and testing accuracy.

10.5 SensorNet Hardware Architecture Design

Implementing hardware architecture for SensorNet faces several challenges such as computational model implementation, efficient parallelism, and managing memory transfers. Following are the objective for the hardware architecture design: consumes minimal power, meets the latency requirement of an application, occupies a small area, needs to be fully reconfigurable, and requires low memory. Also, the design constraints require SensorNet hardware architecture to be reconfigurable because different applications have different requirements. Parameters such as filter shapes, the number of filters in the convolutional layers, sizes of the fully connected, and softmax layers are configurable.

Figure 10.19 depicts SensorNet hardware architecture with implementation details. This architecture is designed based on Algorithm 5 (Page 240) was depicted and explained in Sect. 10.2.1. The main components of SensorNet on hardware consist of the following:

(a) Convolutional Performs convolutional layer operations. Also, this block includes ReLU activation logic.
(b) Max-pooling Performs max-pooling operations.
(c) Fully connected Performs fully connected layer operations. The fully connected block includes ReLU and softmax activation functions. ReLU will be used as the activation for the first fully connected layer, and softmax will be used for the last fully connected layer and will perform the classification task.

Figure 10.19a shows the convolution block. As is shown, the convolution block contains one multiplier, one adder/subtractor, one cache for saving filters, input feature maps and output feature maps, multiplexers, a few registers, and a state machine block. When the convolution operations are done for all the input feature maps, the output feature maps will be saved into the main feature map memory. The input data coming from the sensors are 16-bit two's complement. After performing the convolution, the data will pass to ReLU activation function. The output of ReLU is truncated to 16 bits and saved in feature map memory. Offline training is performed to obtain model weights using Keras. The model weights are converted into fixed-point format and are represented by 16 bits. The floating-point arithmetic is complex and requires more area. Therefore, the use of fixed-point arithmetic will avoid complex multipliers. Figure 10.19b shows the max-pool block that contains some registers and a comparator. The input to the max-pool is feature maps data, which is formed by a convolution block. After max-pooling operations finish, the results will be saved into the main feature map memory. Figure 10.19c shows the fully connected blocks. As is shown, the architecture consists of a serial dot product engine, a dynamic sorting logic for the softmax activation function, ReLU logic, and a state machine for controlling all sub-blocks. Depending on the layer, either ReLU or softmax can be used. After finishing computations for the fully connected layers, the results will be saved into the main feature memory.

10.5.1 Exploiting Efficient Parallelism

Scalability is one of the key features of the proposed SensorNet on hardware. Therefore, SensorNet hardware architecture was designed to be configured to perform convolution operations in parallel if it is needed, especially for fast applications. In deep convolutional neural networks, convolutional layers dominate the computation complexity and consequently affect the latency and throughput. Therefore, for the applications with many input modalities or the applications that need to issue a command very fast, efficient forms of parallelism that exist within convolutional layers must be exploited. There are three main forms of parallelism methods that can be employed in convolutional layers. The basic process for the three tiling methods is shown in Fig. 10.20. The first method, referred to as input channel tiling, convolves multiple input feature channels concurrently for a given feature map. The second method, output channel tiling, performs convolution across multiple output

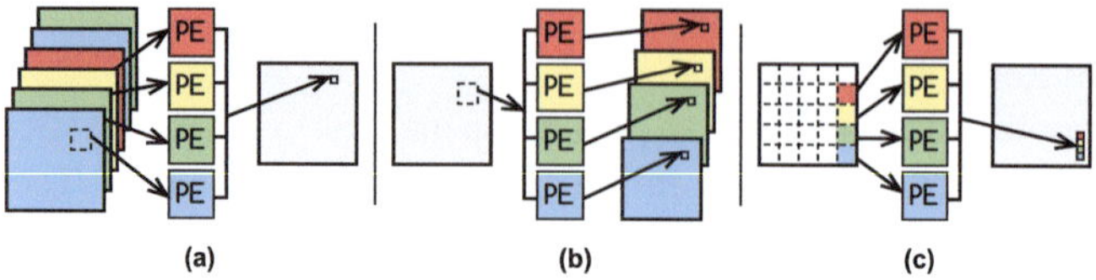

Fig. 10.20 Comparison of different parallel tiling techniques for convolutional layers. Output channel tiling has the least communication contention and inter-core dependency. (**a**) Input channel tiling. (**b**) Output channel tiling. (**c**) Image patch tiling

channels for a given input channel simultaneously. The third method, referred to as image patch tiling, breaks a given input feature channel into patches and performs convolution on the patches concurrently. Usually, output channel tiling provides the best form of parallelism when taking into account I/O memory bandwidth and computational load using the computation to communication (CTC) ratio. Therefore, output channel tiling due to minimal dependency among the parallel cores and minimal communication contention is primarily used in SensorNet.

10.5.2 Hardware Performance Parameters

In practice, hardware performance is evaluated using several metrics, and the use of specific metrics varies based on the application. However, the following are the basic performance parameters that are taken into account for portraying the efficacy of the hardware and to create a point of comparison among similar designs.

Latency In fundamental terms, latency relates to the delay in time due to the cause and effect of some materialistic change in the system. From a hardware perspective, latency is the time taken for a process to generate outputs from a given set of inputs. It is calculated in units of time, i.e., hours, minutes, seconds, microseconds, etc.

Throughput Throughput is directly related to latency. In a practical sense, it is the number of outputs generated per unit of time. For example, if a process is able to execute 10 frames in a second, the throughput for that process will be 10 frames per second. Additionally, it is the inverse of latency.

FPGA Performance The performance metric changes depending on the application. However, for an FPGA, it is usually denoted as the number of operations performed for a given frequency. Furthermore, there are two variants of this performance, i.e., actual performance and peak performance. Actual performance relates to the number of computations executed for the process latency, and the unit is defined as giga operations per second (GOP/S), where the form of the equation is the following:

$$Actualperformance = TotalNumberofComputation/Latency.$$

Peak performance, on the other hand, depicts the maximum capability of the hardware. For a hardware operating with n processing engines with the M number of multiply-accumulate (MAC) units and a frequency of f, the peak performance is given by

$$Peakperformance = 2 * n * M * f.$$

Energy Efficiency The performance of hardware is not always the standard metric to compare among similar architectures. In this case, energy efficiency provides additional insights regarding the performance of the hardware, where the power consumed by the device is also taken into consideration. Energy efficiency for an FPGA is given by

Energy Efficiency = Actual Performance / Total Power.

The unit is GOP/S/W where the total power is calculated in watts.

10.6 Resources

The CNN architecture explored in this chapter has been constructed based on several tools and libraries. In this section, some of the tools have been categorized:

Numpy Library Numpy arrays are standard representations of numerical data in relation to the Python language. Arrays of such structure ensure efficiency and timing economy in terms of large computations. van der Walt et al. [335] introduce the NumPy array and illustrate how to coordinate this with other libraries.

Pandas Library Pandas is a Python library designed to work with structured datasets allowing ease of manipulation and computation of the data. The work in [336] provides detailed design and features of Pandas that serve as a strong complement to the existing Python stack.

TensorFlow TensorFlow is a machine learning module utilized in a variety of environments depending on the application in focus. This maps the dataflow graph neurons across many machines in a cluster including CPU, GPU, and TPU (Tensor Processing Units). The flexible architecture of TensorFlow makes it suitable to be deployed in training and inference tasks of deep neural networks. TensorFlow as a large-scale machine learning tool was introduced and delineated in [337].

Keras API Keras is a deep learning API (Application Programming Interface) written in Python. Keras allows the swift compilation of deep learning layers and complements the machine learning platform of TensorFlow. The combination of Keras API and TensorFlow ensures efficient low-level tensor operations, scaling computation to many devices, and precise computing of the model gradients.

Scikit-Learn Library This library provides modular representations of machine learning algorithms for supervised and unsupervised learning. Along with packages designated to efficiently implement deep learning models, Scikit-learn also allows functions and building blocks designed for data structure manipulation. Pedregosa et al. [338] give a walkthrough regarding the features, packages, and documentation of the library to enhance ease of use, performance, and consistency.

Jupyter Notebook Jupyter notebook is an open-source software designed for interactive data science computing mainly across the Python programming language. It started from the 2014 IPyhton project detailed in [339]. This was developed in the open on GitHub and is a free-to-use platform for algorithmic processes.

10.7 Exercise Problems

Problem 10.1 Consider an input sequence of the following form:

$$A = [1, 0, 1, 0, 2, 0].$$

The input sequence A goes through a 1D max-pooling layer where the stride is of size 2. If valid padding is applied, what will be the shape of the output sequence after max-pooling?

Problem 10.2 Consider an input sequence of the following form:

$$A = [1, 2, 3, 4; 5, 6, 7, 8; 9, 10, 11, 12].$$

The input sequence A goes through a 2D max-pooling layer where the stride is of size [2, 2]. If valid padding is applied, what will be the shape of the output sequence after max-pooling?

Problem 10.3 Consider an input sequence of the following form:

$$A = [1, 0, 1; 0, 2, 0; 1, 0, 1].$$

The input sequence A goes through a 2D convolution layer containing a filter B of the form:

$$B = [1, 0; 2, 2].$$

What will be the output sequence after convolution?

Problem 10.4 Consider the same input sequence and filter in Problem 10.1. The input sequence A goes through a 2D max-pooling layer having a stride B of the form:

$$B = [2, 2].$$

What will be the output sequence after max-pooling?

Problem 10.5 Consider an input sequence of the following form:

$$A = [1, 5, 7, 13, 10].$$

This input vector is connected to a fully connected layer with 2 filters. The filter weight values are the following:

$$A = [9, 1, 0, 0.1, 1; 3, 0.5, 0.6, 0.3, 1].$$

What will be the output sequence after the fully connected operation?

Problem 10.6 Suppose a program is training a deep neural network architecture in a CPU at a frequency of *1.5 GHz*. One pass over the whole training set takes 2 minutes in general. If there is no other dependency on the execution of the program, then approximately how long it will take for the program to complete 100 iterations over the training set?

Problem 10.7 An input vector contains 12 elements. You are asked to create overlapping window frames from this vector where each frame will contain 3 elements. The overlap corresponds to a step size of 1. How many frames will be generated?

Problem 10.8 For the same vector in Problem 10.7, non-overlapping window frames are created where each frame contains 2 elements. How many frames are generated by this process?

Problem 10.9 A raw sensor data sequence contains 1000 entries in total, which is collected via a device that operates at a frequency of *100 Hz*. You need to create non-overlapping window frames from this sequence where size of the frame is denoted by the sampling frequency. How many frames will be created?

Problem 10.10 Consider a deep neural network architecture that consists of one 2D convolution layer, one 2D max-pooling layer, and a fully connected layer, respectively. The input resolution is of shape 10×10. The numbers of filters for convolution and fully connected layers are 3 and 20, respectively. The convolution layer is implemented without zero-padding and with a stride of 1. If the convolution filter and the max-pooling layer have a window of shape [2, 2], what is the total number of parameters for the whole architecture?

Problem 10.11 Use the same deep neural network framework from Problem 10.10 and calculate the number of computations for the network?

Problem 10.12 A convolution layer processes an array of shape [10, 10, 1] and produces an output of shape [5, 5, 3]. If the shape of the convolution filter is [2, 2], what is the stride and the number of filters for the convolution?

Problem 10.13 Consider a hardware that has a frequency of 10 MHz. It executes a program in 10 seconds that has 10^9 computations. If the hardware has only one processing engine consisting of one MAC unit, what is the actual performance of the hardware?

Problem 10.14 For the same hardware as in Problem 10.13, consider that now it has 2 processing engines. If there is no dependency between the processing engines and no additional latency due to the inclusion of another processing engine, what will be the new latency?

Problem 10.15 For the same hardware as in Problem 10.13, the static power for the device is 90 mW, and the dynamic power consumption for the process is 150 mW. What will be the energy efficiency for the design?

Problem 10.16 Loading the Dataset
Download the dataset (PAMAP2) for the physical activity case study from this link https://archive.ics.uci.edu/ml/datasets/PAMAP2+Physical+Activity+Monitoring.

Read in the dataset as .numpy files or in pandas dataframe and report the following:

(1) The number of samples for each subject.
(2) Perform normalization of the data for each subject.

Problem 10.17 Creating the Window Frames
One of the fundamental steps of preparing the time-series dataset for inference is to create windows from the stream of data as described in Sect. 10.2.2. Build a function that employs striding to create window images from the datasets while taking into account of the sampling frequency of the sensors.

Problem 10.18 Generating One-Hot Encoded Labels
In deep neural networks involving activity classification, the loss function used is most often categorical cross-entropy. In order to use this loss function, the labels themselves must be categorized. One-hot encoding is one such way to create categorical labels. Build a function that can take in any stream of labeled data and generate one-hot vectors from it.

Problem 10.19 Obtaining Customized Training, Validation, and Test Sets
Usually, the input data fed to the model is segmented into training, validation, and test sets for convenience of functionality. Create a function that can read in the whole data stream and create segments of 70% training, 10% validation, and 20% test sets.

Problem 10.20 Building the Model
The SensorNet architecture is illustrated in Fig. 10.6. Using the Keras libraries, perform the following:

(1) Build the SensorNet model with the number of filters and parameters defined as in Sect. 10.2.3.
(2) Summarize the model architecture in terms of parameters.
(3) Find out the total number of computations for the architecture generated in (1).

Problem 10.21 Compiling the Model
Use the model built in Problem 10.5 and change the activation functions to tanh. Now, compile the model with 32 batch size, “Adam” optimizer, “categorical_crossentropy” loss, and “accuracy” as the metric for 150 epochs.

Problem 10.22 Consider an input sequence of the following form:

$$A = [1, 0, 1, 0, 2, 0].$$

The input sequence A goes through a 1D convolution layer containing a filter B of the form:

$$B = [0, 1].$$

What will be the output sequence after convolution?

Problem 10.23 Suppose you are given raw data coming from a sensor with a sampling frequency of $100\,Hz$. You are required to make window frames that are 5 seconds long, how many samples will the frames contain?

Problem 10.24 Consider a neural network containing a 2D convolution layer followed by a fully connected layer. If the input is of shape 16×16, where the convolution layer has 5 filters of shape $3\times\ 3$ and the dense layer consists of 10 filters, count the total number of parameters? (Hint: Use Table 10.2.)

Problem 10.25 Let us consider a hardware that has a frequency of 10 MHz. It executes a program in 10 seconds that has 10^9 computations. If the hardware has only one processing engine consisting of one MAC unit, what is the peak performance of the hardware?

Chapter 11
Transfer Learning in Mobile Health

11.1 Introduction

The rapid integration of wearable sensor technologies, along with advanced computational algorithms, has created a unique opportunity for ubiquitous and objective monitoring applications that impact virtually every aspect of modern life. A popular area with great potential to improve people's quality of life is mobile health applications on a daily basis, such as remote health monitoring [340, 341], long-term fitness tracking [342, 343], and fall detection [344]. A core task to support these applications is *activity recognition* with the use of machine learning techniques [345]. However, the scalability of sensor-based activity recognition systems for everyday living scenarios is challenging, mainly due to the following reasons:

- *Cross-user variations.* Machine learning models trained with sensor data collected in a lab setting do not necessarily represent the movement patterns of an unknown user. As you will see in this chapter, the performance of an activity recognition algorithm drops up to 63.7% when the system is adopted by a new user.
- *Spatial uncertainty.* Spatial uncertainty refers to "hardware-induced uncertainty" and/or "software-induced uncertainty" across two sensing platforms. The former includes any cross-platform changes due to variations in hardware, and the latter includes differences in software configurations of the devices. Experiments presented in this chapter show a 60.3% accuracy decline when an activity recognition model trained on one smartwatch is used to detect activities observed using another smartwatch of a different model.
- *Dependency on labeled training data.* Traditional machine learning algorithms rely on large amounts of labeled training data to obtain reliable models. However, manually annotating sensor data is an expensive process, and it has been identified as a major barrier to personalized motion analysis [346]. Therefore, it is infeasible to label sensor data in every new setting.

S. Rafatirad et al., *Machine Learning for Computer Scientists and Data Analysts*,
https://doi.org/10.1007/978-3-030-96756-7_11

The popularization of personal mobile devices has brought in new challenges with respect to generalizability of activity recognition algorithms, mainly because of the spatial uncertainty among various consumer platforms [347]. Consider a practical scenario, where the user replaces or upgrades an old mobile device (e.g., smartphone) with a new one of the same or different model. The user intends to maintain the usability of a motion analysis application installed on the old device, which has a well-trained and accurate activity recognition model for motion analysis. However, the user prefers not to provide additional manual annotations for model re-training on the new device.

It is well-known that the performance of an activity recognition model running on a new device is adversely impacted due to spatial uncertainties, such as sensor biases in low-quality modules, varying sampling frequencies, and the instability of the sampling rate [347]. In particular, the performance decline in F1-score is 34.4% on average for an activity recognition model, when the training dataset was resembled data samples gathered by one smartphone, whereas the test dataset was constructed using another smartphone of a different model (e.g., Samsung Galaxy S3 versus LG Nexus). Figure 11.1 shows an example where the acceleration signals of one subject's walking behavior gathered by two smartphones exhibit different patterns. Such divergence in sensor readouts propagates through the data processing pipeline and leads to a significant accuracy decline in the performance of the machine learning models.

This chapter aims to use transfer learning as a new means for the autonomous development of machine learning models in new settings without collecting any

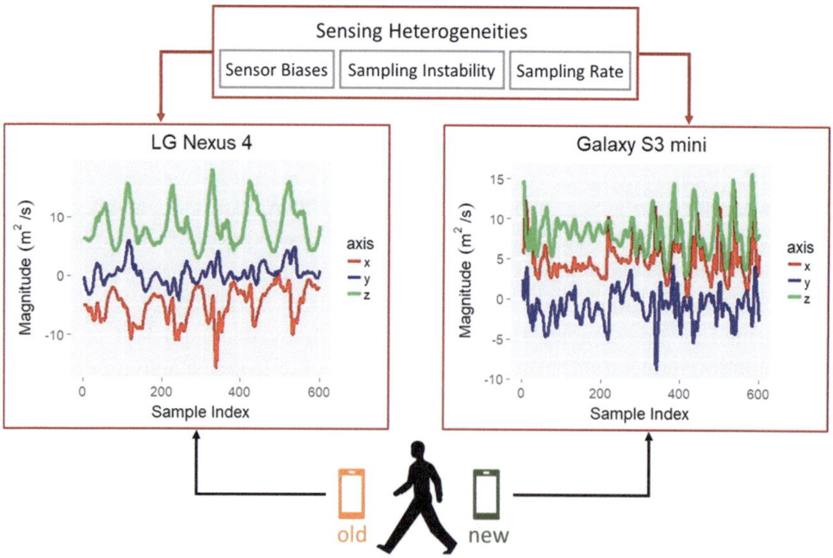

Fig. 11.1 Sensor readings collected from the same subject using two smartphones of different models

labeled data for re-training. To this end, the chapter focuses on introducing an asynchronous knowledge transfer framework, referred to as *TransFall*, to overcome the challenges caused by individual variations and spatial uncertainties for activity recognition, meanwhile, reducing the dependency on the labeled training data. The framework design follows a waterfall-like structure that starts with a two-tier data transformation layer based on marginal distribution matching approaches, followed by a label estimation layer using the kernel method encoded in a weighted least-mean-squares fitting, and ends with a model generation layer using the obtained label set from the previous steps of the transfer learning framework.

11.2 Transfer Learning

The goal in transfer learning is to extend the knowledge from a known domain (e.g., old sensor device, previous user), often referred to as *source*, to an unknown domain (e.g., new sensor device, new user), referred to as *target*.

Teacher/learner architecture is a widely used technique when there is no direct access to the training data in the target domain. Instead, a machine learning model trained on the source domain operates simultaneously with the target learner for label inference [348]. Several variations of the teacher/learner architecture have been developed to perform activity recognition using transfer learning [349]. One study [350] introduced a system-supervised learning approach to send activity labels from the teacher sensor to the learner sensor so that the learner could develop a new model with its own measurements combined with the activity labels received from the source domain. One limitation of such a teacher/learner based approach is the requirement of simultaneous data collection between the teacher and the learner. Furthermore, the activity recognition accuracy of the learner is bounded by the performance of the teacher [348].

Another well-adopted strategy in transfer learning for activity recognition is cluster mapping. In this method, an unsupervised clustering algorithm is first applied to the target data, followed by inferring the corresponding labels of target clusters using the source data through certain similarity measures. Authors in [351] introduced a framework named transELMAR, which utilized a one-step k-means clustering algorithm in conjunction with the extreme learning machine model for cross-mobile activity recognition. This approach attempted to mitigate the adverse impact of different sampling frequencies between the source and the target sensor devices. Another study [352] proposed a cross-subject transfer learning algorithm. It first detected communities on the target dataset using cosine similarity and a threshold learned from the source dataset and then inferred label information through a bipartite mapping between the clusters on the target and the classes on the source. Finally, an adaptive label fusion method was performed on the obtained labels, as well as a separate label set generated by classifiers trained with the source dataset, to create a labeled dataset for the target.

Instead of mapping detection from a label/cluster perspective, another popular approach for label inference across different domains is to reduce the overall discrepancy between the source and the target datasets. Authors in [353] introduced an adaptive approach called importance-weighted least-squares probabilistic classifier (IWLSPC), to overcome the divergence in data distributions across different subjects for activity recognition. IWLSPC combines a least-squares probabilistic model with a sample re-weighting approach, to handle the changes in data distribution between the source and the target datasets. The experimental results demonstrated that IWLSPC achieved the lowest mean mis-classification rate compared to other semi-supervised learning methods.

TransFall employs a similar structure to that of IWLSPC, which utilizes a sample re-weighting technique to transfer the knowledge of activity labels from the source domain to the target domain. However, TransFall performs a two-tier data transformation on both datasets to empirically match the distributions of the two datasets for more accurate label estimation.

Example 11.1 (Transfer Learning)
Problem: Consider you intend to use a public dataset that contains 10 participants' biosignal data and activity labels to perform activity recognition on yourself while you are collecting biosignal data using wearable device that is different from the one the public dataset used. Is this scenario regarded as transfer learning? If yes, point out *source* domain and *target* domain.
Solution: Yes this is a transfer learning problem. Transfer learning aims to extend the knowledge from a known domain to an unknown domain. In this scenario, the known domain (*source* domain) is the public dataset that is labeled and the unknown domain (*target* domain) is the newly collected data without label.

11.3 Problem Statement

Let us refer to the transfer learning problem addressed in this chapter as *Asynchronous Transfer Learning (ATL)*. This section first gives a formal definition of the ATL problem. You will see that this problem can be broken down into two subproblems. This section will also present formulations for these sub-problems. The section concludes by proposing a cascaded transfer learning approach to solve these optimization problems.

The sensing platform used to gather a labeled dataset is referred to as the "*source*" ($\mathcal{S}$), and the other platform used to collect unlabeled data samples is referred to as the "*target*" ($\mathcal{T}$). In the case of the device replacement or upgrade, data collection on the two devices usually does not occur simultaneously. Therefore, the assumption of synchronous data collection between the source device and the target device, which is commonly used in teacher/learner based architectures, cannot hold under this scenario.

11.3.1 Problem Definition

A data sample $x \in \mathcal{R}^d$ is a d-dimensional feature vector extracted from sensor signals, where the ith column variable x_i corresponds to the feature $f_i \in \mathcal{F}$. Each data sample $x \in \mathcal{X}$ is presumed to be drawn from a marginal probability distribution $P(x)$.

In addition to the data samples, the activity recognition task also involves a label space $\mathcal{Y}$ associated with L unique activities. The conditional probability distribution $P(y|x)$ refers to the likelihood of assigning a label $y \in \{a_1, \ldots, a_L\}$ to the observed data sample x. A general assumption adopted in many relevant studies is that the conditional probability distribution $P(y|x)$ remains unchanged for the two datasets, although the marginal distribution of the source dataset differs from that of the target dataset.

The goal of this problem is to perform accurate activity recognition on the target platform, by transforming the knowledge of a labeled dataset gathered by the source platform, without additional ground truth labels for model retraining.

Asynchronous Transfer Learning (ATL) Let X_s be a set of N_s data samples collected on the source platform $\mathcal{S}$, where $x^s \in X_s$ is a d-dimensional variable drawn from the marginal distribution $P_s(x)$; the corresponding label set Y_s is associated to L unique user activities. Furthermore, let X_t be another set of N_t data samples collected asynchronously on the target platform $\mathcal{T}$, where N_t is not necessarily equal to N_s, and $x^t \in X_t$ is drawn from the marginal distribution $P_t(x)$. The same set of activities are observed in X_t. The objective of this problem is to develop an activity recognition model $\mathcal{M} : \mathcal{X} \mapsto \mathcal{Y}$, capable of accurately estimating the corresponding labels for the data samples collected on $\mathcal{T}$.

11.3.2 Problem Formulation

We can tackle the ATL problem described above through two subproblems including *data transformation* and *label estimation*. Here we define each subproblem and formulate them as optimization problems.

With changes on the sensing platform, there is presumed to be a distribution shift in the covariates x between the source dataset X_s and the target dataset X_t, resulting in $P_s(x) \neq P_t(x)$. Therefore, the first task to solve the ATL problem is to address the covariate shift through a data transformation process.

Data Transformation Let $x \in \mathcal{R}^d$ be a d-dimensional variable, and the ith column variable x_i corresponds to the feature $f_i \in \mathcal{F}$. The source dataset X_s is drawn from a marginal distribution $P_s(x), x \in X_s$, whereas the target dataset X_t is drawn from a marginal distribution $P_t(x), x \in X_t$ and $P_s(x) \neq P_t(x)$. The objective of data transformation is to find a mapping function $\phi : \mathcal{P} \mapsto \mathcal{P}$ that transforms $P_t(x)$ to $\tilde{P}_t(x)$ such that $\tilde{P}_t(x) \simeq P_s(x)$.

To minimize the discrepancy between $P_s(x)$ and $P_t(x)$ through a mapping function ϕ, the problem described above can be formulated as follows:

$$\underset{\phi}{\text{Minimize}} \int_x |P_s(x) - \phi(P_t(x))| dx \ . \tag{11.1}$$

After data transformation, the second task in solving the aforementioned ATL problem is to estimate the corresponding labels for the target dataset, X_t. This is an important task for constructing an accurate model on $\mathcal{T}$, because it provides the ground truth information necessary for classification purposes.

Label Estimation Let X_t be a dataset consisting of N_t data samples collected on $\mathcal{T}$. The dataset X_t contains observations associated with L unique activities. A set of labels $\{a_1, \ldots, a_L\} \in \mathcal{Y}$ are used to represent each activity exclusively. The objective of this problem is to find a classification model $f : \mathcal{X} \mapsto \mathcal{Y}$, which can assign a label $\hat{y}_i \in \{a_1, \ldots, a_L\}$ for the data sample $x_i \in X_t$ with minimal mistakes.

To minimize the overall classification error between the true label y_i and the estimated result $\hat{y}_i$ for all $x_i \in X_t$, this problem can be formulated as follows:

$$\underset{f}{\text{Minimize}} \sum_{i=1}^{N_t} |y_i - f(x_i)| \ . \tag{11.2}$$

11.4 TransFall Framework Design

Figure 11.2 shows a waterfall-like structure that addresses the problem of asynchronous transfer learning by integrating the two subproblems discussed previously. Throughout this chapter we refer to this proposed approach as *TransFall*.

TransFall is composed of four computational modules. The first module performs data transformation on column variables along the vertical direction. The second module further transforms the marginal distribution of the covariates along the horizontal direction. These two modules form a two-tier data transformation layer to solve data transformation problem.

The transformed dataset and the obtained weight parameters are passed into the third module for label estimation, by encoding the kernel methods into a weighted least-mean-squares fitting problem to reliably estimate labels as required by label estimation problem. Finally, with the estimated label set $\hat{Y}_t$ for X_t, TransFall can accomplish the goal discussed in ATL Problem by training a classification model

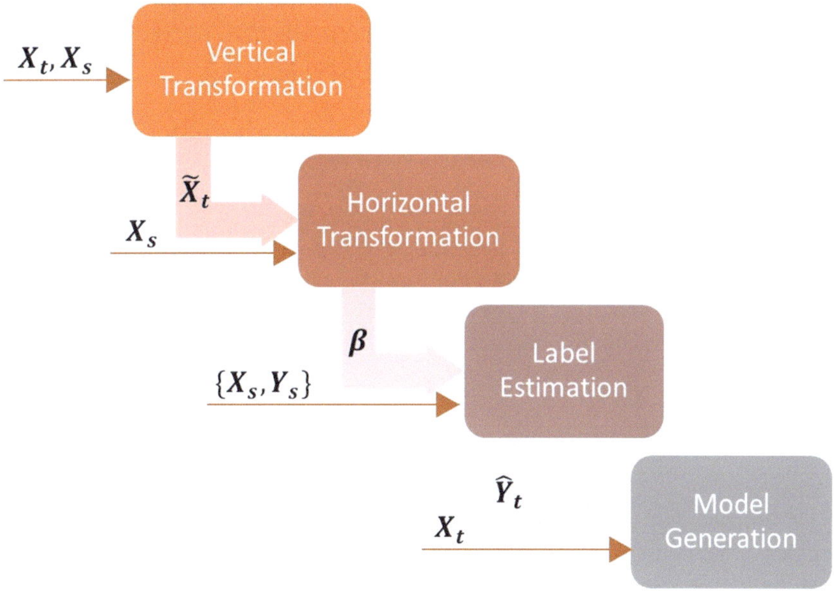

Fig. 11.2 TransFall: a sequential transfer learning design for activity recognition

in a supervised manner to conduct activity recognition on future sensor data in the target platform.

The two-tier data transformation approach in TransFall performs marginal distribution matching through two modules, namely vertical transformation and horizontal transformation. The output of data transformation is fed into the label estimation module, which utilizes a weighted least-mean-squares fitting and kernel methods to label the target data with respect to the source data. Generating a model using labeled training data is a straightforward machine learning task that is eliminated for brevity.

11.4.1 Vertical Transformation

The vertical transformation module aims to match the marginal distributions of individual column variables between X_s and X_t. This goal is accomplished using a naive Bayes approximation approach and a linear transformation.

Due to the nature of multidimensionality in the data sample $x \in \mathcal{R}^d$, the marginal distribution $P(x)$ can be determined by a joint probability distribution $P(x_1, \ldots, x_d)$. The raw signals collected with a sensing platform are converted into vector objects in the feature extraction phase. This process can be viewed as applying a projection of the input signals onto a designated feature space $\mathcal{F}$ [354].

Each feature $f_i \in \mathcal{F}$ is computed independently, and hence, we can use naive Bayes approximation to factorize the joint probability distribution (Eq. (11.3)), where the ith column variable in data sample x corresponds to feature f_i:

$$P(x) = P(x_1, \ldots, x_d) = \prod_{i=1}^{d} P(x_i). \tag{11.3}$$

Therefore, the optimization problem presented in (11.1) can be rephrased as the summation of d subordinate optimization problems for d column variables, as shown below, where $\phi_i \in \Phi$ is a mapping function that converts the original probability distribution of a one-dimensional variable into a different distribution:

$$\underset{\Phi}{\text{Minimize}} \sum_{i=1}^{d} \int_{x_i} |P_s(x_i) - \phi_i(P_t(x_i))| dx_i. \tag{11.4}$$

In practice, however, the true distribution of a random variable is unattainable, and hence the normal distribution is commonly adopted to approximate the marginal distribution of sensor data in activity recognition applications. As a result, each column variable x_i is assumed to be drawn from a normal distribution with mean μ_i and variance σ_i^2, denoted as $x_i \sim \mathcal{N}(\mu_i, \sigma_i^2)$. These two parameters can be empirically computed as follows:

$$\mu_i := \frac{1}{N} \sum_{j=1}^{N} x_{ij} \tag{11.5}$$

$$\sigma_i^2 := \frac{1}{N} \sum_{j=1}^{N} (x_{ij} - \mu_i)^2. \tag{11.6}$$

To minimize the difference of the empirical means and variances between column variables in X_s and X_t, we perform a linear transformation on each dimension $i \in [1, d]$ in X_t with respect to X_s, as shown in Algorithm 6.

The output of the vertical transformation module is the transformed target dataset $\tilde{X}_t$, where $\tilde{x}_i^t \sim \mathcal{N}(\mu_i^s, \sigma_i^{s2})$ for each $i \in [1, d]$. It is worth mentioning that this module can be potentially used in conjunction with standard machine learning algorithms to provide an additional layer of data transformation by replacing the original target dataset, X_t, with the transformed dataset, $\tilde{X}_t$.

Algorithm 6 Vertical transformation

Input: $X_s \in \mathcal{R}^{d \times N_s}$ and $X_t \in \mathcal{R}^{d \times N_t}$
Output: $\tilde{X}_t$
Initialize a $d \times d$ matrix W
Initialize a $d \times 1$ vector $\overline{b}$

for dimension $i \in [1, d]$ **do**
  Compute μ_i^s and σ_i^s of X_i^s
  Compute μ_i^t and σ_i^t of X_i^t
  Solve b_i and w_i for $(b_i + w_i x_i^t) \sim \mathcal{N}(\mu_i^s, {\sigma_i^s}^2)$
  $W_{ii} = w_i$
  $\overline{b}_i = b_i$
end
$\tilde{X}_t = \overline{b}I + W X_t$

Example 11.2 (Vertical Transformation)
Problem: Consider the ith column in X_s and X_t:

$$x_i^s = \{-2, 4, 2, 1, 0\}$$
$$x_i^t = \{-7, 4, 0, -4, 2\}$$

Obtain w_i and b_i.
Solution: To find w_i and b_i, we have

$$\mu_i^s = 1, \sigma_i^s = 2$$
$$\mu_i^t = -1, \sigma_i^t = 4$$

w_i and b_i are calculated as follows:

$$\frac{x_i^t - \mu_i^t}{\sigma_i^t} = \frac{x_i^s - \mu_i^s}{\sigma_i^s}$$
$$\frac{x_i^t + 1}{4} = \frac{x_i^s - 1}{2}$$
$$\frac{1}{2}x_i^t + \frac{3}{2} = x_i^s \sim N(\mu_i^s, \sigma_i^s)$$
$$w_i = \frac{1}{2}, b_i = \frac{3}{2}$$

11.4.2 Horizontal Transformation

The horizontal transformation module aims to further reduce the discrepancies in multivariate variables (e.g., row vectors) between X_s and $\tilde{X}_t$ using an importance sampling technique.

A number of studies [355, 356] have utilized the importance sampling method in the context of covariate shift. This technique aims to solve the covariate shift problem by finding a weight factor β for X_s. The approach assigns a higher weight to those source data samples that are more representative of the target dataset. To this end, Eq. (11.1) in data transformation problem can be revised as follows:

$$\underset{\beta}{\text{Minimize}} \int_x |\beta(x)P_s(x) - P_t(x)|dx \,. \tag{11.7}$$

However, as discussed previously, the real probability distributions $P_s(x)$ and $P_t(x)$ are unknown in practice. To address this limitation, we can use a kernel-based algorithm such as empirical Kernel Mean Matching (eKMM) [354] to find the optimal weight factor β with the use of Reproducing Kernel Hilbert Space (RKHS) technique.

Let $\Phi : \mathcal{X} \mapsto \mathcal{F}$ be a function that maps a vector variable X into a feature space $\mathcal{F}$. The output of the eKMM algorithm is the optimal weight factor β, which can minimize the distance between the empirical means of X_s and $\tilde{X}_t$ on the feature space $\mathcal{F}$, as shown in the following equations:

$$\begin{aligned}
&\underset{\beta}{\text{Minimize}} && \left\| \frac{1}{N_s}\sum_{i=1}^{N_s} \beta_i \Phi(x_i^s) - \frac{1}{N_t}\sum_{j=1}^{N_t} \Phi(x_j^t) \right\|^2 \\
&\text{Subject to} && \beta_i \geq 0,\ i \in [1, N_t] \\
& && \left| \frac{1}{N_t}\sum_{i=1}^{N_t} \beta_i - 1 \right| \leq \epsilon.
\end{aligned} \tag{11.8}$$

The first constraint in (11.8) refers to the non-negative property of probability and the second constraint guarantees that the re-weighted distribution $\beta(x)P_s(x)$ is close to a valid probability distribution that can sum to 1 [354].

In this chapter, we use RKHS technique to solve the optimization problem in (11.8) based on an important property described below [357].

Proposition 11.1 *Given a positive definite kernel k over a vector space $\mathcal{X}$, we can find a Hilbert space $\mathcal{H}$ and a mapping function $\Phi : \mathcal{X} \mapsto \mathcal{H}$, such that*

$$k(x_i, x_j) = \langle \Phi(x_i), \Phi(x_j) \rangle_{\mathcal{H}},$$

where $x_i, x_j \in \mathcal{X}$.

Therefore, with the use of a kernel function that is positive definite on Euclidean space $\mathcal{R}^d$, the optimization problem in (11.8) can be solved without explicitly defining the mapping function Φ. For this purpose, we use Gaussian kernel as follows:

$$k(x_i, x_j) = \exp -\frac{\|x_i - x_j\|^2}{2\sigma^2}.$$

Therefore, the objective function in (11.8) can be rephrased as shown in (11.9):

$$\underset{\beta}{\text{Minimize}}\ \frac{1}{2}\beta^\top K\beta - \kappa^\mathsf{T}\beta, \tag{11.9}$$

where the kernel matrix K and the kernel expansion κ are given by:

$$\begin{aligned} K_{ij} &:= k(x_i^s, x_j^s) \\ \kappa_i &:= \frac{N_s}{N_t}\sum_{j=1}^{N_t} k(x_i^s, x_j^t). \end{aligned} \tag{11.10}$$

As a result, the optimal β can be determined by solving (11.9) with the constraints listed in (11.8) using quadratic programming. Similar to the vertical transformation module, the horizontal transformation module also has the potential to be coupled with existing machine learning algorithms that support the sample re-weighting.

Example 11.3 (Horizontal Transformation)
Problem: Consider X_i and X_j:

$$x_i = \{-2, 1, 4, 0.2, 0\}$$
$$x_j = \{-1, 0, 2, 0.5, 1\}$$

Given $\sigma = 0.3$, determine $k(x_i, x_j)$.
Solution: In order to obtain $k(x_i, x_j)$, one can use Gaussian kernel function as below:

$$\begin{aligned} k\left(x_i, x_j\right) &= \mathrm{e}^{-\frac{\|x_i - x_j\|^2}{2\sigma^2}} \\ &= e^{-\frac{2.66^2}{2\times 0.3^2}} = 8.5 \times 10^{-18} \end{aligned}$$

11.4.3 Label Estimation

Given the transformed target dataset $\tilde{X}_t$ and the weight factor β, which approximates the distribution of $\tilde{X}_t$ using the source dataset X_s, the label estimation module intends to estimate the label set $\hat{Y}_t$ for $\tilde{X}_t$ in preparation for training an activity recognition model for the target platform using $\{X_t, \hat{Y}_t\}$, as described in label estimation problem.

We note that the ordering of the data samples in X_t and $\tilde{X}_t$ are identical because the vertical transformation module only works on the column variables of X_t rather than the rows. Therefore, the label set $\hat{Y}_t$ estimated for $\tilde{X}_t$ can be directly used to label the original target dataset X_t.

The optimal solution f for (11.2) should minimize the overall estimation errors between $f(x_i)$ and y_i for all $x_i \in \tilde{X}_t$. However, the true label $y_i \in Y_t$ is unknown for the target dataset. To leverage the label information of the source dataset, we rewrite the equation in (11.2) using X_s, by encoding the weight factor β in the objective function as follows:

$$\underset{f}{\text{Minimize}} \sum_{i=1}^{N_s} |y_i - \beta(x_i) f(x_i)|,$$

where $x_i \in X_s$.

We rewrite this optimization problem using a weighted least-mean-squares (LMS) fitting technique with a 2-norm regularization term as shown below. The LMS technique is a popular approach for parameter estimation in linear models [354, 358].

$$\underset{f}{\text{Minimize}} \sum_{i=1}^{N_s} \beta_i (y_i^s - f(x_i^s))^2 + \lambda \|f\|^2. \tag{11.11}$$

However, the optimal function f in (11.11) is not necessarily a linear model. Therefore, we convert (11.11) into a new form using a linear model, based on the representer theorem described below [359].

Proposition 11.2 *Given a kernel k and the corresponding RKHS $\mathcal{H}$, for a function $L : \mathcal{R}^n \mapsto \mathcal{R}$ and a non-decreasing function $\Omega : \mathcal{R} \mapsto \mathcal{R}$, if the optimization problem can be expressed as:*

$$\underset{f \in \mathcal{H}}{\mathit{Minimize}}\ L(f(x_1), \ldots, f(x_n)) + \Omega(\|f\|^2_{\mathcal{H}}),$$

then the optimal solution can be expressed as:

$$f^* = \sum_{i=1}^{n} \alpha_i k(x_i, \cdot),$$

Furthermore, if Ω is strictly increasing, all solutions have this form.

Utilizing this theorem, Eq. (11.11) can be rewritten as follows:

$$\underset{\alpha}{\text{Minimize}} \sum_{i=1}^{N_s} \beta_i \left(y_i^s - \sum_{j=1}^{N_s} \alpha_j k(x_j^s, x_i^s) \right)^2 + \lambda \left\| \sum_{j=1}^{N_s} \alpha_j k(x_j^s, \cdot) \right\|^2,$$

which can be written (after extension) as shown in (11.12):

$$\underset{\alpha}{\text{Minimize}} \ (Y_s - K\alpha)^\top \overline{\beta} (Y_s - K\alpha) + \lambda \alpha^\top K \alpha, \tag{11.12}$$

where K represents the kernel matrix defined in (11.10), and $\overline{\beta}$ is a $N_s \times N_s$ diagonal matrix of β. If K and $\overline{\beta}$ are full rank matrices, the optimal solution for α can be directly derived using the following equation [354]:

$$\alpha = (\lambda \overline{\beta}^{-1} + K)^{-1} Y_s. \tag{11.13}$$

Because the label set in activity recognition often contains multiple activity classes, we use a one-to-all approach for label estimation, by first solving L optimal linear models α^m for all activity labels, and then combining L corresponding estimations of the data sample $x_i^t \in \tilde{X}_t$, to make the final prediction.

$$\hat{y}_i^t = \underset{m}{\text{argmax}} \sum_{j=1}^{N_s} \alpha_j^m k(x_j^s, x_i^t). \tag{11.14}$$

After obtaining the label set $\hat{Y}_t$, the last step to solve the ATL problem described earlier is to train an activity recognition model on the target platform. Given the training set $\{X_t, \hat{Y}_t\}$, a variety of classification algorithms can be used to accomplish this task in a supervised manner, and hence the last module, model generation, is not the focus of our framework design.

Example 11.4 (Label Estimation)
Problem: Consider two row vectors in source domain x_1 and x_2:

$$x_1 = \{0.2, 0, 1\}$$

(continued)

Example 11.4 (continued)

$$x_2 = \{0.3, -1, 0.8\}$$

$$y = \{1, 2\}$$

$$\beta = \{0.5, 1\}$$

Given $\lambda = 0.001, \sigma = 0.3$, determine α.
Solution: According to (11.14), α is obtained as follows:

$$\begin{aligned}
\alpha &= \left(\lambda\bar{\beta}^{-1} + K\right)^{-1} Y_s \\
&= \left(0.001 \times \begin{bmatrix} 0.5 & 0 \\ 0 & 1 \end{bmatrix}^{-1} + \begin{bmatrix} k(x_1, x_1) & k(x_1, x_2) \\ k(x_2, x_1) & k(x_2, x_2) \end{bmatrix}\right)^{-1} \begin{bmatrix} 1 \\ 2 \end{bmatrix} \\
&= \left(\begin{bmatrix} 0.002 & 0 \\ 0 & 0.001 \end{bmatrix} + \begin{bmatrix} 0 & 0.003 \\ 0.003 & 0 \end{bmatrix}\right)^{-1} \begin{bmatrix} 1 \\ 2 \end{bmatrix} \\
&= \begin{bmatrix} 714 \\ -143 \end{bmatrix}
\end{aligned} \tag{11.15}$$

11.5 Validation Approach

We conducted experiments on three publicly available datasets [347, 360], as shown in Table 11.1. We designed two tasks, *label estimation* and *activity recognition*, to evaluate the performance of TransFall in three scenarios including *cross-platform*, *cross-subject*, and *hybrid*. The cross-platform scenario was further separated into four sub-scenarios, regard to different settings of the source and the target for knowledge transfer.

In this chapter, we will use the notations shown in Table 11.1 to refer to various datasets, transfer learning scenarios, and comparison approaches. Before presenting the actual results, let us present experimental settings, and details of the comparison approach and algorithms against which the performance of TransFall is compared.

11.5.1 Overview of the Datasets

Three publicly available datasets obtained using a variety of mobile devices are used to assess the performance of TransFall on various data sources. For clarification, this

Table 11.1 Notations and abbreviations used in experimental analyses

	Notation	Description
Dataset	Phone	Dataset gathered by 8 smartphones from 9 subjects
	Watch	Dataset gathered by 4 smartwatches from 5 subjects
	HART	Dataset gathered by one smartphone from 30 subjects
Cross-platform scenario	P2P-S	Same-model phone-to-phone
	P2P-D	Different-model phone-to-phone
	W2W-S	Same-model watch-to-watch
	W2W-D	Different-model watch-to-watch
Comparison group	NN	Using nearest neighbor algorithm following the naive approach
	DT	Using decision tree algorithm following the naive approach
	LR	Using logistic regression algorithm following the naive approach
	SVM	Using support vector machine algorithm following the naive approach
	Upper	Machine learning model trained with ground truth data
	IWLSPC	Comparison framework introduced in [353]

chapter uses the term "*individual data*" to refer to the data samples gathered by a mobile device with a subject.

The first dataset *Phone* [347] was obtained using a variety of smartphones with different models. Sensor data was collected from 9 subjects for 6 daily activities (i.e., biking, sitting, standing, walking, climbing upstairs, and climbing downstairs). This dataset was provided in the form of raw signals, and we extracted 8 widely used statistical features from the 3D acceleration signals using a sliding window of 2 seconds. For each subject and each smartphone device, the number of data samples ranged from 944 samples to 2283 samples, with an average of 1526 samples.

The second dataset *Watch* [347] was obtained using 4 smartwatches with 2 different device models in the same experiment as the previous one. During the data collection, each participant was asked to wear 4 smartwatches on both arms and perform 6 daily activities as above-mentioned. The same process was performed on sensor signals as for the first dataset, to extract 8 features on the time domain, including the *mean*, *standard deviation*, *root-mean-squares*, *maximum*, *minimum*, and *pairwise correlation* among the three uni-axial signals.

The third dataset *HART* [360] contained sensor data regarding 6 daily activities (i.e., standing, sitting, lying, walking, climbing upstairs, and climbing downstairs) collected from 30 subjects. This dataset contained feature representations of signals. We first performed a correlation-based feature selection combined with the best-first-search strategy to reduce the dimensionality of the original feature space. The integrated feature selection functions in the WEKA (Waikato Environment for Knowledge Analysis) platform [361] were used to accomplish this task, which resulted in 27 most salient features out of 561 original features. The total number of data samples was 10,411. For each subject, the number of data samples ranged from 284 to 413, with an average of 347.

The individual data sizes vary across different devices, in particular for the Phone dataset. This is due to the instability of sensing platforms in practical use, which results in some devices having more missing samples than others.

11.5.2 Cross-Domain Transfer Learning Scenarios

Two individual datasets were first selected as "source" and "target." At the beginning of one trial, no more than 50% of the data were randomly selected from the source and the target, where the data from the source had true labels, and the objective of the label estimation task was to estimate the corresponding labels for the data from the target. In the activity recognition task, the data selected from the target, together with the obtained labels from the first task, were fed into a machine learning model as the training set, to perform activity recognition on the remaining data from the target.

Three scenarios of source-and-target pairs were tested in the experiments discussed in this chapter. These scenarios are discussed next.

For cross-platform transfer learning, the *Phone* and *Watch* datasets were used, because they are gathered by multiple devices with different models. Four sub-scenarios were designed with respect to different settings of the source and the target for knowledge transfer. These sub-scenarios are described below:

1. *Same-Model Phone-to-Phone*. This scenario contained 34 individual data files collected from 9 participants. In each trial, the source data and the target data were gathered by two smartphones with the same device model.
2. *Different-Model Phone-to-Phone*. This scenario contained 35 individual data files collected with 9 participants. In each trial, the source data and the target data were gathered by two smartphones with different device models.
3. *Same-Model Watch-to-Watch*. This scenario contained 10 individual data files collected from 5 participants. In each trial, the source data and the target data were gathered by two smartwatches with the same device model.
4. *Different-Model Watch-to-Watch*. This scenario contained 10 individual data files collected with 5 participants. In each trial, the source data and the target data were gathered by two smartwatches with different device models.

For cross-subject transfer learning, the source data and the target data were collected from different subjects but using the same device. All three datasets were used in this scenario, and the evaluation results were averaged over all the devices for each dataset.

The hybrid transfer learning scenario was the combination of the previous two scenarios, where both the device and the subject of the source data differ from that of the target data. For this purpose, *Phone* and *Watch* datasets were used in this scenario because they represented both cross-platform and cross-subject cases.

Example 11.5 (Cross-Domain Transfer Learning Scenarios)
Problem: Determine which sub-scenarios does Example 11.1 belong.
Solution: Scenario in Example 1.1 is a hybrid scenario since it contains cross-platform (i.e., source domain and target domain from different subjects) and cross-subject scenarios (i.e., source domain and target domain use different device model).

11.5.3 Comparison Approach and Performance Metrics

We compared the performance of TransFall against other approaches including four popular machine learning algorithms and another transfer learning algorithm. These algorithms are briefly described as follows:

- **Naive**: naive approach trained an existing machine learning algorithm using the source data and then directly applied the model to the target data to estimate the corresponding labels. Four standard machine learning algorithms, including nearest neighbor (NN), decision tree (DT), logistic regression (LR), and linear support vector machine (SVM), were tested following this approach to provide baseline measures for comparison.
- **Upper**: the upper bound performance of label estimation was naturally 100% because it involves gathering ground truth labels in "target." The upper bound of activity recognition, however, was estimated by training a machine learning model with the true labels of the training dataset.
- **TransFall**: the proposed framework consists of a two-tier data transformation (vertical and horizontal) layer and a label estimation layer.
- **IWLSPC**: a cross-subject learning framework for activity recognition [353], referred to as importance-weighted least-squares probabilistic classifier (IWL-SPC).

Two separate metrics were adopted for performance evaluation of the two tasks. A description of those metrics is as follows:

- **Labeling accuracy**: this refers to the correct labeling rate of all the data samples been labeled in the first task.
- **Classification accuracy**: the classification accuracy on a test set was computed to evaluate the performance of the trained model for activity recognition.

11.5.4 Choice of Classification Model

This section discusses the empirical choice of a machine learning model for the activity recognition task. The objective of this task is to investigate the effectiveness of transfer learning on the performance of the downstream machine learning models. Four popular machine learning algorithms were examined for their performance in activity recognition. Those algorithms included nearest neighbor (NN), decision tree (DT), logistic regression (LR), and linear support vector machine (SVM).

In each trial, no more than 50% of the data samples in one data file were randomly selected as the training set, and the remaining data were used as the test set to estimate the classification accuracy of each model. Ten rounds of such trial were performed on each individual data file. Overall, logistic regression achieved the best performance, with an average accuracy of 92% on the three datasets. For this reason, logistic regression was used as the machine learning classifier in the activity recognition task presented in this chapter.

11.6 Results

11.6.1 Cross-Platform Transfer Learning Results

This scenario refers to the case when source data and target data are gathered by different devices asynchronously. We further divide this scenario into four sub-cases as shown in Table 11.1.

Figure 11.3 shows the results of label estimation using different approaches in two cross-platform transfer learning cases including P2P-S and P2P-D. In this figure, the red central mark on each box indicates the mean value of the labeling accuracy, the bottom and top edges indicate the 25th and 75th percentiles, respectively, and the outliers are denoted using plus symbols.

The labeling accuracy is higher in P2P-S case than that of the P2P-D case. This result is consistent with general expectation because the source device and the target device are of different models for the P2P-D case, which results in a higher level of diversity between the two domains due to differences such as sampling frequency and platform configuration.

Comparing to the other five approaches shown in Fig. 11.3, TransFall achieves the highest labeling accuracy on average, with a correct labeling rate of 0.88 and 0.79 in the two phone-to-phone transfer learning cases. The increase in the labeling accuracy of TransFall compared to other approaches is more than 2.7% in the P2P-S case and more than 6.3% in the P2P-D case.

For activity recognition, a logistic regression model was empirically chosen to carry out activity recognition after label information transfer. Intuitively we expect the classification accuracy to be consistent with the labeling accuracy because the quality of the training dataset is determined by the precision of

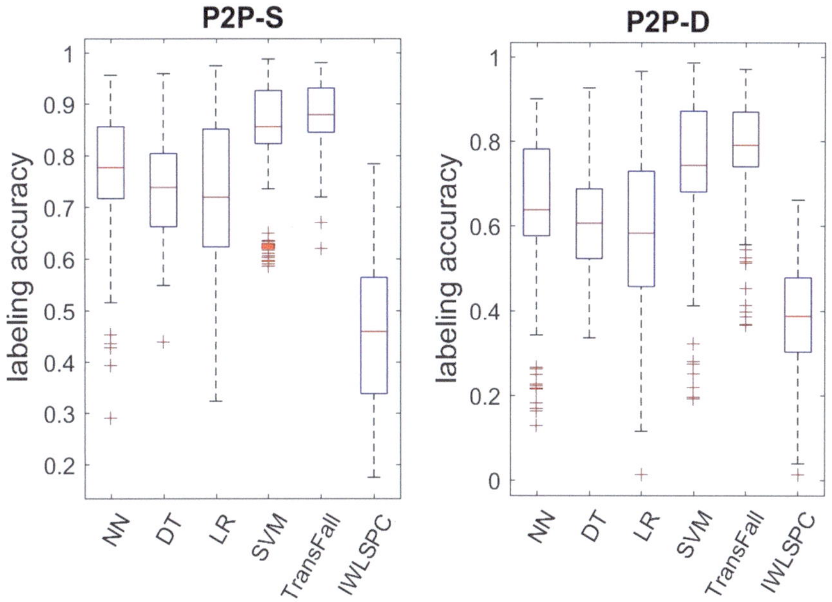

Fig. 11.3 Results of label estimation in phone-to-phone transfer learning scenario

labeling the target data in the previous task. By perming this analysis, we can see that TransFall still achieves better performance in activity recognition than other approaches. Its classification accuracy is 88.4% and 76.6% in the two phone-to-phone transfer learning scenarios. Moreover, the performance improvement of the machine learning model trained by TransFall is more than 7.7% compared to the machine learning models trained by other approaches on the smartphone data.

11.6.2 Cross-Subject Transfer Learning Results

In this scenario, source data and target data are collected from two subjects using the same type of mobile device.

Figure 11.4 shows the results of label estimation on datasets "Phone" and "Watch" separately. TransFall achieves a labeling accuracy that is 6.9% better than the other approach on the "Phone" dataset and approximately 9.5% better than the other algorithms on the "Watch" dataset.

After the label estimation is done, we have a training dataset that can be used for model creation. Given label sets obtained using different approaches, we can examine the performance of activity recognition model trained. The empirical upper bound of classification accuracy is 92.1% on Phone dataset and 88% on Watch dataset. TransFall still has the best performance in activity recognition compared

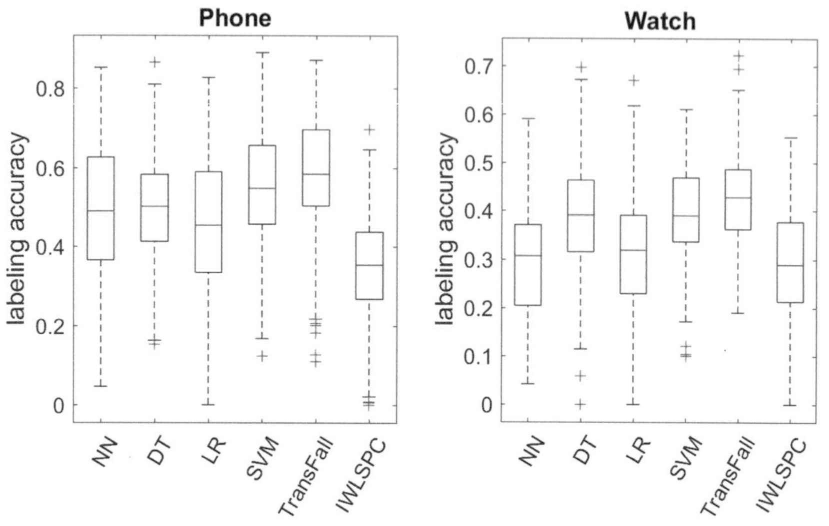

Fig. 11.4 Results of label estimation in cross-subject scenario on datasets "Phone" and "Watch"

to other approaches. TransFall achieves an accuracy that is 7.4% higher than other techniques on Phone dataset and 20.2% higher than other techniques on Watch dataset.

11.6.3 Hybrid Transfer Learning Results

The hybrid scenario combines a cross-platform scenario and a cross-subject scenario. As a result, the discrepancies between source data and target data are more significant than that of the previous two scenarios.

Figure 11.5 shows the results of label estimation on Phone dataset and Watch dataset. Overall, the performance of all the approaches has an obvious decline in the hybrid scenario compared to the previous two scenarios. This result is mainly caused by increased uncertainty in the target dataset with respect to the source dataset. Nevertheless, TransFall achieves the best performance comparing to other approaches in this scenario, with an accuracy increase of 41.9% on average on Phone dataset, and 26.4% on average on Watch dataset.

11.6.4 Transformation Module Analysis

One can further investigate the potential of using the individual transformation module of TransFall in conjunction with standard machine learning algorithms,

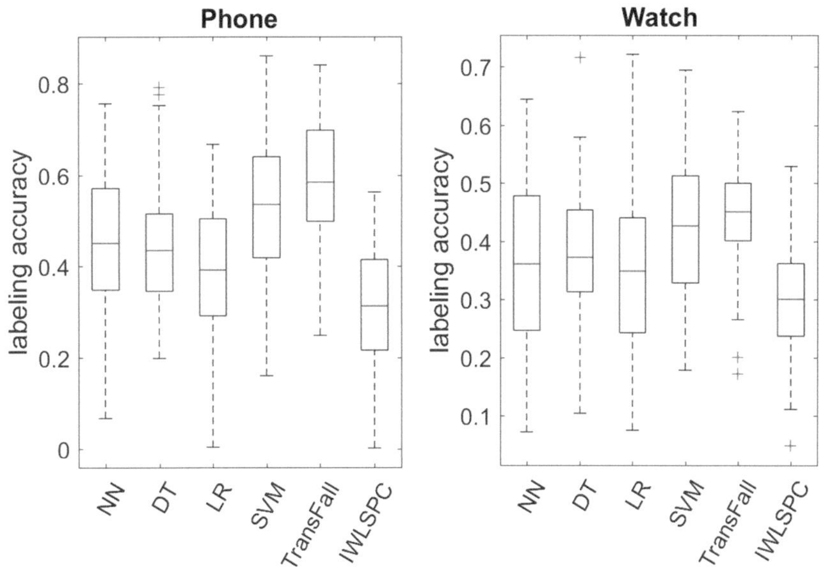

Fig. 11.5 Results of label estimation in hybrid scenario on "Phone" and "Watch" datasets

to reduce the differences in data distributions between the source and the target. The results of label estimation in three scenarios are shown in Fig. 11.6, where the blue bars refer to the results of applying the approaches mentioned in Table 11.1, the circle marks indicate the results of performing vertical transformation (VT) module on target data before using a machine learning algorithm to estimate the corresponding labels, and the square marks refer to the results of performing horizontal transformation (HT) module on source data to obtain the weight factor that can be used in the process of IWLSPC.

In general, utilizing VT module as an additional layer for data transformation can steadily improve the labeling accuracy of an existing machine learning algorithm in the cross-platform scenario, in which Phone dataset is used for validation. The performance improvement is 9.5–20.7%, with an average of 17.4% over the five algorithms. In the cross-subject scenario where HART dataset is used for validation, although adding VT module reduces the labeling accuracy of DT, it still achieves an accuracy increase of 3.8–11% for the other four algorithms. These results demonstrate the potential of combining VT module in TransFall with existing machine learning algorithms to diminish the discrepancy between the source data and target data in the presence of covariate shift.

Unlike VT module that performs a linear transformation on target data independently to machine learning models, HT module determines a weight factor for source data that cannot directly work in conjunction with any existing machine learning algorithms. In this experiment, we combine HT module with the probabilistic least-squares fitting model used in IWLSPC, to investigate the potential of

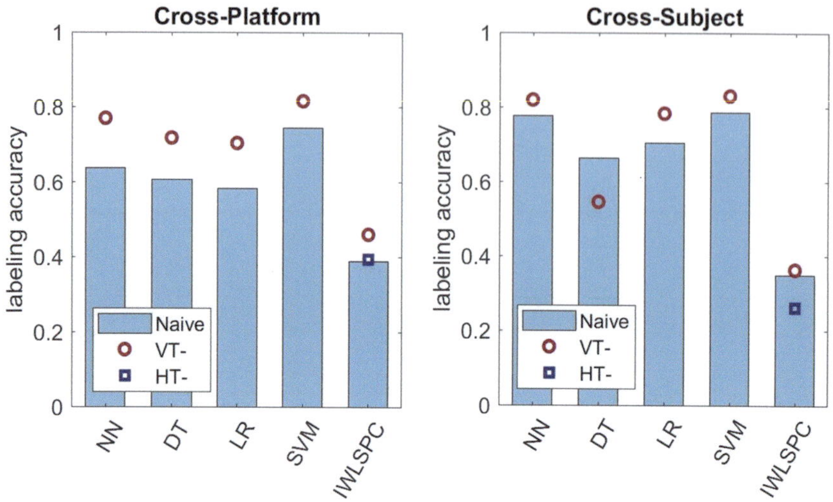

Fig. 11.6 Results of label estimation using individual data transformation module in conjunction with existing machine learning algorithms in cross-subject and cross-platform learning scenarios

utilizing HT module with other weighted label estimation algorithm. However, the results shown in Fig. 11.6 do not support this hypothesis.

11.6.5 Parameter Examination

Different values were examined for the two parameters, σ and λ, for the label estimation task in the three scenarios. Parameter σ is used in Gaussian kernel to scale the input, and λ is used in the least-mean-squares fitting to weigh the regularizer term. In this test, when the value of σ changes (left plot in Fig. 11.7), λ is set to 0.001; when the value of λ changes (right plot in Fig. 11.7), σ is set to 0.3.

When the value of σ increases from 0.01 to 0.3, the performance of TransFall improves by 9.3% in cross-platform scenario, 2.2% in cross-subject scenario, and 26.7% in hybrid scenarios. However, the changes in labeling accuracy tend to be different among the three scenarios along with the continuous increase of σ value, as the accuracy gradually increases in the cross-platform scenario but drops in the other two scenarios.

In the right plot in Fig. 11.7, the performance of TransFall is relatively stable in cross-platform and cross-subject scenarios, when λ is less than 0.72; but the accuracy drops by 3% and 8.6%, respectively, when λ increases to 10. In hybrid scenario, the performance of TransFall fluctuates slightly along with the changes of λ, with a standard deviation of 0.009 in the labeling accuracy. To validate the

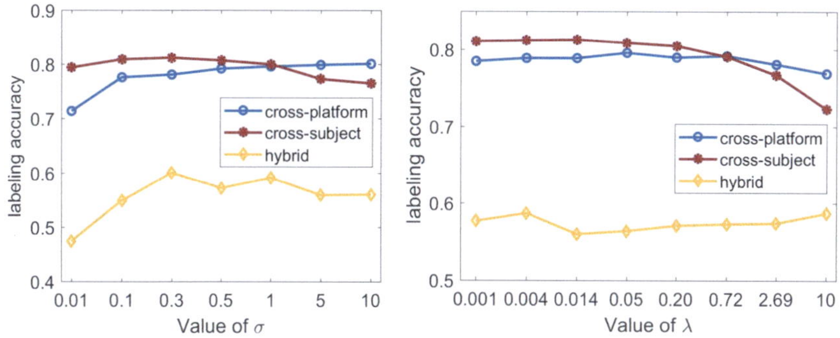

Fig. 11.7 Results of label estimation in three scenarios with different parameter values in TransFall

performance of TransFall in other experiments discussed in this section, σ is set to 0.3 and λ is set to 0.004.

11.7 Exercise Problems

Problem 11.1 What are scalability challenges of machine learning algorithm design in mobile health systems?

Problem 11.2 Explain spatial uncertainty and provide two examples of such uncertainties.

Problem 11.3 What is the goal of a transfer learning algorithm? Explain the terms *source* and *target* in the context of transfer learning.

Problem 11.4 What is a teacher/learner approach to transfer learning?

Problem 11.5 Discuss similarities and differences between TransFall IWLSPC.

Problem 11.6 What is *importance sampling* and how is it used in the design of TransFall?

Problem 11.7 What is the goal of *label estimation* in TransFall? Can we do transfer learning without this label estimation module?

Problem 11.8 Prove that the overall complexity of TransFall is cubic in the size of the source dataset.

Problem 11.9 In this chapter, we used classical machine learning algorithms such as nearest neighbor, decision tree, logistic regression, and support vector machine classifiers for activity recognition. What are the advantages and disadvantages of using deep neural networks instead of these classical algorithms for activity recognition and transfer learning?

Problem 11.10 We performed phone-to-phone transfer learning in this chapter. Use a similar strategy to obtain results for watch-to-watch transfer learning. The two new scenarios that you will be focusing on are denoted as W2W-S and W2W-D in Table 11.1. Plot and discuss your results.

Problem 11.11 Using the labeled training data that you obtain in Problem 11.10, train a machine learning model using logistic regression, similar to what we did in this chapter. Plot your results and compare the performance of TransFall against the competing algorithms. Also, discuss the difference between the performance of TransFall for phone-to-phone transfer learning versus that of watch-to-watch transfer learning.

Problem 11.12 We showed the results of cross-subject transfer learning on two datasets, "Phone" and "Watch" in this chapter. In this problem, we want to perform transfer learning for label estimation on HART dataset. Plot and discuss your results.

Problem 11.13 Using the labeled training data that you obtain in Problem 11.12, train a machine learning model for activity recognition on HART dataset. Plot your results and compare the performance of TransFall against the competing algorithms on the HART dataset. Also, discuss the difference between the performance of TransFall on "Phone" and "Watch" datasets compared to HART.

Problem 11.14 We discussed the labeling accuracy of various approaches for hybrid transfer learning in this chapter. Continue the analysis presented in that section to calculate activity recognition accuracy of all the five competing algorithms on the two datasets under consideration, namely "Phone" and "Watch" datasets. Plot and discuss your results.

Problem 11.15 We performed analysis using the vertical transformation module of TransFall in conjunction with standard machine learning algorithms for cross-subject and cross-platform scenarios in this chapter. Design an experiment for hybrid scenario and obtain the performance results. Plot and discuss your results.

Chapter 12
Applied Machine Learning for Computer Architecture Security

12.1 Introduction

The security of a computer system can be compromised at the computer architecture level through various types of attacks such as by executing malicious applications to infect the system or deploying microarchitectural Side-Channel Attacks (SCAs) to infer confidential information [286, 290, 362–364]. The rapid development of information technology has made malware and microarchitectural SCAs serious threats to the security of modern computer systems.

The recent proliferation of computing devices in mobile and Internet-of-Things (IoT) domains further exacerbate these threats calling for effective security countermeasure solutions. There exist some important factors influencing the security vulnerability of embedded systems and IoTs including the limited energy and resources available, the low computational capacity, and a significant number of computing nodes in the network. These devices are connected over the Internet network, which drastically increases the necessity of providing advanced security mechanisms to protect the integrity and confidentiality of the authenticated users' information.

12.1.1 Malware

Malware, a broad term for any type of malicious software, is a piece of code designed by cyber attackers to infect the computing systems without user consent. Though varied in classes and functionalities, malware primarily serves harmful purposes such as providing a remote control for an attacker to use an infected machine, stealing sensitive information, unauthorized data access, destroying files, running intrusive programs on devices to perform Denial of Service (DoS) attack, and disrupting essential services to perform financial fraud.

S. Rafatirad et al., *Machine Learning for Computer Scientists and Data Analysts*,
https://doi.org/10.1007/978-3-030-96756-7_12

Malicious software infections have plagued companies, organizations, and individual users for many years, and are significantly growing stealthier and increasing in numbers. Given the different malicious purposes and characteristics, malware can be classified into various categories. One could refer to a group of malware that shares many important characteristics as belonging to a single malware family. Below, we present brief descriptions of common types of malicious software attacks that are widely analyzed in prior malware research in computer security.

Virus It is a program that infects the executable software and, when run, causes the malicious pattern to spread and malfunction other executables. Virus typically refers to a malicious program that has the ability to replicate itself and attaches itself to a legitimate program or document while relying on the host program to get activated and affects the functionality of the systems when get executed. Viruses may further perform other malicious activities including creating a Backdoor malware for later malicious activities.

Trojan In computing, a Trojan, a.k.a Trojan Horse, is a program that appears harmless and legitimate while performing malicious operations in the background to damage and infect the computer systems such as downloading and installing other malware on the target system, modifying the user's system settings, or manipulating the host files without the user authorization. Unlike Viruses, a Trojan horse is not able to replicate itself, nor can it propagate without an end user's assistance. Hence, cyber attackers deploy social engineering techniques to persuade the end-users to execute the Trojan programs, which can occur through clicking on the email attachment or downloading the free program. Afterward, the malware is transferred to the user's computing device and the malicious code can run and carry out any malicious activities designed by the cyber attackers.

Rootkit A Rootkit is a malicious program that has the ability to hide its presence from the user system while enabling continued privileged access to the host system. Rootkits can prevent a malicious process from being visible in the system's list of processes, or keep its files from being read. They allow this concealment and remain undetected by modifying the host's operating system. Rootkit techniques are applied by various malicious software, at both user-mode and kernel mode. At user-mode, Rootkits take actions by instrumenting Application Programming Interface (API) calls, whereas at kernel-level, Rootkits interfere with operating system organizations as a device driver or a kernel module to hide the information about the victim's computer system.

Backdoor A Backdoor exploits method of bypassing normal authentication procedures in computing systems by avoiding the traditional security mechanisms. Backdoors can install themselves as part of an exploit to take advantage of the system's weaknesses or vulnerabilities. Once a system has been compromised by one of the malware attacks, one or more Backdoors may be installed on the victim computing system to facilitate the access of the cyber attacker to the target system in the future. Backdoors can also be installed prior to the main malicious activities allowing the cyber attackers entry without the consent of the user.

Worm A computer Worm is a self-replicating malicious software that duplicates itself to spread from one system to another. Worm typically spreads itself over a network of computers, relying on security failures on the target computer. Another characteristic that distinguishes Worms from other forms of malware is that they are standalone software that is able to operate and propagate autonomously, without the requirement of being activated by a host file on the target computer. Worms often deploy the invisible parts of an operating system to the user to exploit the computer system vulnerability. Once the computer system is infected by this attack, the Worm can perform different malicious activities to make the system more vulnerable and to degrade overall system performance.

Ransomware It is a recent malicious cyber-attack in which the user's information on the target system is locked usually through encryption algorithms while a payment is demanded from the user to restore access. The files are still available on the target computer but inaccessible by the user. After receiving the payment from the victim, the ransomed data is unlocked and decrypted. It is notable that the goal of Ransomware is mostly fraud, and unlike other types of malware, the victim is usually notified that an exploit has occurred and is given instructions on how to recover from the attack. While initially targeting individuals, recent ransomware cyber-attacks have been tailored toward larger groups such as businesses and organizations with the purpose of demanding larger payouts.

Blended Threat A.k.a. hybrid malware that combines the characteristics of two or more types of malware to build a more powerful and sophisticated malware attack. It can cause harm to the infected system or network as they propagate, using multiple infection methods while exploiting the systems' vulnerabilities to create various points of entry subject to attack.

Example 12.1 (Characterization of Malware Attacks)
Problem: Categorize different classes of malware.
Solution: This example characterizes various malicious software types comprising the security of computer systems and highlights important characteristics of each malware attack type.

(continued)

Example 12.1 (continued)

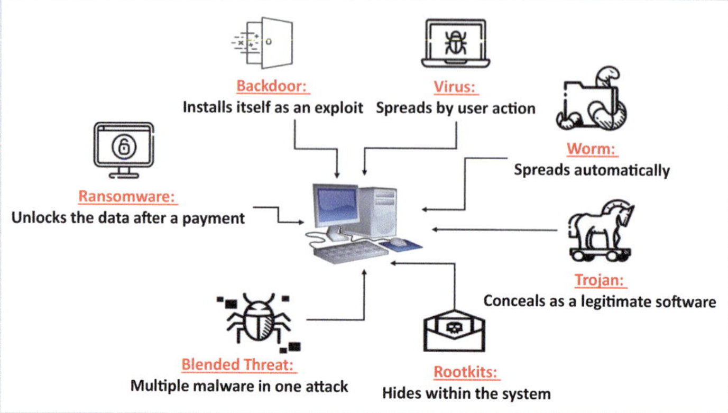

12.1.2 Microarchitectural Side-Channel Attacks

Microarchitectural side-channel attacks have posed serious threats to the security of modern computing systems. Such attacks exploit side-channel vulnerabilities stemming from fundamental performance-enhancing components such as cache memories, out-of-order execution, and speculative execution units. Recently, cache-based SCAs have shown powerful capabilities of stealing users' critical information such as secret keys of cryptographic applications within the same processing core or cross-core residency of victim applications. Below, an overview of some of the recently proposed microarchitectural SCAs is presented. Furthermore, Fig. 12.1 illustrates the working principles of three important cache-based SCAs discussed in this section including Flush+Reload, Prime+Probe, and Flush+Flush attacks.

Flush+Reload Flush+Reload attack exploits the weakness of page de-duplication and monitors memory access lines in shared pages. This attack flushes out the victim's data in the cache and waits for the victim's execution. The attacker then reloads data by accessing them and measures the accessing time. If accessing time is shorter, it infers the data is accessed by the victim; else, it has not been accessed by the victim [365].

Flush+Flush In this type of SCA, the setup and first stage are the same as Flush Reload. In the second stage, instead of reloading the shared memory blocks, the adversary still flushes the blocks. If the victim fetches a block into the cache, then flushing it will take a longer time than when it is out of the cache. So Flush has the same effect as Reload. Besides, a single Flush operation can serve as Check for the current round as well as Set for the next round [366].

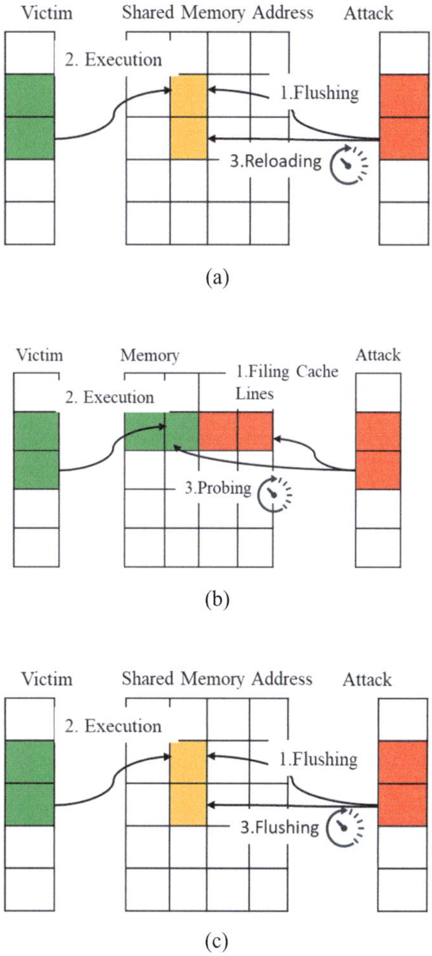

Fig. 12.1 Working principle of three cache-based side-channel attacks. (**a**) Flush+Reload. (**b**) Prime+Probe. (**c**) Flush+Flush

Prime+Probe This attack targets L1/L3 data caches. Prime Probe attack consists of two stages: Prime and Probe. In the Prime stage, the attacker builds an eviction set which is a group cache sets causing potential conflict with victims and fills cache with the eviction sets. Next, the attacker waits for victim execution and then re-accesses the eviction sets (Probe stage). If the accessing time is long enough, it means the victim accessed the data; else, the victim does not access the data [367].

Spectre Speculative execution is used in commercial processors to boost performance by executing the next execution path predicted by control flow [368]. When CPU waits for data coming from the memory or disk, the current register state is stored, and then the speculative instruction is executed. Once data arrives, the processor validates the correctness of the executed speculative path. If the prediction is right, the performance can be improved significantly; otherwise, the execution time is equal to being idle during the waiting period. And if the predicted execution path is wrong, executed results will still be kept. Recent Spectre attack takes exploit speculative execution by locating the instructions firstly and tricking the CPU into speculative and erroneous execution of this instruction sequence which results in information leakage.

Meltdown This type of attack exploits out-of-order execution of microprocessors to leak secret information from user space and kernel space of the same process and other processes [369]. Memory protection mechanism is provided in all processors through an operating system that prevents a user-level program from reading data from kernel space or from any other user-level program. However, side effects generated by out-of-order execution make it possible to bypass this protection mechanism. Even an unprivileged user level program can read all main memory with Meltdown attack which is a two-step attack. In the first step, meltdown bypasses the memory isolation by executing unprivileged instruction out-of-order. In the second step, it performs cache-based side-channel attack to observe footprints of accessed data from the cache.

12.2 Challenges Associated with Traditional Security Mechanisms

As mentioned in the prior section, due to the exceedingly challenging task of detection of new variants of malicious applications and emerging microarchitectural side-channel attacks, and to keep on combating the increase in malicious cyber-attacks, there is an urgent need to develop intelligent countermeasures to enhance the security of the system and protect legitimate users from these threats. Malware detection techniques are typically classified into two categories including signature-based detection and anomaly-based detection methods. The former looks for signatures of known malware traces to detect malicious patterns and the latter models the normal structure/behavior of programs or systems and attempts to detect deviations from this predefined detection model. Existing traditional malware detection methods such as Anti-Virus (AV) tools are mostly signature-based detection and semantics-based anomaly detection techniques. These are all considered software-based solutions, which often incur significant computational overheads to the system, making them unfit for computing devices, especially those with limited available computational and memory resources such as embedded systems and IoT devices [290]. In particular, these software-based malware detection methods like AV software pose several drawbacks. First, they rely on static signature-based detection in order to detect malicious patterns of the infected application. Such a detection mechanism searches for suspicious byte patterns in the program, whereas an attacker can deceive AV software by programming and crafting malware in such a way that its signature appears as benign software. Second, AV software is prone to misuse like any other software that can ultimately compromise protection if exploited. Third, AV software tools are slow and resource hungry. They mostly impose significant complexity and computational overheads on the system. Conditions become even worse for metamorphic viruses, as effective detection of such attacks is an NP-complete problem.

In addition to the aforementioned shortcomings of traditional malware detection mechanisms, the emergence of new malware threats requires patching or updating the software-based malware detection solutions (such as off-the-shelf anti-virus software) that need a vast memory footprint and a large amount of hardware resources, which is not feasible for emerging computing systems, especially in embedded mobile and IoT devices. The emerging embedded systems, which account for a wide range of applications, are often highly resource-constrained, challenging the conventional software-based methods traditionally deployed for detecting and containing malware in general-purpose computing systems. In addition to the complexity and cost (computing and storage), software-based malware detection methods mostly rely on the static signature analysis of running programs, requiring continuous software updates in the field to remain accurate in capturing emerging malware. This is not affordable for embedded systems which are limited in computing and communication bandwidth. Moreover, most of these advanced analysis techniques are architecture-dependent, i.e., dependent on the underlying hardware. Hence, this makes existing traditional software-based malware detection techniques hard to apply effectively to emerging embedded computing devices.

12.3 Deployment of Hardware Performance Counters for Computer Architecture Security

The complexity of today's computing systems has tremendously increased compared to the prior systems. Hierarchical cache subsystems and pipeline, non-uniform memory, simultaneous multithreading, and out-of-order execution have a significant impact on the performance of modern processors. Performance monitoring is an essential feature of a microprocessor. Access to the performance monitoring hardware is usually provided in the form of hardware performance counters. Recent studies have demonstrated that malicious activities at the hardware level ranging from application-based malware to microarchitecture side-channel attacks can be effectively recognized by classifying anomalies in the low-level feature spaces such as microarchitectural events collected by Hardware Performance Counters (HPCs) registers using Machine Learning (ML) techniques.

HPCs are special purpose registers embedded inside modern microprocessors to monitor and capture different microarchitectural events. The primary purpose of HPC is to analyze and tune the architectural level performance of running applications. There exist numerous low-level features (e.g., bus-cycles, instructions, cache-misses, branch instructions, branch-misses, cache references, etc.) that are captured by HPC registers using different monitoring tools such as *Perf*, Intel V-Tune, and AMD uProf. Recent works have proposed to utilize the HPCs for securing the hardware systems against both malware (application execution based attacks) and side-channel attacks. Hardware-based detectors offer fast online detection, efficiency in resource utilization, and invulnerability from getting infected by attackers which make them suitable for mitigating newer threats. These features are collected by profiling the applications (benign vs. malicious programs) on the target system to build an extensive dataset that will be used to train and test the ML models used for hardware-based malware and side-channel attacks detection.

Example 12.2 (Hardware Performance Counter Events Description)
Problem: Describe different hardware performance counter events in Intel processor related to the caches.
Solution: As mentioned earlier, functional components in a processor deliver different HPC events that are considered as the applications' signature that are left on underlying processor architecture. In this example, briefly describe each of the following hardware performance counter events that could be monitored from applications execution and used in applied ML for computer architecture security:

- *Cache-misses:* The number of misses in the last level cache references.
- *L1-dcache-loads:* The number of retired memory load operations.
- *Branch-misses:* The number of branch instructions that are mispredicted.
- *Cache references:* The total number of last level cache references.
- *L1-icache-load-misses* The number of instruction misses in the first level instructions cache.
- *iTLB-load-misses:* The number of misses in instruction TLB (Translation Lookaside Buffer) during load operations.
- *L1-icache-load-misses* The number of instruction misses in the first level instructions cache.
- *iTLB-load-misses:* The number of misses in instruction TLB during load operations.
- *L1-dcache-load-misses:* The number of cache lines transferred into the first level cache (L1) from DRAM main memory.

While HPCs are finding their way in various processor platforms from high-performance to low-power embedded processors and IoT devices, they are limited in the number of microarchitectural events that can be captured simultaneously. This is mainly due to the limited number of physical registers on the processor chip because of the high price. A variety of processor platforms such as Intel, ARM, and AMD includes HPCs on its processors. The number of physical registers present on each core usually ranges from 2 to 8. For example, the number of counter registers in the Intel Ivy-bridge and Intel Broadwell CPUs is limited to only four per processor core, meaning that only four HPCs can be captured simultaneously. In addition, Intel SandyBridge and Haswell architectures both have total 8 general-purpose counters per core. This limitation can be mitigated by multiplexing performance counters, but at the cost of accuracy degradation.

These registers are easily programmable across all platforms and are able to count a variety of low-level events such as cache memories access and misses, TLB hits and misses, branch mispredictions, and core stalls of the chip that are used for various optimization targets such as performance, energy-efficiency, and security enhancement. In particular, HPCs are programmed to issue an interrupt when a counter overflows or even be set to start the counter from the desired value.

12.4 Application of Machine Learning for Computer Architecture Security Countermeasures

Machine learning algorithms have been extensively deployed to enhance the security of computing systems. Machine learning-based countermeasures are trained by low-level microarchitectural features and continuously learn by analyzing the HPC data to identify the malicious patterns by accurately detecting malware and protecting the processor architecture against information leakage caused by emerging SCAs.

12.4.1 Feature Selection: Key Microarchitectural Features

Determining the most significant low-level features is an important step for developing effective ML-based countermeasures. As there exist numerous microarchitectural events (e.g., +100 in Intel Xeon), each of them representing a different functionality, collecting all features leads to high-dimensional data. Furthermore, processing raw dataset involves computational complexity and induces delay. In addition, incorporating irrelevant features would result in lower detection accuracy and performance for the ML classifiers. This poses two research questions. First, which low-level features are relevant to be employed to detect and classify a class of malicious attacks? Second, how to perform feature reduction of collected data to alleviate unnecessary computational overheads? Hence, to perform an efficient ML-based security countermeasure to detect security attacks with minimal overhead, a minimal set of HPCs is determined that can effectively represent the application behavior.

Feature selection methods are capable of enhancing the performance of learning process, decreasing the computational complexity, building better generalizable ML models, and reducing the required storage and memory on the computing systems. In the process of hardware-level malware and SCAs detection using machine learning techniques, the feature selection step intends to identify a set of critical HPCs that can effectively represent the attack application behavior. For effective run-time malware detection in embedded and resource-limited systems having a limited number of available HPCs, feature reduction even plays a more crucial role in which it determines the minimal set of critical HPCs that can effectively represent the malware class behavior and are feasible to collect in a single run even on low-end processors with few HPCs to perform HMD with minimal overhead avoiding multiple runs [290].

Therefore, instead of accounting for all captured features, irrelevant features need to be identified and removed using a feature reduction algorithm and a subset of HPC events is selected that represents the most important features for classification. The selected features are then supplied to each ML-based detector. The detector attempts to find a correlation between the feature values and the application behavior to predict the benign or attack type. Principle Component Analysis (PCA), correlation attribute evaluation, Information Gain Ratio, and Fisher Score (FS) are some well-known feature selection techniques that have been used in ML-based security countermeasures.

12.5 ML for Hardware-Assisted Malware Detection: Comparative Analysis

To address the traditional malware detection shortcomings, Hardware-assisted Malware Detection (HMD), by employing low-level features captured by HPCs, has emerged as a promising solution. HMD solutions reduce the latency of the detection process by order of magnitude with small hardware overhead [290]. Demme et al. [286] was the first study that proposed to deploy HPCs for malware detection and demonstrated the effectiveness of using ML models for hardware-based malware detection. They showed high detection accuracy results for Android malware by applying complex ML algorithms like Artificial Neural Network (ANN) and K-Nearest Neighbor (KNN). Tang et al. [362] further discussed the feasibility of complex unsupervised learning on low-level features to detect buffer overflow attacks that incur large overhead and sophisticated analysis. Ozsoy et al. [363] used sub-semantic features to detect malware using Logistic Regression (LR) and ANN algorithms. Moreover, they suggested changes in microprocessor pipeline to detect malware in a truly real-time nature which increases the overhead and complexity.

The research in [290] proposed ensemble learning techniques for effective run-time hardware-assisted malware detection and improved the performance of HMD by accounting for the impact of reducing the number of HPC features on the performance of malware detectors. In [370], a machine learning-based HMD is proposed that uses various traditional classifiers that requires 8 or more features to achieve high accuracy which makes it less suitable for online malware detection. In addition, recent work in [7] proposed a two-stage machine learning-based approach for run-time malware detection in which in the first level classifies applications using a multiclass classification technique into either benign or one of the malware classes

(Virus, Rootkit, Backdoor, and Trojan). In the second level, to have a high detection performance, the authors deploy an ML model that works best for each class of malware and further apply effective ensemble learning to enhance the performance HMD.

In this section, we describe an effective hardware-assisted approach that takes advantage of machine learning algorithms to classify malware and benign programs. In this research, we deploy HPCs information to construct a vector of microarchitectural features by profiling malware and benign applications. We take advantage of the HPC registers to collect execution traces for various microarchitectural events by executing malware and benign applications in an isolated environment (details will be discussed in Sect. 12.5.1). The profiling process shows that if two different programs are executed on a processor, they generate different performance counter traces providing a unique opportunity to detect the behavior of the running application using effective machine learning-based predictive modeling.

Example 12.3 (Sample Malware and Benign HPC Traces)
Problem: Demonstrate the effectiveness of HPCs in classifying malicious and benign programs.
Solution: As an example to demonstrate the effectiveness of using HPC traces for malicious pattern detection, below we illustrate the trace of *branch instructions* and *branch-misses* features for normal and malware applications. As seen, the malware traces are significantly different from benign applications for both examined HPC features. Using this observation, malware can be distinguished from normal applications by its different HPC values. This further highlights the goal of hardware-assisted malware detection that learns malware behavior with the aid of machine learning methods based on microarchitectural features captured by a limited number of available HPCs from various applications.

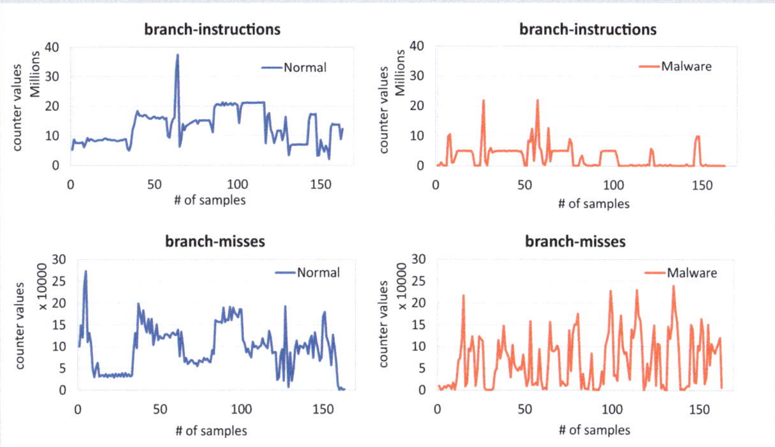

12.5.1 Experimental Setup and Data Collection

This section provides the details of the experimental setup and data collection process. The applications (both malware and benign) are executed on an Intel Xeon X5550 machine running Ubuntu 14.04 with Linux 4.4 Kernel. In order to extract the HPC information, we deployed *Perf* tool available under Linux. *Perf* provides rich generalized abstractions over hardware-specific capabilities exploiting *perf-event-open* function call in the background which can measure multiple events simultaneously. For this problem, we have executed more than 100 benign and malware applications for HPC data collection. Benign applications include MiBench [371] and SPEC2006 [372], Linux system programs, browsers, and text editors. For malware applications, Linux malware is collected from virustotal.com [373]. Malware applications include Linux ELFs, python scripts, Perl scripts, and bash scripts which are created to perform malicious activities.

Figure 12.2 depicts a general overview of the data collection process and proposed run-time HMD framework. It is primarily composed of various stages including feature extraction, feature reduction, and ML classifiers implementation. In this framework, HPC information is captured by executing all applications in Linux Containers (LXC) which is an isolated environment [374]. We have extracted 44 low-level CPU events available under *Perf* tool using static performance monitoring approach where we can profile applications several times measuring different events each time. In other words, since Intel Xeon processor hosts only 4 HPC registers physically available [375], we can only measure 4 events at a time. As a result, multiple runs are required to fully capture all events. We divide 44 events into 11 batches of 4 events and run each application 11 times at a sampling time of 10 ms to gather all microarchitectural events. Running malware inside the container can contaminate the environment which may negatively impact subsequent data collection. As a result, to ensure that there is no contamination in the collected data due to the previous run, the container is destroyed after each run.

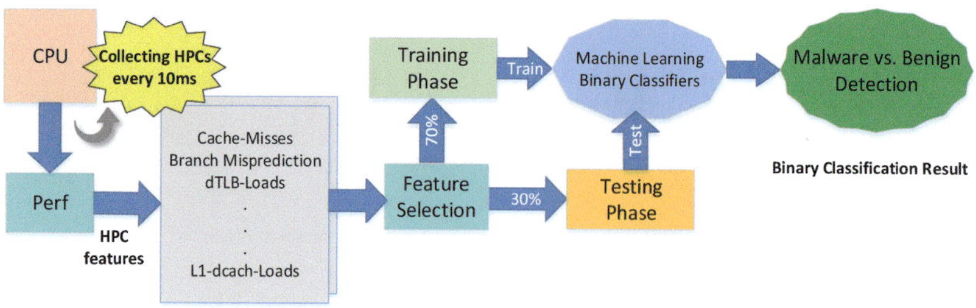

Fig. 12.2 Overview of the hardware-assisted malware detection using ML

12.5.2 Feature Selection and ML Classifiers Implementation

Training ML classifiers involves profiling the incoming application with *Perf* tool and extracting low-level feature values for each training program, reducing the extracted features to the most vital HPCs, and developing a learning model from the training data. It is important to note that the input variables in our classifiers are the HPCs extracted every 10 ms interval from running applications, and the output variable is the class of an application. In order to validate each of the utilized ML classifiers, a standard 70%–30% dataset split for training (known applications) and testing (unknown applications) is followed. To ensure a non-biased splitting, 70% benign-70% malware application for training and 30% benign-30% malware applications for testing are used. In order to identify the most important HPC features for effective malware detection, we applied a two-level feature selection as described below.

(a) Manual Feature Reduction We first analyze the collected HPC events manually and exclude events which are obviously not related to the target variable (malware behavior). Out of all 44 events, there are certain features provided by Linux kernel that is included as software events under *Perf*. We exclude a total of 12 events from the final selected features list. These events are alignment-faults, context-switches, cpu clock, cpu migrations, emulation-faults, major-faults, dummy, minor-faults, page-faults, and task-clock. In addition, events like cpu-cycles and ref-cycles do not represent uniqueness in terms of program phase. Thus, they are also excluded from the final list. Hence, a total of 12 events are removed using manual feature reduction.

(b) Algorithmic Feature Reduction Following the manual approach, we deploy *Correlation Attribute Evaluation* to rank 32 remaining features under WEKA. Correlation evaluation algorithm calculates Pearson correlation between each attribute and class as follows:

$$\rho(i) = \frac{cov(X_i, C)}{\sqrt{(var(X_i)var(C))}}, \tag{12.1}$$

where ρ is the Pearson correlation coefficient. X_i is an input dataset of any performance counter event i. C is an output dataset contains different classes, "Malware" or "No Malware" in our case. Value of i represents any feature out of 32 features and $cov(X_i; C)$ measures covariance between input dataset and output dataset. The $var(X_i)$ and $var(C)$ also measure variance of both input and output dataset, respectively. Based on the ranking of ρ, the top 16 features are selected for analysis. This algorithm finds correlation coefficient for all 32 features as per above equation in which the branch instruction has the highest value of ρ than other features. Equation (12.1) can be elaborated further as shown below:

Table 12.1 Hardware performance counters in order of importance

Rank	HPC feature	Rank	HPC feature
1	branch instructions	9	dTLB-stores
2	branch loads	10	iTLB-loads
3	iTLB-load-misses	11	L1-icache-load-misses
4	dTLB-load-misses	12	branch-load-misses
5	dTLB-store-misses	13	branch-misses
6	L1-dcache-stores	14	LLC-store-misses
7	cache-misses	15	node-stores
8	node-loads	16	L1-dcache-load-misses

$$\rho(i) = \sum_{k=1}^{n} \frac{(x_{k,i} - \bar{x_i})(c_k - \bar{c})}{\sqrt{\sum_{k=1}^{n}(x_{k,i} - \bar{x_i})^2 \sum_{k=1}^{n}(c_k - \bar{c})^2}}, \tag{12.2}$$

where i represents the index of feature ($i = 1, ..32$); k is the number of input values; $x_{(k,i)}$ is kth value in input dataset for feature i; and c_k is kth value in output dataset. The mean of input for feature i is denoted by $\bar{x_i}$, and that for the output data by $\bar{c}$. Based on the ranking of ρ, the top 16 features are selected for analysis. This algorithm finds correlation coefficient for all 32 features as per above equation. We list the top 16 features with the highest correlation coefficient value. These reduced features are described in Table 12.1. These events have a mixture of branch-related events representing core behavior and cache related events representing memory behavior in which the branch instruction has the highest value of ρ than other features.

12.5.3 Evaluation Results of ML-Based Malware Detectors

Detection Accuracy To evaluate the detection accuracy of our malware classifiers, we calculate the percentage value of samples that are correctly classified. Table 12.2 shows the accuracy results of various ML classifier used for malware detection. We have implemented five different ML classifiers and calculated their accuracy in classifying malware and benign applications. The accuracy of malware detection with two sets of hardware performance counters (8 and 4) are reported. It is notable that the OneR classifier performs well even after feature reduction. The reason that OneR classifier is not affected by feature reduction and shows almost constant accuracy results are that it only selects one performance counter (branch instructions) to predict the malware behavior.

Hardware Implementation As mentioned earlier, the software implementation of ML classifiers for malware detection is slow in the range of tens of milliseconds which is an order of magnitude higher than the latency needed to capture the

Table 12.2 Accuracy results of ML-based malware detectors

ML classifier	8HPCs	4HPCs
BayesNet	85	81
OneR	81	81
MLP	85	81
JRIP	83	84
J48	82	82

malware at run-time. Therefore, we develop a hardware implementation of the machine learning-based malware detectors. To this aim, we use Vivado High-Level Synthesis (HLS) compiler to develop the Hardware Description Language (HDL) implementation of the classifiers and deploy it on Xilinx Virtex 7 FPGA. FPGA (Field Programmable Gate Array) is a target in our study, as few modern microprocessors have on-chip FPGAs available for programmable logic implementation. Such arrangement makes it feasible to implement reprogrammable low-level malware detection logic (ML model) which can detect malware by reading the CPU hardware performance counters through a shared memory bus. As a result, when it comes to choosing the ML classifiers for hardware implementation, the accuracy of an algorithm is not the only parameter in decision-making. Design area and response time (latency) overhead of ML classifiers also play a key role in selecting the cost-efficient hardware solution. While complex algorithms such as neural networks can deliver high accuracy, they will also add significant overhead in terms of hardware implementation cost. Also, given their complexity, they can be slow in detecting malware.

For synthesizing ML classifier models, we assume that the logic of fetching counter values periodically from CPU is already implemented and is the same for all studied classifiers. Hence, we are excluding data fetching logic for latency, area, and power estimation. Moreover, we assume that vectors with all HPC events are already available at the input to classifier. Every model is treated as a black box which accepts HPC event vector of size 8 and outputs binary value "malware" or "not malware." During high-level synthesis latency and area/utilization of all ML classifiers are collected. To collect the total power estimation of the implemented model, IP core is synthesized in Vivado. Power estimation is collected for 100 MHz clock attached to the IP core. Power estimation contains both static power and dynamic power consumption of digital logic. Results of estimated latency, area, and power are shown in Table 12.3. The latency unit is in terms of the number of clock cycles required to classify the input vector. Area unit is the total number of utilized LUTs, FFs, and DSP units inside Virtex 7 FPGA. The unit for power consumption is Watt.

To evaluate the efficiency of machine learning-based malware detectors in Fig. 12.3, we present the results comparison in terms of accuracy over the unit of hardware area for various implemented ML classifiers. The reason for considering such metric is that area and accuracy comparison is putting more emphasis on silicon area budget. We use the ratio of Accuracy over Area to list down ML

Table 12.3 Hardware implementation results of the ML-based malware detectors

ML classifier	Latency	Power	Area overhead
BayesNet	14	0.445	6794
OneR	1	324	1258
MLP	302	1.03	36,252
JRIP	4	0.436	1504
J48	9	0.436	1801

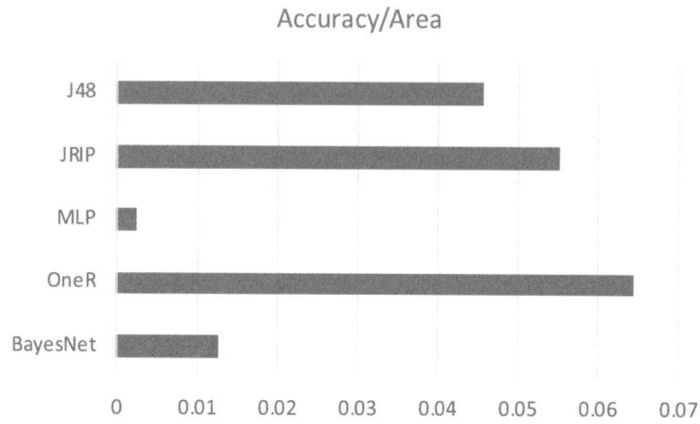

Fig. 12.3 Efficiency analysis (Accuracy/Area) of implemented ML-based malware detectors

classifiers which require small area and yet can predict with high accuracy. Classifier with a higher ratio is considered better than with a lower ratio. As the results show, rule-based and tree-based classifiers are performing significantly better in terms of accuracy per area compared to highly accurate but complex Bayesnet and MultiLayerPerceptron classifier.

12.6 ML for Microarchitectural SCAs Detection: Comparative Analysis

The detection work in [376] monitors HPCs trace of both victim and attack processes and compares the effectiveness of three ML classifiers: neural network, decision tree C4.5, and K-nearest neighbors. The work in [377] proposes a detection system containing one analytic server and one or more monitored computing devices to detect SCAs, including Spectre and Meltdown. The analytic server receives HPCs data from monitored devices and identifies suspicious core activity. Once detected, application level monitors will be deployed on the computing devices and take corrective actions as soon as they find suspicious application activity. Recent work [378] uses cache latency to build cache occupancy of victims and attacks. Based on the cache occupancy relation of the two processes, SCAs can be deduced.

Chiappetta et al. [287] collected HPC features for building the ML classifiers and compared three different attacks scenarios including finding a correlation between victims and attacks, building supervised machine learning models based on HPCs from victims and attacks, and detecting anomalies by validating attack HPCs as samples and other processes as outliers. This work shows that tweaking the attack

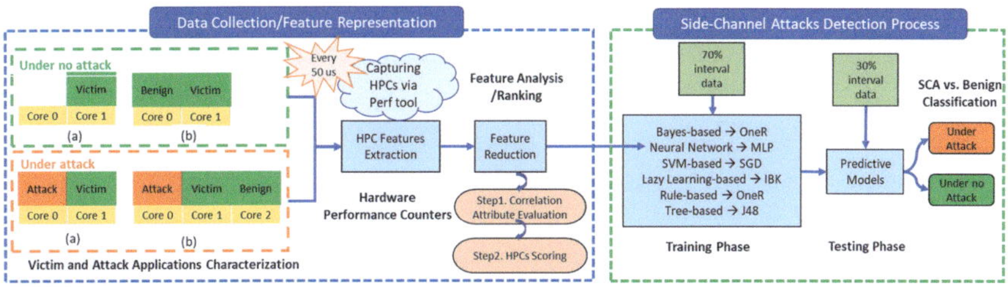

Fig. 12.4 Overview of the real-time SCAs detection methodology based on victim application HPCs

application can easily result in bypassing the detection mechanism indicating that profiling attack applications and building the ML models based on the attacks HPCs can lead to potential vulnerabilities in the detection system.

Similarly, in [379] the authors presents CloudRadar which aims at detecting cross-VM side-channel attacks by deploying HPC patterns. The work in [380] proposes an online detection of Spectre by monitoring microarchitectural features using time-series classification.

In this section, we first present the motivation for using victims' HPC data for SCAs detection. Then we follow similar steps in malware for collection and feature selection. Lastly, the machine learning-based detector shown in Fig. 12.4 will be introduced. As shown, the detector is comprised of different steps such as data collection, feature reduction, training phase, testing phase. First, for feature extraction, the "under no attack" and "under attack" HPC data will be collected within (a) isolated scenario, and (b) non-isolated scenario. The "isolated" environment refers to the case that a computer only processes victim applications, whereas the "non-isolated" environment denotes that a computer system processes victim applications on one core while benign applications are being executed on the rest of the cores. Then the importance of HPCs is evaluated and only 4 most prominent HPCs are selected for ML-based detection. Next, the trained models will be employed in the testing phase.

12.6.1 Detection Based on Victim Applications' HPCs Data

Current SCAs intentionally cause influence victim applications' cache or branch predictor by flushing/priming cache, mistraining branch predictors and then observe accessing time of the cache sets, which changes caching victims' data and microarchitectural behaviors of victim applications. This also provides the opportunity of detecting SCAs by observing the alteration in microarchitectural behaviors.

Example 12.4 (Sample Victim Application with SCA)
Problem: Explain the effectiveness of HPCs in detecting side-channel attacks.
Solution: As an example to show the effectiveness of using HPC traces for the purpose of side-channel attacks detection, the following graph indicates that there exists a clear difference between the behavior of victim under attack and victim application under no attack conditions. In this motivational case study, the HPC traces of L1 HIT for the tested victim application under no attack (RSA) and under L3 Flush Reload attack (RSA with FR) are illustrated. It can be observed that the L1 HIT of VA shows a significantly different trend compared to that of victim under no attack condition. This observation highlights the effectiveness of using HPCs data of only victim applications (excluding the impact of attack applications' HPCs) for detecting the behavior of SCAs.

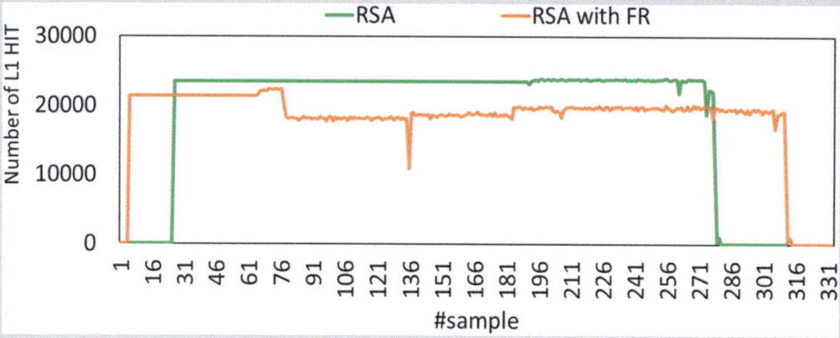

12.6.2 ML Classifiers Implementation

Following a similar process introduced in Sect. 12.5.1, HPCs for victim under attack and victim under no attacks are collected. All data is split into 70% and 30% for training and testing. Then, the importance of HPCs is evaluated with *Correlation Attribute Evaluation* detailed in Sect. 12.5.2.

For the purpose of a thorough analysis of various types of ML classifiers, NavieBayes, MLP, SGD, IBK, OneR, and J48 ML algorithms are deployed as our final classification models. The rationale for selecting these machine learning models are: firstly, they are from different branches of ML: Bayes-based, neural network, SVM-based, lazy learning, rule-based, and tree-based techniques covering a diverse range of learning algorithms which are inclusive of model both linear and non-linear problems; secondly, the prediction model produced by these learning algorithms can be a binary classification model which is compatible with the SCA detection problem in our work. As mentioned before, only four HPCs can be

Table 12.4 The collected HPC features and their importance ranking

Ranking	HPC name	Ranking	HPC Name
1	L1 HIT	9	L1 MISSES
2	UOPS_RETIRED	10	BRANCHES MISPREDICTED
3	BR_NONTAKEN_CONDICTIONAL	11	L2 HIT
4	ALL BRANCHES RETIRED	12	TAKEN_INDIRECT_NEAR_CALL
5	INST_RETIRED_ANY	13	L3 HIT
6	L2 MISSES	14	ITLB_MISSES
7	BR_TAKEN_CONDITIONAL	15	DTLB_STORE_MISSES
8	L3 MISSES	16	DTLB_LOAD_MISSES

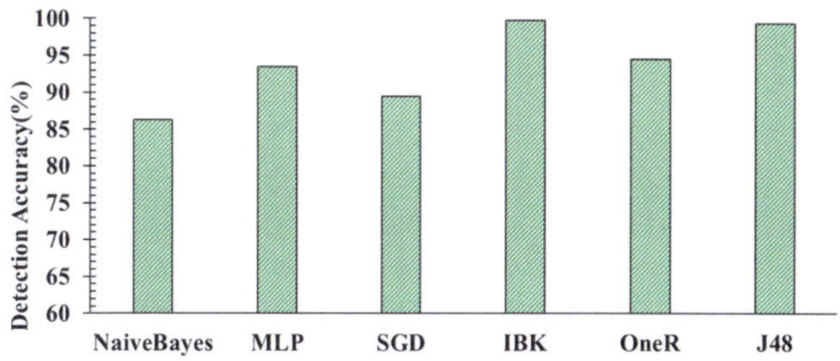

Fig. 12.5 Prediction accuracy comparison among various machine classifiers

collected for most processors at once due to a limited number of registers for storing them. Hence, reducing the number of HPCs required for ML models is important to eliminate the need for multiple runs. For this purpose, various numbers of HPCs from 16 to 4 (16, 12, 8, and 4 selected based on the ranking in Table 12.4) are examined to evaluate the influence of reduced HPCs on classification accuracy and highlight the importance and motivation of using a lower number of HPCs (only 4) for effective real-time SCA detection.

12.6.3 Evaluation Results of ML-Based SCAs Detectors

In this section, we present the detection accuracy and efficiency (F-Measure/latency) that can be used for evaluating ML-based SCAs detector.

Detection Accuracy As shown in Fig. 12.5, the detection accuracy of the six different ML classifiers for 30% split testing accuracy results ranges from 85% to 99%. NaiveBayes and SGD result in significantly lower detection accuracy compared with the rest four classifiers, less than 90%. IBK and J48 give above 99% detection accuracy, indicating they are more effective in capturing SCAs.

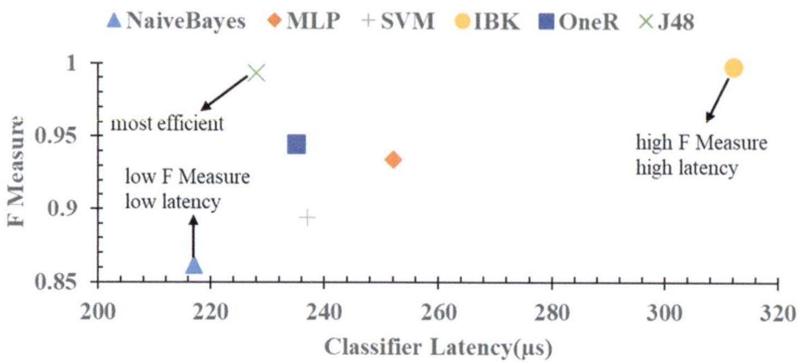

Fig. 12.6 Efficiency comparison among various classifiers

Efficiency Analysis: F-measure vs. Latency Though important, detection accuracy and F-measure are not the only evaluation metrics used in analyzing the effectiveness of machine learning algorithms used for attack detection. As a result, for the purpose of comprehensive analysis, here we further evaluate the efficiency of ML classifiers in terms of F-measure vs. Latency. As mentioned in Sect. 1.7.5, F-measure is a more comprehensive evaluation metric over accuracy (percentage of correctly classified samples) since it takes both the precision and the recall into consideration. Moreover, latency here represents the duration that predictive models take for delivering the "under no attack" or "under attack" results.

To accordingly account for both SCAs detection rate and cost of ML classifiers, in Fig. 12.6 we compare the detection rate over a computational latency (F-measure/Latency) for various ML classifiers. We use F-measure over latency to identify the SCA detectors that require small cost and yet can detect the maliciousness of program with high accuracy and performance. A classifier with a higher ratio is considered a more efficient detector than the classifier with a lower ratio. As shown in Fig. 12.6, a clear trade-off is seen between detection rate and latency achievable for real-time hardware-assisted SCAs detection. The ML classifiers such as IBK achieves high detection rate, but also higher computational overhead. The techniques such as NaiveBayes, MLP, SGD, IBK, OneR, and J48 show relatively smaller timing costs with high SCAs F-measure. For highly resource-constrained embedded systems, techniques such as J48 provide the smallest computational overhead, while achieving an F-measure of close to 0.993 on average. Clearly, the results show trade-offs between F-measure and latency. Therefore, it is important to compare ML classifiers for effective SCAs detection by taking all these parameters into consideration.

12.7 Exercise Problems

Problem 12.1 Explain three major shortcomings of software-based malware detection techniques in securing the modern computer systems.

Problem 12.2 What is hardware performance counter and how it could be used to detect emerging security threats?

Problem 12.3 Write functions to read in a CSV dataset and perform a train-test split. Description: Create a Jupyter Notebook to read the data file named in the same folder "4HPCSsmallersamplenolabels.csv" into Pandas's dataframe, then use sklearn library to do a train-test split at 70% train—30% test to split the dataset into training and testing sets. Save both training and testing datasets to disk for later use. Name them "4HPCSsmallersamplenolabels_train.csv" and "4HPCSsmallersamplenolabels_test.csv", respectively.

Problem 12.4 For this problem, use the saved training dataset from Problem 3 (named "4HPCSsmallersamplenolabels_train.csv") to train a binary classifier from a list of machine learning algorithms below to better realize each model' detection accuracy and save all models to disk. Make sure to do a train-test split to ensure model evaluation is based on the validation dataset. (a) Decision Tree, (b) Random Forest, (c) KNN, (d) SVM

Problem 12.5 Use the test dataset from Problem 3 to test on the pre-trained Random Forest model in Problem 4 to output confusion matrix. In this problem, read in the test dataset from Problem 3 named "4HPCSsmallersamplenolabels_test.csv" to predict on the previously trained and saved Random Forest model from Problem 4. Use true label and predicted label to obtain the confusion matrix. Plot the confusion matrix.

Problem 12.6 This problem is focused on the application of the Information Gain feature selection method for malware detection problems to identify the most critical features. Using the provided binary classification dataset implement the Information Gain feature selection algorithm and show the ranking of each feature to narrow down to the top 4 suitable features for malware detection.

Problem 12.7 In this problem, using the reduced binary classification dataset with 4HPC features from Problem 6, implement Random Forest algorithm for malware detection and report the precision, recall, and F-measure results. Please note that the top 4 features identified by Information Gain feature selection method in the previous example should be used to train the ML model.

Problem 12.8 For this problem, using the reduced binary classification dataset with 4HPC features from Problem 3, implement K-Nearest Neighbor algorithm for malware detection and compare its precision, recall, and F-measure results with the Random Forest classifier in Problem 12.7.

Problem 12.9 Refer to the hardware performance counter-based binary dataset of side-channel attack and benign application, named as "binary.csv". Choosing one of the feature evaluation algorithms in Scikit-Learn library to evaluate the importance of all provided 16 HPC events for SCA detection. Specify the importance score of each HPC and plot a bar chart for it.

Problem 12.10 Refer to the hardware performance counter-based multiclass dataset of side-channel attack and benign application, named as "multiclass.csv". Choosing one of the feature evaluation algorithms in Scikit-Learn library to evaluate the importance of all provided 16 HPC events for SCA detection. Specify the importance score of each HPC and plot a bar chart for it.

Problem 12.11 Use the 16 HPCs from the HPCs importance ranking performed in Question 9, and classify benign or SCAs samples using a SVM binary classifier. Report the ML implementation results including (a) Training accuracy, (b) Testing accuracy, and (c) False positive rate.

Problem 12.12 In this problem, implement an SVM-based SCA detector using the 4 most important events determined in Problem 11. Compare the training accuracy, testing accuracy, false positive rate with the SVM classifier built with 16HPCs from Problem 12.11.

Problem 12.13 Repeat the task described in the Problem 12.12 using the 4 least important HPC events. Compare the implementation results including training accuracy, testing accuracy, false positive rate. Provide your conclusion.

Problem 12.14 In this problem, implement a MultiLayer Perceptron-based SCA detector using the 4 most important events determined in Problem 12.9. Compare the training accuracy, testing accuracy, false positive rate with the SVM classifier built from Problem 12.12.

Problem 12.15 Measure and compare the execution time of SVM and MLP built with the 4 most prominent HPCs based on the binary file in Problems 12.12 and 12.14.

Chapter 13
Applied Machine Learning for Cloud Resource Management

13.1 Introduction

The continuous increase in the volume of data due to the rise of social media, Internet-of-Things (IoT), and multimedia has produced an overwhelming flow of data referred to as big data. In order to efficiently process such massive data, scale-out architecture has gained interest as a promising solution that is designed to provide a massively scalable computer architecture. Recent improvements in networking, storage, energy-efficiency, and infrastructure management have made Cloud (the best example of scale-out architecture) a preferable approach to responding to the new computing challenges.

Cloud computing is a significant paradigm shift in service for enterprise applications and has become a powerful scale-out architecture to perform large-scale computing. Today, cloud computing platforms host a wide range of applications such as scientific computing workloads, latency-sensitive web services, machine learning, large-scale distributed data analytics, etc. The advantages of cloud computing include parallel processing, security, scalable data storage, and virtualized environment.

13.1.1 Challenge of Diversity

Virtualization is a process of resource sharing and isolation of underlying hardware to increase computer resource utilization, efficiency, and scalability. However, from the application perspective, as different applications have different characteristics, we are experiencing that one architectural configuration fits all does not provide the best performance and energy efficiency for every application. Therefore, the cloud service providers offer a wide range of cloud configuration choices such as Virtual Machine (VM) instances with a variety of CPUs, memory, disk, and

S. Rafatirad et al., *Machine Learning for Computer Scientists and Data Analysts*,
https://doi.org/10.1007/978-3-030-96756-7_13

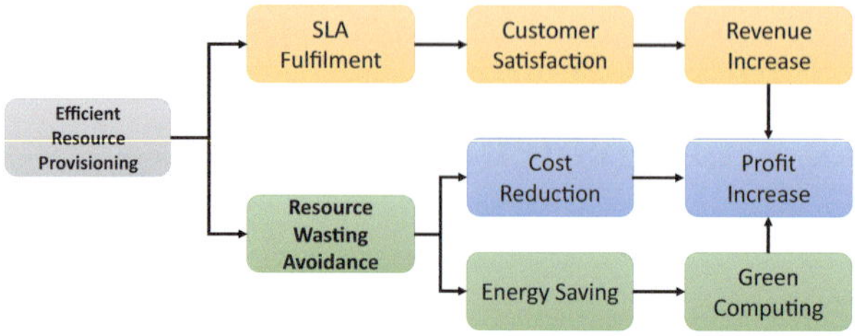

Fig. 13.1 Impact of resource provisioning on the cloud's aspects

network configurations, and also customized VMs for analytics applications. At the same time, the advancements of hardware architecture designs lead to datacenters with diverse hardware systems. This hardware diversification at different levels is marking the beginning of an era of super-heterogeneous datacenters.

The super-heterogeneous datacenter makes determining the best cloud configuration for a given application by brute-force search expensive and exhaustive. Choosing the right cloud configuration is essential, as a non-optimal configuration results in more cost for the same performance target as different analytic jobs have diverse behaviors and resource requirements. As Fig. 13.1 illustrates, efficient resource provisioning impacts three different aspects of the cloud. It fulfills service-level agreements (SLA) and meets cloud customers' requirements. It guarantees cloud obligations to its users. It also prevents resource waste, thereby reducing energy consumption and operational cost. The reduction of energy consumption leads to a decrease in carbon emission, which facilitates green computing. Hence, energy-aware resource provisioning is also important for reducing cost and for increasing revenue that improves the profit of cloud providers.

A more challenging problem is that the behavior and resource requirements of applications running on the cloud vary during different phases of execution. Each application faces various phases of execution, each with different memory and processing requirements. Based on the top-down methodology, three major phases can be identified in an application, namely I/O-bound phase, memory-bound phase, and compute-bound phase. These phases are different in terms of their microarchitectural behavior, therefore requiring different processing and memory resources for performance and energy-efficiency optimization. For instance, the compute-bound phase requires more cores, higher core frequency, and higher DRAM bandwidth.

Figure 13.2 illustrates the microarchitectural differences between those three phases. The micro-op (μop) queue of an out-of-order processor is used to abstract the microarchitectural behavior. The op queue is classified into four broad categories: Retiring, Front-end bound, Bad speculation, and Back-end bound. Out of these categories, only the Retiring is classified as "useful work," while the rest prevents the workload from utilizing the full core bandwidth. In addition to μop queues, C0 (active state residency of processor) is a metric that can be used to differentiate among phases. As the figure shows, the main difference between

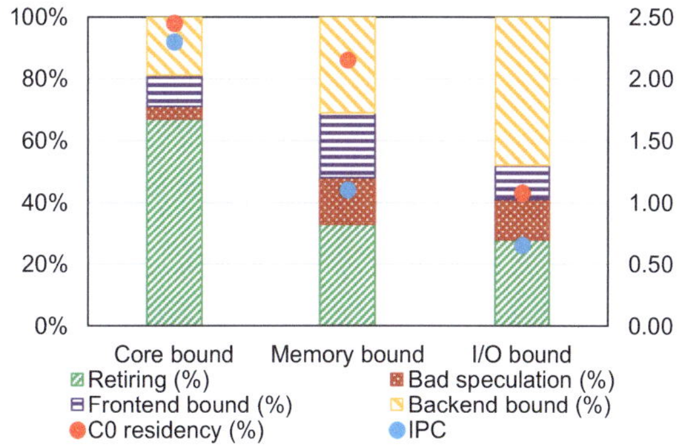

Fig. 13.2 Microarchitectural break-down of workloads for different phases

memory-bound and I/O-bound is C0 residency. This can be explained as follows: in I/O intensive phase, the core is waiting for I/O; hence, the core changes its state to save power. Therefore, C0 residency drops.

There are thousands of applications running at the same time in a cloud, and each requires different processing and memory resources to be allocated at different phases of run-time. It is, therefore, necessary that the resource management system identifies those phases at run-time to be able to allocate resources accordingly. Hence, this makes existing traditional reactive resource allocation methodologies achieve a sub-optimal performance gain and hard to apply effectively to emerging cloud computing services.

13.2 Modern Resource Provisioning Systems: ML Comes to the Rescue

Challenges described in the previous section have motivated several researchers to devise a new resource management methodology for the cloud. A resource provisioning system facilitates various services, including resource efficiency, security, fault tolerance, and monitoring, to achieve the performance goals while maximizing the utilization of available resources in the cloud. Researchers have utilized machine learning solutions to overcome the challenge of applications' diversity and heterogeneity of resources in the cloud and they were successful to significantly improve the utilization.

Several machine learning-based resource provisioning systems have been proposed for the cloud such as PARIS, Quasar, CherryPick, and Ernest. We will discuss their details later in this section. Table 13.1 summarizes the recent works and differentiates them from each other. In the system column, after the name of each system, we have provided the name and the conference in which the research has been published. Moreover, in this table, proactive means to act before a significant

Table 13.1 Comparison of state of the art

System	Target	Complexity	Accuracy	Proactive	Dynamic	Domain	Cost aware
ProMLB	Performance/cost, Fairness	High	High	Yes	Yes	Big data	Yes
DAC (ASPLOS'18)	Performance	High	High	No	No	In-memory	No
CherryPick (NSDI'17)	Performance	Low	Low	No	No	Big data	No
MeNa (IISWC'17)	Performance/cost	Low	Low	No	No	Broad	Yes
HCloud (ASPLOS'16)	Cost	Medium	Medium	No	Yes	Scale-out	Yes
Ernest (NSDI'16)	Performance	Medium	High	No	No	Big data	No
Heracles (ISCA'15)	Performance	Low	Medium	No	Yes	Latency-critical	No
Quasar (ASPLOS'14)	Performance	Medium	Medium	No	Yes	Scale-out	No
REF (ASPLOS'14)	Fairness	low	Low	No	Yes	Broad	No
Paragorn (ASPLOS'13)	Performance	Medium	Low	No	Yes	Scale-out	No

change occurs in the behavior of an application and influences the performance of the system.

As we mentioned, the RPS attempts to meet the user performance requirements and provider efficiency in terms of multiple aspects such as load balancing among servers, minimum number of active hosts, and least response time, to avoid service-level agreement violations in the cloud platform. Hence, RPSs or schedulers to fulfill their objectives must have two main tasks:(1) instance initialization and (2) periodic monitoring of applications.

During the instance initialization stage, when an instance is created and submitted to a scheduler, the scheduler profiles the application and based on the application's behavior determines the resources required for meeting its SLA. Machine learning can be used in this stage to identify the application's characteristics and determine its basic requirements. After that, the scheduler allocates the instance to a host in the infrastructure.

During the periodic monitoring stage, the scheduler monitors the application's behavior to guarantee the SLA all the time. In the case that application's behavior changes, the scheduler attempts to reschedule and migrate the instance to a new host to provide required resources to meet the SLA agreement. In this stage, machine learning can be leveraged to first detect the behavior change, and secondly to model the performance, cost, or even the energy of the application for determining the best instance that can cope with the change of the application's requirements.

Figure 13.3 shows how a general ML-based RPS works. First, the system monitors the application and extracts its microarchitectural and system-level information. Then based on the current behavior and server configuration, it may predict the performance of the application to make sure that the performance of the application will not be degraded. If the RPS identifies a performance degradation, then by leveraging optimization techniques, it determines another suitable configuration and host for the application.

One of the most popular RPS proposed so far is Quasar [381] that leverages machine learning and collaborative filtering to quickly determine which applications can be co-scheduled on the same machine without destructive interference. CherryPick [382] is another successful system that leverages Bayesian Optimization

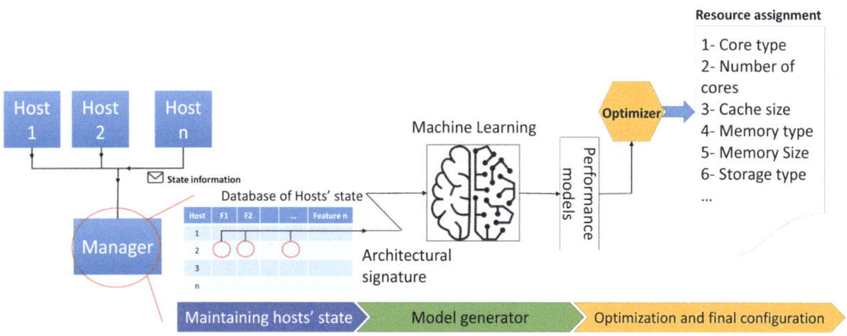

Fig. 13.3 ML-based resource provisioning system

and Regression technique to build performance models for various applications to distinguish the close-to-the-best configuration. Ernest [383] uses common machine learning kernels and statistical techniques for selecting the optimal configuration on the cloud. PARIS [384] is another ML-assisted system that uses Random Forest for predicting performance from the application's microarchitectural behavior to find the best VM type configuration.

There are several works that have focused on other aspects of resource provisioning such as energy-efficient resource provisioning for cloud. Zhang et al. [385] provided a control-theoretic solution to the dynamic capacity provisioning problem that minimizes the total energy-cost while meeting the performance objective in terms of task scheduling delay. Guevara et al. [386] studied how heterogeneous platforms bring energy efficiency for cloud applications. Guenter et al. [387] proposed an automated server provisioning system that aims to meet workload demand while minimizing energy consumption in datacenters. Altomare et al. [388] developed a system for energy-aware allocation of virtual machines on cloud physical nodes.

Delimitrou and Kozyrakis [389], Kousiouris et al. [390], Delimitrou and Kozyrakis [391] were proposed to address QoS-aware, performance-aware, and cost-aware scheduling and resource allocation. REF [392] is a method to provide a fair set of resources for each user at cloud. Kulkarni et al. [393], and Bolt[394] are other works related to heterogeneity and security. Kousiouris et al. [390] proposed to use a two-layer service in cloud to translate high-level application parameters (workload and QoS based on Service Level Agreement) to resource level attributes. Their work did not consider any performance model to select the optimum configuration. Also, they have not considered the cost efficiency.

There are other systems that adaptively allocate resources based on feedback. RightScale [395] creates additional VM instances when a load of an application crosses a threshold for EC2. YARN [396] decides resource needs based on requests from the application. Other systems have explicit models to inform the control system, e.g., [397]. Wrangler [398] identifies overloaded nodes in map-reduce clusters and delays scheduling jobs on them. Interference is creating challenge in accurate performance estimation. In recent works [399] and [400] explore placing applications on particular resources to reduce interference, by co-scheduling applications with disjoint resource requirements[399]. However, users requesting VM types in cloud services like Amazon EC2 cannot usually control what applications are co-scheduled. None of these studies have focused on the influence of system parameters such as memory or storage on the performance and cost in the cloud. In this chapter, we cover all parameters to shed light on all aspects.

13.3 Applications of Machine Learning in Resource Provisioning Systems

RPS is a configuration tuning methodology that automatically adjusts the hardware configuration assigned to a VM in a proactive manner in order to dynamically optimize a target metric such as energy consumption, execution time, or even the cost of a given program on a given heterogeneous cluster of servers. A general RPS usually should consist of four major components: a predictor to be able to act proactively, an estimator that can model the target metric, an explorer to optimize and search for the best solution, and a decision maker to put all information together from different components and make the final decision about the resource allocation. In the following, we elaborate more about each component and discuss how a machine learning technique can be utilized to improve the functionality of that component.

Figure 13.4 illustrates the block diagram of a modern RPS. RPS server must maintain a database of per-host state and update it on each monitoring interval. RPS may predict the next phase of an application and its microarchitectural signature based on the current and previous states. Given the predicted signature and corresponding server configuration, RPS can estimate the target metric. Next, RPS would search for the best platform and configuration that minimizes or maximizes the target metric for a given application. There is a searching component (explorer) in the RPS to automatically search for the configuration that achieves the optimal target metric. Overall, the estimator component of RPS relies on the results of the predictor, and the explorer component selects the best configuration from the outcome of the estimator.

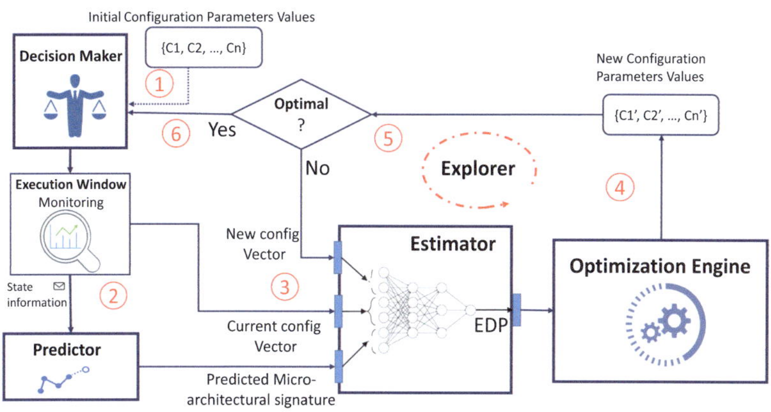

Fig. 13.4 Components of a modern RPS

13.3.1 Monitoring and Prediction of Applications' Behavior

Online Monitoring

RPS must run a monitoring agent on each host to be able to continuously monitor each host's state. The monitoring agent should extract information such as architectural information and resource utilization of application during each execution monitoring period (window) and report it periodically to the manager server. The window can be defined as a sufficient amount of time that a VM on a particular physical server can be maintained without migration. The architectural information can be collected through profiling tools such as the Intel Performance Counter Monitor tool (PCM), DSTAT, Perf, Intel VTune, and AMD uProf. Architectural information includes memory and processor behavior. Although the reports of the power consumption of CPU and memory can be gathered for energy and power modeling. The key features that can be easily collected as a representation of an application's behavior (or signature of an application) are architectural features. Example 13.1 shows few system-level and microarchitectural-level features.

Example 13.1 (Sample Features of Application's Signature)
Problem: Discuss different classes of application's signature on different system components.
Solution: One can see various types of microarchitectural information can be used for monitoring the state of a server in the cloud.

1. **Memory related:** Available virtual, physical, and shared memory, the cache and buffer space, memory bandwidth utilization, etc.
2. **Disk related:** Ratio of free to total disk space, storage bandwidth utilization, etc.
3. **Network related:** Bytes sent and received, etc.
4. **CPU related:** L2, and Last Level Cache (LLC) hits ratio, instruction per cycle (IPC), core C0 state residency, CPU idle time, system time, user time, etc.

The dimensionality reduction technique can be used to reduce the number of features and it is called the Principal Component Analysis (PCA). It is a powerful technique that arises from linear algebra and probability theory. In essence, it computes a matrix that represents the variation of your data (covariance matrix/eigenvectors), and ranks them by their relevance (explained variance/eigenvalues).

Phase Prediction

As an effective RPS needs to be proactive to determine an optimal server configuration for the running application, act before a significant change occurs in the behavior of an application and influences the performance of the system, it should be equipped with a phase predictor.

The goal of the phase predictor in the RPS is to predict the future behavior of the application based on the current and previous behavior of that application. The accurate prediction is important as other components of RPS depend on the predicted phase of the application. Based on the accurate prediction, RPS can assign an appropriate resource to the application in advance before the performance is degraded or resource utilization drops.

In this section, we briefly introduce few simple ML techniques such as time-series neural network, hidden Markov model, and K-nearest neighborhood that can be used as a predictor in any RPS. Each ML technique has its own trade-offs. We also explain the ensemble method that uses a combination of multiple learning algorithms to obtain better predictive performance than could be obtained from any of the constituent learning techniques alone.

Time-Series Neural Network

Time-series neural network or TSNN is an eager learning technique. The training of TSNNs can be done offline. The time-series neural network module is based on a non-linear autoregressive network with an exogenous inputs network. The following formula shows the basic function to predict the future behavior. A simple way to determine the architecture of TSNN is Grid Search for reaching the highest possible accuracy.

$$Y(t) = F(Y(t-1), Y(t-2), \ldots, Y(t-n)).$$

Hidden Markov Model

The hidden Markov model (HMM) is another eager technique employed for effective prediction. The HMM is used extensively for performance modeling and performance-prediction analysis, where the HMM can predict the future state of a target system based on its current state. In reality, as the relationship between the observed time and the observed state is not one to one, a group of probability distributions for two stochastic processes are involved, called the HMM. In an HMM, the states are not observable, but when we visit a state, an observation is recorded that is a probabilistic function of the state.

K-Nearest Neighbors (KNN) Regression

KNN is a lazy learning technique that does not require training. Suppose the dataset has m samples that each sample x_i is described by n input variables and an output variable y_i such as $x_i = \{x_{i1}, \ldots, x_i n | y_i\}$. The goal is to learn a mapping function F: x –> y known as a regression function that captures and models the relationship between input variable x and an output variable y. The KNN regression estimates the function by taking a local average of the dataset. Locality is defined in terms of the k samples nearest to the estimation sample. As the performance of KNN algorithm strongly depends on the parameter k, finding the best values of k is essential. A large k value decreases the effect of noise and minimizes the prediction losses. However, a small k value allows simple implementation and efficient queries.

Ensemble Method

Ensemble learning is a branch of machine learning, which is used to improve the accuracy and performance of general ML predictors. Using ensemble learning enables the RPS to use both eager learning techniques and lazy learning technique that does not require training. Using lazy learning technique enables RPS to be more flexible and have better accuracy for unknown applications. A common ensemble method is Bagging, or Bootstrap Aggregation. It is a statistical prediction technique where a future state of an application is estimated from voting of prediction results of few models. Each model is exploited to make a prediction and the results are voted to give a more robust and generalized prediction. If the prediction of all ML techniques is different from each other, then the voter can select the current state as the predicted result.

13.3.2 Using ML for Performance/Cost/Energy Estimation

In this section, it has been shown how a machine learning technique can be leveraged to formulate the performance, cost, or energy efficiency of different applications in a scale-out environment. The first part of this section is devoted to performance modeling. The second part is to formulate the dependency of the price that subscribers must pay for utilizing different server configurations. Then the developed models to formulate the performance improvement of applications with respect to the baseline hardware configuration will be presented. These models can be exploited by the optimizer in the next step to select the most performance- or cost-efficient server configuration for a given application.

Example 13.2 (Performance Modeling)

Problem: Describe the performance model space for graph analytic workloads.

Solution: Here, we present the performance model space for the graph analytic workload from Flink framework.

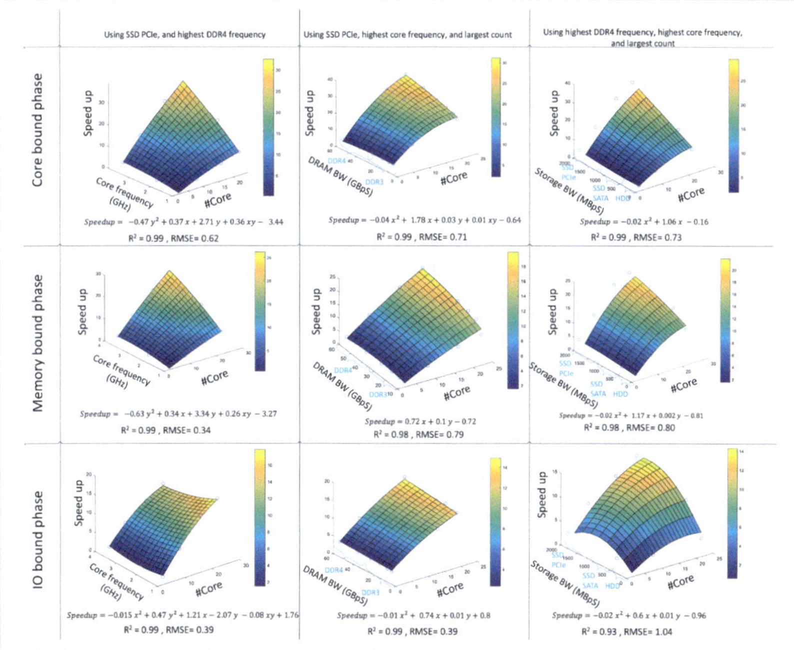

One of the most important components of RPS is to dynamically generate a performance model for each phase of applications. This leads to a more accurate model and helps the optimizer to select the best configuration. Example 13.2 is presented to illustrate that each application has a different performance model depending on its phase and the server platform. Example 13.2 shows a subset of generated performance models for three different phases of graph analytic application in Flink framework. In each sub-figure, X represents the number of cores and Y stands for the other parameter. If we model the performance of applications as a convex function of servers' parameters such as core count and core's frequency, then a generic performance model can be developed. However, this generic model has to be adopted for each application.

The simplest way to formulate the server performance is to use the product of processor performance and the number of processor in each server. There are several parameters that can be configured in a server such as the number of cores, core frequency, DRAM bandwidth and capacity, storage bandwidth, and network parameters. As the performance does not scale linearly with its parameters , a non-linear modeling is required. To clarify this concept, we demonstrate a simple generic performance model example, generated by regression technique, and it has been used in Example 13.3:

Example 13.3 (Binomial Equation as a Generic Performance Model)
Problem: How to model the speedup of a server based on different system components?
Solution: We present an equation that can be used as a performance model:

$$Speedup = \alpha_1 x^2 + \beta_1 y^2 + \alpha_2 x + \beta_2 y + \omega xy + \gamma,$$

where $x, y \in \{core, freq, DRAMBW, DRAMcap, StorageBW\}$ and $x \neq y$.

The advantage of generating an accurate performance model of applications at each phase of their execution for various types of servers is to improve the server selection. An appropriate resource provisioning will decrease the execution time of the subscriber's job, increases the resource utilization of scale-out infrastructure and eventually brings economic benefits for both subscriber and provider. This is important because performance improvement in datacenters translates into millions of dollars revenue per year for cloud provider and also it decreases the cost for subscriber and makes cloud services more attractive for the end users.

To model the cost of a server, first, the parameters that influence the pricing of a server must be analyzed. The goal of cost modeling is to establish a relationship between the performance of applications and the cost of a server running those applications. The server price can be determined as a function of server configuration as follows:

$$C_{server} = C_{processor} + C_{memory} + C_{disk} + C_{network}.$$

The per-server costs usually include configurable DRAM, configurable processor, disk, and network costs. Any machine learning technique (the simplest one is regression technique) can be used to derive a cost equation for storage, memory, and processor.

To evaluate the accuracy of the predictor and the estimator, we use equation of the RMSE (Root Mean Squared Error):

$$RelativeRMSE = \sqrt{\frac{1}{N}\sum_{n=1}^{N}\left(\frac{p_i - a_i}{a_i}\right)^2} \times 100, \tag{13.1}$$

where N is the number of samples, and p_i and a_i are the predicted and actual values of the sample, respectively. We want the % relative RMSE to be as low as possible. RMSE is a standard metric in regression, which is sensitive to scalability. For example, an RMSE of 1 s in runtime prediction is not acceptable if the actual runtime is 2 s but can be acceptable if the actual runtime is 1000 s. Expressing the error as a percentage of the actual value solves this issue.

13.3.3 Explore and Optimize the Selection

The purpose of the Explorer component is to find the best configuration that minimizes or maximizes the target metric. Therefore, it is an optimization problem in which the optimization variables are the server's configuration, and the cost function is the target metric. In order to solve this optimization problem, Explorer should use an optimization engine (OE). Popular optimization algorithms such as genetic algorithm (GA), imperialist competitive algorithm (ICA), Discrete Particle Swarm Optimization (PSO), and Bat algorithm can be used for this purpose. There exist other techniques for navigating complex configuration spaces, e.g., recursive random search and pattern search. Random recursive search has shown to be ineffective since it is more likely to be locked in local optima, preventing the system from achieving maximum efficiency. Pattern search typically suffers from slow local (asymptotic) convergence rates and is not suitable to be used in the RPS.

Figure 13.5 describes a simple configuration search procedure, which consists of six steps, as shown in Fig. 13.4.

Algorithm 1 Explorer functionality

Input: Architectural signature of application, List of available servers
Output: An optimal configuration
Result: Minimized EDP

```
Optimal_EDP = ∞
Conf ← Get_Random(List of servers)
Sign ← Architectural Signature
E ← ANN_Estimate(Conf,Sign)
while iteration ≤ Threshold do
    Opimal_EDP← OE_Assimilation(E_EDP,Opimal_EDP)
    Conf' ← OE_Im_compet.(OE_Revolu.(E_Conf))
    E ← ANN_Estimate(Conf',Sign)
return E_Conf
```

Fig. 13.5 Explorer functionality

The configuration vector is as follows:

$$Conf = \{c_1, c_2, \ldots, c_{10}\}, \tag{13.2}$$

where $Conf$ is the configuration vector and c_i is the value of the i_{th} configuration parameter such as number of sockets, number of cores, core frequency, cache size, memory capacity, memory frequency, number of memory channels, storage capacity, storage speed, and network bandwidth. The estimator estimated the target metric based on the signature of application and server configuration, e.g., $Sign$ and $Conf$. The target metric model is described by:

$$Target\_metric = f(Conf, Sign). \tag{13.3}$$

Note that $f(Conf, Sign)$ is a data model, which means it can be modeled by a machine learning model such as neural network.

In step 1, RPS runs the application on a random platform (random configuration) and starts monitoring its microarchitectural behavior for N execution windows. Then, RPS extracts the architectural information to predict the next phase of the application using the predictor component. In step 3, RPS inputs the initial values of the configuration parameters and architectural signature of the next phase of application to the estimator. In step 4, RPS passes the estimated target metric and configuration parameter values to the OE. Note that the configuration parameters are randomly selected. In this way, OE generates a new set of configuration parameter values. In step 5, these configuration parameter values are fed to the estimator again to estimate a new target metric's value and then, it will be passed to the OE again to check if the stop condition is satisfied or not. Steps from 3 to 5 are repeated a number of times until the optimum configuration is found. The next step (6) includes sending the optimal configuration to the decision maker. All steps are repeated for the next window of operation until the execution of the application finishes.

For a given application and workload, our goal is to find the optimal or a near-optimal server configuration that simultaneously satisfies the performance requirements with minimal operational cost. For this purpose, we use Bounded Knapsack algorithm to solve the aforementioned optimization problem.

Bounded Knapsack Solution

If the optimization is constrained by a Budget, then Bounded Knapsack algorithm is a quick and simple solution to select the optimal server configuration. The bounded knapsack problem can be implemented using dynamic programming:

Maximize $\Sigma_{i=1}^{n} Perf_i \times Conf_i$,
Subject to $\Sigma_{i=1}^{n} Cost_i \times Conf_i \leq Budget$ and $\min_i \leq Conf_i \leq \max_i$,

where $Conf_i$ represents the number or the value of parameter i, and $\min_i$ and $\max_i$ are the minimum and the maximum available resource for parameter i. Also, $Cost_i$ presents the cost corresponding to $Conf_i$. Similarly, $Perf_i$ presents the performance improvement corresponding to $Conf_i$. The result of this optimization is the recommended configuration for decision maker. Budget is a constraint that the user provides. The result of solving the optimization problem is a set of configurations such as the number of sockets, number of nodes, the number of cores, core frequency, memory capacity, memory bandwidth, storage capacity, and storage bandwidth. The OE recommends this configuration to the decision maker. It is the responsibility of decision maker to decide about the action that needs to be taken for scaling the current platform to make it as close as the recommended configuration for the targeted VM.

13.3.4 Decision Making

After finding the optimum configuration, it is time for decision maker component to determine which actions can be executed for efficient resource allocation in the cloud infrastructures. RPS usually makes a decision based on a defined policy. As the cloud gets more complicated, the complexity of developing a policy to handle the resource management increases too. Therefore, designers and researchers are replacing the old school policy making by latest machine learning techniques such as reinforcement learning, e.g., Q-learning. Reinforcement learning enables the system to find the best policy for the resource allocation tasks by itself.

Example 13.4 shows few reallocation actions that can be considered by decision maker.

Example 13.4 (Reallocation Actions' Samples)
Problem: List different actions that can be taken by decision maker.
Solution: (1) Increase CPU frequency. (2) Decrease CPU frequency. (3) Increase CPU core. (4) Decrease CPU core. (5) Add allocated storage. (6) Remove allocated storage. (7) Increase memory capacity. (8) Decrease memory capacity. (9) Migrate VM to different node.

Changing the CPU frequency can be performed by Dynamic Voltage Frequency Scaling (DVFS). The first escalation level ("change VM configuration") can be done locally on a host and can change the host resources. This operation can be performed by hot-(un)plugging of resources (memory and cores). For example, mounting storage or adding memory to the VM is more lightweight than migrating VMs. If there is no appropriate VM available on the host, the decision maker should create a new VM on an appropriate host or migrate the VM to a host that has enough available resources.

Considering Migration Time

The decision maker must take into account the live migration time overhead for deciding whether it is beneficial to migrate the VM or not. Therefore, a performance model of live migration is required to estimate the time that it takes to migrate a VM from host A to host B and resume the job. Performance modeling of live migration involves three main factors: the size of VM memory (V_{Mem}), the memory dirtying rate (D), and network transmission rate (J). Live VM migration achieves negligible application downtime by iteratively pre-copying the pages dirtied at the previous round of transmission.

Resource Sharing Consideration
Another aspect to consider is resource sharing. If running multiple VMs on a host is required, decision maker can use Resource Elasticity Fairness (REF) [392] in order to allocate the resources among VMs, as co-scheduling multiple VMs on a single server could result in interference. REF is a fair allocation mechanism, which satisfies three game-theoretic properties (sharing incentives (SI), envy-freeness (EF), and Pareto efficiency (PE)) using Cobb–Douglas utility function. For that purpose, RPS can begin with the space of possible allocations. Then it adds constraints to identify allocations with the desired properties as follows:

Suppose multiple VMs share a server with R types of hardware resources. Let $x_i = \{x_{i1},\ldots,x_{iR}\}$denote i-th VM's hardware allocation. Further, let $u_i(x_i)$ denote i-th VM's utility. The following equation defines utility within the Cobb–Douglas preference domain:

$$u_i(x_i) = a_{i0} \sqcap_{r=1}^{R} x_{ir}{}^{a_{ir}}.$$

The parameters a_i=$\{a_{i1},...,a_{iR}\}$ quantify the elasticity with which a VM demands a resource. Let C_r denote the total capacity of resource r in the system. RPS can find fair multi-resource allocations given Cobb–Douglas preferences with the following feasibility problem for N virtual machines and R resources:
Find x subject to:

(1) $u_i(x_i) \geq u_i(x_j)$ $\quad i, j \in$[1,N]
(2) $\frac{a_{ir}}{a_{is}}\frac{x_{is}}{x_{ir}} = \frac{a_{jr}}{a_{js}}\frac{x_{js}}{x_{jr}}$ $\quad i, j \in$[1,N]; $r, s \in$[1,R]
(3) $u_i(x_i) \geq u_i$(C/N) $\quad i \in$[1,N]
(4) $\sum_{i=1}^{N} x_{ir} \leq C_r$ $\quad r \in$[1,R]

Where C/N=$\{C_1/N,\ldots,C_R/N\}$. In this formulation, the four constraints enforce EF, PE, SI, and capacity. The outcome of applying Cobb–Douglas utility function is a fair resource allocation among multiple VMs running on a server.

13.4 Security Threats in Cloud Rooted from ML-Based RPS

We discussed that maximizing utilization, i.e., sharing of resources, is key to achieving cost efficiency in the cloud. However, it also opens the door for security and privacy vulnerabilities. In particular, these resources will be shared among different users due to the multi-tenancy capability of hosts in the cloud, which facilitate a platform for performing a wide range of resource sharing-based attacks, including transient execution attacks, rowhammer attacks, distributed side-channel attacks and distributed denial of service (DDoS) attacks, data leakage exploitation, and attacks that pinpoint target VMs in a cloud system. Mounting such attacks is trivial once the attacker is co-located with the victim. Therefore, the biggest challenge of attacks that exploit resource sharing in cloud environments is co-location.

Unfortunately, RPSs can become a blind spot and vulnerability that can be exploited to solve the co-location challenge of resource sharing-based attacks. In particular, adversarial attacks against machine learning models can be adopted to force RPSs to co-locate the attacker with the victim. These attacks work by adding specially crafted perturbations to the input data, i.e., an image, of machine learning models to manipulate their outcome. However, pixels of an image can be easily manipulated independently without changing the appearance of the image since images have high entropy. In contrast, adding adversarial perturbations to attack programs has different challenges since the attacker needs to ensure that the adversarial perturbations do not alter the malicious payload.

These attacks not only show that current RPSs can be exploited to facilitate a wide range of attacks by solving the co-location challenge in the cloud but also highlight a serious need for new techniques to be invented that guarantee security. Specifically, although there is a large number of defense classes that were developed against computer vision-based adversarial attacks, these defenses are limited in defending against such attacks to RPSs. In particular, these defenses assume that the attacker has a budget, i.e., the maximum amount of perturbation that can be added to an image without changing its content. This is important in the computer vision domain since the goal of adversaries is to perturb an image to fool a specific machine learning classifier but can still be classified correctly by a human. However, for program perturbations, there is no such budget/constraint, allowing the attacker to have an unlimited degree of freedom to add perturbations without the risk of increasing the possibility of being detected.

After co-locating the adversarial VMs with the targeted victim, attackers face two more challenges, namely detection and migration. Specifically, the RPS job does not stop after the instance initialization phase, i.e., the initial deployment of a VM on a suitable host. It has been shown that periodic monitoring after the initial deployment can be unitized to improve both security and performance. For security, the trace information can be used to detect attacks based on a computational anomaly. For performance, the trace can be utilized to detect performance degradation due to resource contention or behavioral change of the running application and migrate the VM to a different host. However, attacks can evolve to bypass such detection as well as avoiding VM migration.

13.4.1 Adversarial Machine Learning Attack to RPS

Adversaries are rarely interested in a random service running on a public cloud. They need to pinpoint where the target resides in a practical manner to be able to perform denial of service (DoS) attacks, resource freeing attack (RFA), or side channel (SC) attacks. This requires a launch strategy for co-location and a mechanism for co-residency detection. The attack is practical if the target is located with high accuracy, in reasonable time and with modest resource costs. Performing any distributed attack requires a number of prerequisite steps detailed in the following sections.

Finding Physical Hosts Running Victim Instances

To perform any attacks based on co-location, the attacker needs several VM launching strategies to achieve co-residency with the victim instance, which is impractical and not feasible. A pre-condition for the attack is that the malicious VMs reside on the same physical host as victim VMs, as side-channel and RFA attacks are performed locally. The first and main step is thus to find the location of physical hosts running victim VMs. Several placement variables such as datacenter region, time interval, and instance type are important to achieve co-residency. These variables may be different among IaaS clouds. However, the application type is considered an effective factor in the placement strategy. Let $P(m_{mali})$ be the probability of instance m_{mali} to be co-resident with instance victim m_{vici}. The value of P will be raised by increasing the number of launched attack instances. To make sure that both attacker and victim VMs achieve co-resident placement, the adversary can perform co-residency detection techniques such as network probing. The attacker can also use data mining techniques to detect the type and characteristics of a running application in the victim VM by analyzing interferences introduced in the different resources to increase the accuracy of co-residency detection.

Evasion from Detection and Migration

There are various techniques to detect an attack in a virtualized environment. For example, as side-channel attacks are very fine-grained attacks, the detection of such attacks requires high-resolution information, mainly provided by Hardware Performance Counters (HPCs). The HPCs are a set of special-purpose registers built into modern microprocessors to capture the trace of hardware-related events such as LLC load misses, branch instructions, branch misses, and executed instructions while executing an application. Those events are basically used to profile a program behavior for optimization purposes and are available for any application in the user space. These events are also used in the detection of abnormal behaviors in computer systems. We distinguish two different methods of detection: (1) signature based and

(2) threshold based. Signature-based approaches create the signature of the attack based on received information from HPCs and compare the behavior of the system with the generated signature to identify any eventual malicious activity. On the other hand, threshold-based approaches utilize the HPCs trace to flag anomaly resource utilization that goes beyond a pre-specified threshold.

Overview of Attack

In any resource sharing-based attack, the instance initialization phase of RPS plays a preventive role in such attacks by avoiding a malicious instance to be co-localized with a targeted victim instance on the same physical machine. This phase of the VM placement algorithm is undocumented for security reasons by cloud service providers. However, it is possible that an adversary can bypass the instance initialization phase and get co-located by victims with high probability. Moreover, periodic monitoring and rescheduling are leveraged to mitigate co-residency attacks when the attack is detected. However, it is still possible to disguise the malicious behavior of the adversary's VM and remain on the same host with the victim and avoid the migration.

Figure 13.6 shows the overview of Adversarial Machine Learning Attack to RPS. The attacker can use a fake trace generator (FTG) to wrap it around an adversary application in order to first get *co-located* with a targeted victim and subsequently evade detection and migration. At instance initialization phase, the FTG goal is to perform a trace mimicry task that will mimic the behavior of the targeted victim application to increase the chance of co-location. If the desired server has not been assigned to the adversary VM (co-residency with the victim can be detected by network probing), then the adversary VM terminates, and a new VM must be reinstantiated as the cost of forcing RPS to migrate the adversary VM to another host that may have the victim is high. After co-residency, FTG will change its mode to constantly generate a new specialized trace that changes the behavior of the adversary application to evade not only detection but also migration.

To create FTG, attackers can use the concept of adversarial sample in machine learning where they can add specially crafted perturbations to an input signal to fool the machine learning models. In particular, their goal is to change the model's output

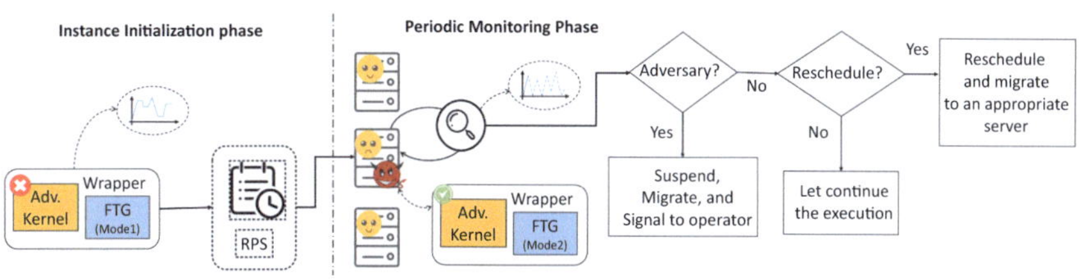

Fig. 13.6 Overview of adversarial machine learning attack to RPS

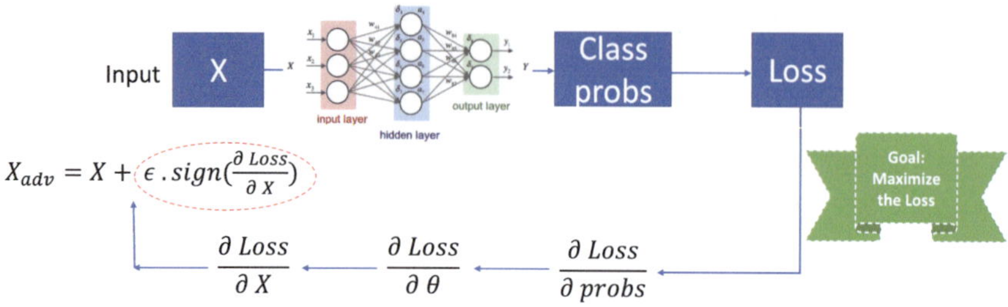

Fig. 13.7 Fast gradient sign method

decision to a specific output, i.e., output that is similar to the targeted victim output. For this purpose, the attack should be performed in two phases. First, they need to reverse engineer the RPS using a machine learning model to have access to how the RPS makes decisions. Second, they should utilize the reversed engineering results to craft a specialized FTG; an FTG that will add perturbations to the adversary's application trace to force the RPS to co-locate it with a targeted victim. This perturbation is the trace that must be generated by FTG. The adversarial perturbation can be calculated based on the gradient loss, as shown in Fig. 13.7, similar to the Fast Gradient Sign Method and is given by:

$$x^{adv} = x + \epsilon sign(\nabla_x L(\theta, x, y)).$$

13.4.2 Isolation as a Remedy

After co-location, a wide range of attacks can be mounted due to the lack of strictly enforced resource isolation between co-scheduled instances, as shown in Fig. 13.8.

In order to study the effects of resource isolation, we enforce several resource partitioning and isolation techniques. We employ core isolation (thread pinning to physical cores), to constrain interference context switching. We employ the Cache Allocation Technology (CAT) available in Intel chips to isolate the LLC. The size of cache partitions can be changed at runtime by reprogramming MSR registers. We also use the outbound network bandwidth partitioning capabilities of Linux's traffic control. We employ the `qdisc` to enforce bandwidth limits. To perform DRAM bandwidth partitioning, RPS can monitor the DRAM bandwidth usage of each application using Intel PCM to co-locate jobs on the same machine where it can accommodate their aggregate peak memory bandwidth usage.

Figure 13.9 shows the impact of isolation techniques on the effectiveness of attacks that we have implemented on a local cluster. The numbers presented here are

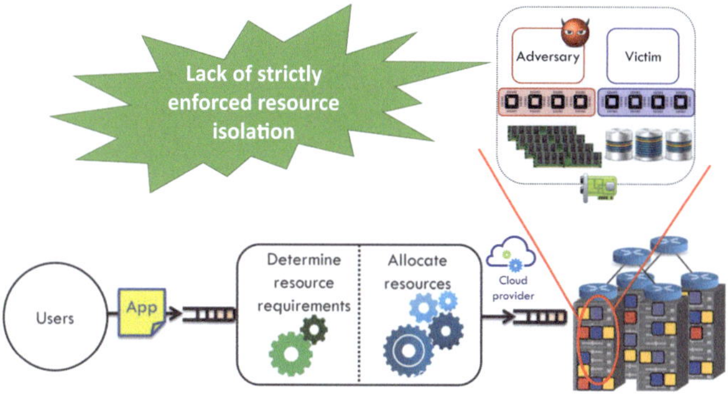

Fig. 13.8 Security threats in the cloud

based on a controlled experimental setup, and the purpose is to provide an insight on this issue.

As expected, when no isolation is used, attacks have a significantly higher success rate. As a result, introducing thread pinning benefits since it reduces core contention. The dominant resource usage of each application determines which isolation technique benefits it the most. Thread pinning mostly benefits workloads bound by on-chip resources, such as L1/L2 caches and cores. Adding network bandwidth partitioning lowers success rate for DoS attacks. It primarily benefits network-bound workloads, for which network interference conveys the most information for detection of co-residency. Cache partitioning has the most dramatic reduction in success rate of SCA, as they are LLC-bound applications. Enforcing core isolation is also sufficient to degrade the success rate of RFAs. Finally, memory bandwidth isolation further reduces success rate by 10% on average, benefiting jobs dominated by DRAM traffic. It is more effective on DoS and RFA and has less impact on SCAs.

The number of co-residents also affects the extent to which isolation helps. The more co-scheduled applications exist per machine, the more isolation techniques degrade success rate, as they make distinguishing between co-residents harder. Improving security using isolation, however, comes at a performance penalty of 32% on average in execution time, as threads of the same job are forced to contend with one another. Alternatively, users can overprovision their resource reservations to avoid performance degradation, which results in a 47% drop in utilization. This means that the cloud provider cannot leverage CPU idleness to share machines, decreasing the cost benefits of cloud computing.

Our analysis highlights a design problem with current datacenter platforms. Traditional multicores are prone to contention, which will only worsen as more cores are integrated into each server, and multi-tenancy becomes more pronounced. Existing isolation techniques are insufficient to mitigate security vulnerabilities, and techniques that provide reasonable security guarantees sacrifice performance or cost

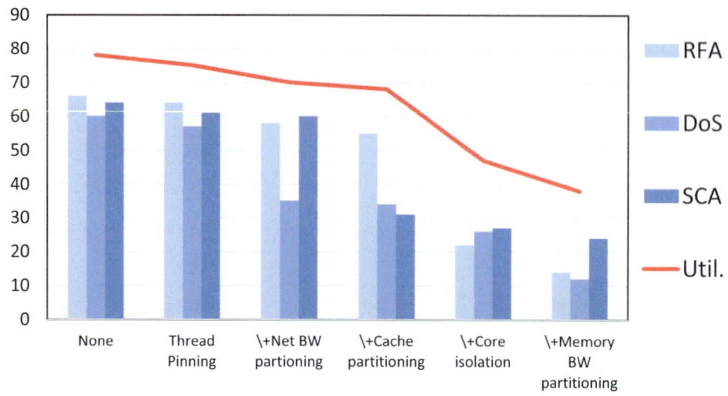

Fig. 13.9 Attacks' success rate and CPU utilization (Util) with isolation techniques

efficiency, through low utilization. This highlights the need for new techniques to be invented that guarantee security at high utilization for shared resources.

13.5 Exercise Problems

Problem 13.1 In this problem, using the link provided below provide the description of the following CPU metrics: CPI Rate, IPC, Back-End Bound, Port Utilization, Bad Speculation, Cache Bound, Front-End Bound, Hardware Event Count, L1 Bound, Memory Bound, DRAM Bound, Retire Stalls, and Retiring.
URL: https://software.intel.com/content/www/us/en/develop/documentation/vtune-help/top/reference/cpu-metrics-reference.html

Problem 13.2 In this problem, using the application's trace dataset (model_knn.csv), implement a K-nearest neighborhood model to predict the EDP of the application based on K steps.

Problem 13.3 In this problem, find the root mean squared error rate for different k values of the previous problem.

Problem 13.4 In this problem, sweep the number of previous steps (K) from 1 to 25 and draw the graph of accuracy based on K for the KNN model developed in the previous problem and find the best K.

Problem 13.5 In this problem, using the application's signature dataset provided for you (model.csv), implement a simple fully connected neural network to classify applications into the following categories: I/O bound or CPU bound. Use *pandas* library to read CSV file, *sklearn* library for processing the dataset, and *Keras* library for implementing a sequential model with dense layers. Use 15% of data for testing and 15% of data for validation.

Problem 13.6 By using grid search (changing the number of dense layers and the number of neurons in each layer), find the minimum number of neurons required in the hidden layer of your neural network trained for the previous problem to reach an accuracy of more than 90% on the testing set.

Problem 13.7 By using grid search (changing the number of epoch and batch size), find the best number of training steps for the previous problem to reach an accuracy of more than 90% on the testing set.

Problem 13.8 In this problem, by using the application's signature dataset (model_knn.csv) implement Random Forest algorithm for the classification task.

Problem 13.9 Using Random Forest model generated in the previous problem, show the ranking of each feature to select the most suitable feature for application classification.

Problem 13.10 Train new random forest with only the two most important variables that you found in the previous problem. Report the ML implementation results including (a) recall and (b) F1-score.

Problem 13.11 Visualize the application's signature dataset with PCA algorithm, understanding the dimension reduction for visualization.

Problem 13.12 Explain how isolation can mitigate the security risk in the cloud? List pros and cons.

Problem 13.13 What is Resource Elasticity Fairness (REF)? Name three game-theoretic properties that REF can satisfy.

Problem 13.14 What types of microarchitectural information can be used for monitoring the state of a server in the cloud? Name four tools that can collect system-level or low-level architectural information for performance analysis.

Problem 13.15 What are the major components of a modern machine learning-based resource provisioning system? What are the main tasks of a resource provisioning system?

References

1. S. Shukla, S. Manoj, and S. Rafatirad, "Rnn-based classifier to detect stealthy malware using localized features and complex symbolic sequence," in *IEEE International Conference On Machine Learning And Applications (ICMLA)*, 2019.
2. ——, "On-device malware detection using performance-aware and robust collaborative learning," in *Design and Automation Conference (DAC)*, 2021.
3. C. Bandi, S. Salehi, R.Hassan, S. Manoj, H. Homayoun, and S. Rafatirad, "Ontology-driven framework for trend analysis of vulnerabilities and impacts in iot hardware," in *IEEE International Conference on Semantic Computing (ICSC)*, 2021.
4. C. R. Boyd, M. A. Tolson, and W. S. Copes, "Evaluating trauma care: The triss method. trauma score and the injury severity score," *The Journal of Trauma*, vol. 27, p. 370–378, 1987.
5. J. Tang, R. Liu, L. Y. L., Y. F. Hu, M. J. Shao, and X. G. Meng, "Application of machine-learning models to predict tacrolimus stable dose in renal transplant recipients," *Scientific reports*, vol. 42192, p. 7, 2017.
6. Y. Zhan, D. Hu, Y. Wang, and X. Yu, "Semisupervised hyperspectral image classification based on generative adversarial networks," *IEEE Geoscience and Remote Sensing Letters*, 2017.
7. H. Sayadi, H. Makrani, S. Dinakarrao, T. Mohsenin, A. Sasan, S. Rafatirad, and H. Homayoun, "2smart: A two-stage machine learning-based approach for run-time specialized hardware-assisted malware detection," in *Design, Automation Test in Europe Conference Exhibition (DATE'19)*, March 2019, pp. 728–733.
8. S. Shukla, G. Kolhe, S. M. P D, and S. Rafatirad, "Stealthy malware detection using RNN-based automated localized feature extraction and classifier," *IEEE International Conference on Tools with Artificial Intelligence (ICTAI)*, 2019.
9. A. Borkar, M. Hayes, and M. Smith, "A novel lane detection system with efficient ground truth generation," *IEEE Trans. Intell. Transp. Syst.*, pp. 365–374, 2012.
10. H. Audibert and J. Ponce, "General road detection from a single image," *IEEE Trans. Image Process.*, pp. 2211–2220, 2010.
11. F. E. Harrell, *Regression Modeling Strategies*. Springer, 2015.
12. E. Alpaydin, *Introduction to Machine Learning*, 2nd ed. MIT Press, 2009.
13. S. Marsland, *Machine Learning: An Algorithmic Perspective*, 2nd ed. Chapman & Hall/CRC, 2009.
14. S. Kumar, S. Mishra, and S. Gupta, "Short term load forecasting using ANN and multiple linear regression," in *Int. Conf. on Computational Intelligence Communication Tech.*, 2016.

S. Rafatirad et al., *Machine Learning for Computer Scientists and Data Analysts*,
https://doi.org/10.1007/978-3-030-96756-7

15. P. Wang, R. Ge, X. Xiao, M. Zhou, and F. Zhou, "hMuLab: a biomedical hybrid MUlti-LABel classifier based on multiple linear regression," *IEEE/ACM Transactions on Computational Biology and Bioinformatics*, vol. PP, no. 99, pp. 1–1, 2016.
16. "Multiple linear regression," Online, http://www.stat.yale.edu/Courses/1997-98/101/linmult.htm.
17. M. A. M. Marinho and et.al., "Array interpolation based on multivariate adaptive regression splines," in *IEEE Sensor Array and Multichannel Signal Processing Workshop (SAM)*, 2016.
18. S. Crino and D. E. Brown, "Global optimization with multivariate adaptive regression splines," *IEEE Trans. on Systems, Man, and Cybernetics*, vol. 37, no. 2, pp. 333–340, Apr 2007.
19. K. Barbé and H. Gosselin, "An ARMA time series approach for analyzing long memory dynamics in measurements," in *IEEE Int. Instrumentation and Measurement Technology Conf.*, 2016.
20. "Auto-regressive moving average," Online, https://www.youtube.com/watch?v=Aw77aMLj9uM&t=1514s.
21. B. Saha, S. Poll, K. Goebel, and J. Christophersen, "An integrated approach to battery health monitoring using Bayesian regression and state estimation," in *IEEE Autotestcon*, 2007.
22. "Bayesian linear regression," Online, https://www.youtube.com/watch?v=d1iIUtnDngg&t=275s.
23. "Bayesian linear regression," Online, http://web.cse.ohio-state.edu/~kulis/teaching/788_sp12/scribe_notes/lecture5.pdf.
24. by Douglas C. Montgomery, G. C. Runger, and N. F. Hubele, *Engineering Statistics*, 5th ed. Wiley, 2010.
25. J. Wakefield, *Bayesian and Frequentist Regression Methods*. Springer, 2013.
26. C. P. Prathibhamol, K. V. Jyothy, and B. Noora, "Multi label classification based on logistic regression (MLC-LR)," in *Int. Conf. on Advances in Computing, Communications and Informatics*, 2016.
27. P. K. Dalvi, S. K. Khandge, A. Deomore, A. Bankar, and V. A. Kanade, "Analysis of customer churn prediction in telecom industry using decision trees and logistic regression," in *Symp. on Colossal Data Analysis and Networking*, 2016.
28. "Logistic regression," Online, https://www.stat.cmu.edu/~cshalizi/uADA/12/lectures/ch12.pdf.
29. "Logistic regression," Online, https://www.youtube.com/watch?v=-Z2a_mzl9LM.
30. M. Qian, "Application of CORDIC algorithm to neural networks VLSI design," in *Conference on Computational Engineering in Systems Applications*, 2006.
31. B. Gisutham, T. Srikanthan, and K. V. Asari, "A high speed flat CORDIC based neuron with multi-level activation function for robust pattern recognition," in *IEEE Int. Workshop on Computer Architectures for Machine Perception*, 2000.
32. W. Bian and X. Chen, "Neural network for non-smooth, non-convex constrained minimization via smooth approximation," *IEEE Transactions on Neural Networks and Learning Systems*, vol. 25, no. 3, pp. 545–556, March 2014.
33. Y. Liu, J. A. Starzyk, and Z. Zhu, "Optimized approximation algorithm in neural networks without overfitting," *IEEE Transactions on Neural Networks*, vol. 19, no. 6, pp. 983–995, June 2008.
34. "Delta rule," Online, https://en.wikipedia.org/wiki/Deltarule.
35. P. D. Heermann and N. Khazenie, "Classification of multispectral remote sensing data using a back-propagation neural network," *IEEE Trans. on Geoscience and Remote Sensing*, vol. 30, no. 1, pp. 81–88, Jan 1992.
36. C. R. Gent and C. P. Sheppard, "Predicting time series by a fully connected neural network trained by back propagation," *Computing Control Engineering Journal*, vol. 3, no. 3, pp. 109–112, May 1992.
37. "Back propagation neural network," Online, http://neuralnetworksanddeeplearning.com/chap2.html.

38. "Back propagation neural network," Online, https://page.mi.fu-berlin.de/rojas/neural/chapter/K7.pdf.
39. D. Masters and C. Luschi, "Revisiting small batch training for deep neural networks," *arXiv preprint arXiv:1804.07612*, 2018.
40. J. Dean, G. Corrado, R. Monga, K. Chen, M. Devin, M. Mao, M. Ranzato, A. Senior, P. Tucker, K. Yang *et al.*, "Large scale distributed deep networks," *Advances in neural information processing systems*, vol. 25, pp. 1223–1231, 2012.
41. D. P. Kingma and J. Ba, "Adam: A method for stochastic optimization," *arXiv preprint arXiv:1412.6980*, 2014.
42. I. Goodfellow, Y. Bengio, and A. Courville, "Deep learning," 2016, book in preparation for MIT Press. [Online]. Available: http://www.deeplearningbook.org
43. J. Heaton, *Artificial Intelligence for Humans, Volume 3: Deep Learning and Neural Networks*. Heaton Research Inc., 2015.
44. Y. H. Chen, T. Krishna, J. S. Emer, and V. Sze, "Eyeriss: An energy-efficient reconfigurable accelerator for deep convolutional neural networks," *IEEE Journal of Solid-State Circuits*, vol. PP, no. 99, pp. 1–12, 2016.
45. S. Lawrence, C. L. Giles, A. C. Tsoi, and A. D. Back, "Face recognition: a convolutional neural-network approach," *IEEE Transactions on Neural Networks*, vol. 8, no. 1, pp. 98–113, Jan 1997.
46. M. Bianchini and M. Gori, "Theoretical properties of recursive neural networks with linear neurons," *IEEE Trans. on Neural Networks*, vol. 12, no. 5, pp. 953–967, Sep 2001.
47. R. J. Williams and D. Zipser, "A learning algorithm for continually running fully recurrent neural networks," *Neural Computation*, vol. 1, no. 2, pp. 270–280, June 1989.
48. S. Nie, H. Zhang, X. Zhang, and W. Liu, "Deep stacking networks with time series for speech separation," in *IEEE Int. Conf. on Acoustics, Speech and Signal Processing*, 2014.
49. L. Deng, X. He, and J. Gao, "Deep stacking networks for information retrieval," in *IEEE Int. Conf. on Acoustics, Speech and Signal Processing*, 2013.
50. J. Lee, J. H. Chang, and J. Sohn, "On using parameterized multi-channel non-causal wiener filter-adapted convolutional neural networks for distant speech recognition," in *Int. Conf. on Electronics, Information, and Communications*, 2016.
51. N. Li, S. Takaki, Y. Tomiokay, and H. Kitazawa, "A multistage dataflow implementation of a deep convolutional neural network based on FPGA for high-speed object recognition," in *IEEE Southwest Symposium on Image Analysis and Interpretation (SSIAI)*, 2016.
52. Y. Zhou and J. Jiang, "An FPGA-based accelerator implementation for deep convolutional neural networks," in *Int. Conf. on Computer Science and Network Technology*, 2015.
53. F. Ikhwantri, N. Habibie, A. R. Syulistyo, Aprinaldi, and W. Jatmiko, "Learning semantic segmentation score in weakly supervised convolutional neural network," in *Int. Conf. on Computers, Communications, and Systems*, 2015.
54. M. Pazouki, S. S. Allaei, M. H. Pazouki, and D. P. F. Múller, "Adaptive learning algorithm for RBF neural networks in kernel spaces," in *Int. Joint Conf. on Neural Networks*, 2016.
55. C.-Y. Lee and L.-C. Liao, "Recognition of facial expression by using neural-network system with fuzzified characteristic distances weights," in *IEEE Int. Conf. on Fuzzy Systems*, 2008.
56. H. Romsdorfer, "Speech prosody control using weighted neural network ensembles," in *IEEE Int. Workshop on Machine Learning for Signal Processing*, 2009.
57. N. B. Karayiannis and G. W. Mi, "Growing radial basis neural networks: merging supervised and unsupervised learning with network growth techniques," *IEEE Transactions on Neural Networks*, vol. 8, no. 6, pp. 1492–1506, Nov 1997.
58. C. Li, Q. Chen, and Y. Zhang, "A fault location method based on the waveform discriminator means and rbf neural networks," in *IEEE Asia-Pacific Power and Energy Engineering Conf.*, 2016.
59. K. Meng, Z. Y. Dong, D. H. Wang, and K. P. Wong, "A self-adaptive rbf neural network classifier for transformer fault analysis," *IEEE Transactions on Power Systems*, vol. 25, no. 3, pp. 1350–1360, Aug 2010.

60. X. Yali and J. Changsheng, "Trajectory linearization control of an aerospace vehicle based on rbf neural network," *Journal of Systems Engineering and Electronics*, vol. 19, no. 4, pp. 799–805, Aug 2008.
61. X. Chen, X. Liu, Y. Wang, M. J. F. Gales, and P. C. Woodland, "Efficient training and evaluation of recurrent neural network language models for automatic speech recognition," *IEEE/ACM Transactions on Audio, Speech, and Language Processing*, vol. 24, no. 11, pp. 2146–2157, Nov 2016.
62. "Recurrent neural networks," Online, https://www.youtube.com/watch?v=iX5V1WpxxkY&t=3174s.
63. "Recurrent neural networks," Online, https://www.youtube.com/watch?v=Ukgii7Yd_cU&t=610s.
64. J. T. Chien and Y. C. Ku, "Bayesian recurrent neural network for language modeling," *IEEE Transactions on Neural Networks and Learning Systems*, vol. 27, no. 2, pp. 361–374, Feb 2016.
65. H. Palangi, L. Deng, Y. Shen, J. Gao, X. He, J. Chen, X. Song, and R. Ward, "Deep sentence embedding using long short-term memory networks: Analysis and application to information retrieval," *IEEE/ACM Transactions on Audio, Speech, and Language Processing*, vol. 24, no. 4, pp. 694–707, April 2016.
66. S. Hochreiter and J. Schmidhuber, "Long short-term memory," *Neural Computation*, vol. 9, no. 8, pp. 1735–1780, Nov 1997.
67. "Lstm for image captioning," Online, https://cs224d.stanford.edu/reports/msoh.pdf.
68. I. Steinwart and A. Christmann, *Support Vector Machines*. Springer, 2008.
69. V. Vapnik, *Estimation of Dependencies Based on Empirical Data*. Springer, 2006.
70. S. Boyd and L. Vandenberghe, *Convex Optimization*. Cambridge University Press, 2004, ch. 5.
71. H. Xue, S. Chen, and Q. Yang, "Structural regularized support vector machine: A framework for structural large margin classifier," *IEEE Transactions on Neural Networks*, vol. 22, no. 4, pp. 573–587, April 2011.
72. "Support vector machine," Online, http://cs229.stanford.edu/notes/cs229-notes3.pdf.
73. "Support vector machine," Online, https://www.youtube.com/watch?v=_PwhiWxHK8o.
74. https://www.benfrederickson.com/multidimensional-scaling/.
75. Y. Tan, W. Liu, and Q. Qiu, "Adaptive power management using reinforcement learning," in *International Conference on Computer-Aided Design (ICCAD)*, 2009, pp. 461–467.
76. W. Liu, Y. Tan, and Q. Qiu, "Enhanced q-learning algorithm for dynamic power management with performance constraint," in *Design, Automation and Test in Europe (DATE)*, March 2010, pp. 602–605.
77. S. Yue, D. Zhu, Y. Wang, and M. Pedram, "Reinforcement learning based dynamic power management with a hybrid power supply," in *30th IEEE International Conference on Computer Design (ICCD)*, Sept 2012, pp. 81–86.
78. C. J. Watkins and P. Dayan, "Q-learning," *Machine learning Journal*, vol. 8, no. 3-4, pp. 279–292, May 1992.
79. M. L. Littman, "Value-function reinforcement learning in markov games," *ACM Cognitive Systems Research*, vol. 2, no. 1, pp. 55–66, Apr 2001.
80. A. Gosavi, "A reinforcement learning algorithm based on policy iteration for average reward: Empirical results with yield management and convergence analysis," *Machine Learning Journal*, vol. 55, no. 1, pp. 5–29, Apr 2004.
81. N. Mastronarde and M. van der Schaar, "Online reinforcement learning for dynamic multimedia systems," *IEEE Tran. on Image Processing*, vol. 19, no. 2, pp. 290–305, Feb 2010.
82. ——, "Online reinforcement learning for dynamic multimedia systems," *IEEE Tran. on Image Processing*, vol. 19, no. 2, pp. 290–305, Feb 2010.
83. R. S. Sutton and A. G. Barto, *Reinforcement Learning: An Introduction*. Cambridge, MA: MIT Press, 1998.
84. S. Shalev-Shwartz *et al.*, "Online learning and online convex optimization," *Foundations and trends in Machine Learning*, vol. 4, no. 2, pp. 107–194, 2011.

85. A. Rakhlin, K. Sridharan, and A. Tewari, "Online learning: Random averages, combinatorial parameters, and learnability," in *Advances in Neural Information Processing Systems*, 2010, pp. 1984–1992.
86. K. Crammer, O. Dekel, J. Keshet, S. Shalev-Shwartz, and Y. Singer, "Online passive-aggressive algorithms," *Journal of Machine Learning Research*, vol. 7, no. Mar, pp. 551–585, 2006.
87. E. Hazan, A. Rakhlin, and P. L. Bartlett, "Adaptive online gradient descent," in *Advances in Neural Information Processing Systems*, 2008, pp. 65–72.
88. O. Dekel, R. Gilad-Bachrach, O. Shamir, and L. Xiao, "Optimal distributed online prediction using mini-batches," *The Journal of Machine Learning Research*, vol. 13, pp. 165–202, 2012.
89. N. Cesa-Bianchi, A. Conconi, and C. Gentile, "A second-order perceptron algorithm," *SIAM Journal on Computing*, vol. 34, no. 3, pp. 640–668, 2005.
90. J. Langford, L. Li, and T. Zhang, "Sparse online learning via truncated gradient," in *Advances in neural information processing systems*, 2009, pp. 905–912.
91. J. Duchi and Y. Singer, "Efficient online and batch learning using forward backward splitting," *The Journal of Machine Learning Research*, vol. 10, pp. 2899–2934, 2009.
92. Y. Nesterov, "Primal-dual subgradient methods for convex problems," *Mathematical programming*, vol. 120, no. 1, pp. 221–259, 2009.
93. L. Xiao, "Dual averaging methods for regularized stochastic learning and online optimization," 2010.
94. E. Hazan, "Introduction to online convex optimization," *arXiv preprint arXiv:1909.05207*, 2019.
95. B. McMahan, "Follow-the-regularized-leader and mirror descent: Equivalence theorems and l1 regularization," in *Proceedings of the Fourteenth International Conference on Artificial Intelligence and Statistics*. JMLR Workshop and Conference Proceedings, 2011, pp. 525–533.
96. J. A. Hartigan and M. A. Wong, "Algorithm as 136: A k-means clustering algorithm," *Journal of the royal statistical society. series c (applied statistics)*, vol. 28, no. 1, pp. 100–108, 1979.
97. S. Guha, N. Mishra, R. Motwani, and L. o'Callaghan, "Clustering data streams," in *Proceedings 41st Annual Symposium on Foundations of Computer Science*. IEEE, 2000, pp. 359–366.
98. L. Kaufman and P. J. Rousseeuw, "Clustering large applications (program clara)," *Finding groups in data: an introduction to cluster analysis*, pp. 126–163, 2008.
99. M. R. Ackermann, M. Märtens, C. Raupach, K. Swierkot, C. Lammersen, and C. Sohler, "Streamkm++ a clustering algorithm for data streams," *Journal of Experimental Algorithmics (JEA)*, vol. 17, pp. 2–1, 2012.
100. E. Schubert, J. Sander, M. Ester, H. P. Kriegel, and X. Xu, "DBSCAN revisited, revisited: why and how you should (still) use DBSCAN," *ACM Transactions on Database Systems (TODS)*, vol. 42, no. 3, pp. 1–21, 2017.
101. J. Ren and R. Ma, "Density-based data streams clustering over sliding windows," in *2009 Sixth international conference on fuzzy systems and knowledge discovery*, vol. 5. IEEE, 2009, pp. 248–252.
102. I. Ntoutsi, A. Zimek, T. Palpanas, P. Kröger, and H.-P. Kriegel, "Density-based projected clustering over high dimensional data streams," in *Proceedings of the 2012 SIAM international conference on data mining*. SIAM, 2012, pp. 987–998.
103. S. Wold, K. Esbensen, and P. Geladi, "Principal component analysis," *Chemometrics and intelligent laboratory systems*, vol. 2, no. 1–3, pp. 37–52, 1987.
104. P. Comon, "Independent component analysis, a new concept?" *Signal processing*, vol. 36, no. 3, pp. 287–314, 1994.
105. Y. Rubner, L. J. Guibas, and C. Tomasi, "The earth mover's distance, multi-dimensional scaling, and color-based image retrieval," in *Proceedings of the ARPA image understanding workshop*, vol. 661, 1997, p. 668.
106. M. Balasubramanian, E. L. Schwartz, J. B. Tenenbaum, V. de Silva, and J. C. Langford, "The isomap algorithm and topological stability," *Science*, vol. 295, no. 5552, pp. 7–7, 2002.

107. L. K. Saul and S. T. Roweis, "An introduction to locally linear embedding," *unpublished. Available at:* http://www.cs.toronto.edu/~roweis/lle/publications.html, 2000.
108. D. A. Ross, J. Lim, R.-S. Lin, and M.-H. Yang, "Incremental learning for robust visual tracking," *International journal of computer vision*, vol. 77, no. 1–3, pp. 125–141, 2008.
109. C. Wang and Y. M. Lu, "The scaling limit of high-dimensional online independent component analysis," *Journal of Statistical Mechanics: Theory and Experiment*, vol. 2019, no. 12, p. 124011, 2019.
110. M. H. Law and A. K. Jain, "Incremental nonlinear dimensionality reduction by manifold learning," *IEEE transactions on pattern analysis and machine intelligence*, vol. 28, no. 3, pp. 377–391, 2006.
111. S. Schuon, M. Durković, K. Diepold, J. Scheuerle, and S. Markward, "Truly incremental locally linear embedding," in *CoTeSys 1st International Workshop on Cognition for Technical Systems*, 2008.
112. D. W. Scott, *Multivariate density estimation: theory, practice, and visualization*. John Wiley & Sons, 2015.
113. A. Zhou, Z. Cai, L. Wei, and W. Qian, "M-kernel merging: Towards density estimation over data streams," in *Eighth International Conference on Database Systems for Advanced Applications, 2003.(DASFAA 2003). Proceedings*. IEEE, 2003, pp. 285–292.
114. M. Kristan, A. Leonardis, and D. Skočaj, "Multivariate online kernel density estimation with gaussian kernels," *Pattern Recognition*, vol. 44, no. 10-11, pp. 2630–2642, 2011.
115. Y. Cao, H. He, and H. Man, "Somke: Kernel density estimation over data streams by sequences of self-organizing maps," *IEEE transactions on neural networks and learning systems*, vol. 23, no. 8, pp. 1254–1268, 2012.
116. Y. Zheng, J. Jestes, J. M. Phillips, and F. Li, "Quality and efficiency for kernel density estimates in large data," in *Proceedings of the 2013 ACM SIGMOD International Conference on Management of Data*, 2013, pp. 433–444.
117. B. Han, D. Comaniciu, Y. Zhu, and L. S. Davis, "Sequential kernel density approximation and its application to real-time visual tracking," *IEEE Transactions on Pattern Analysis and Machine Intelligence*, vol. 30, no. 7, pp. 1186–1197, 2008.
118. D. Pokrajac, A. Lazarevic, and L. J. Latecki, "Incremental local outlier detection for data streams," in *2007 IEEE symposium on computational intelligence and data mining*. IEEE, 2007, pp. 504–515.
119. G. S. Na, D. Kim, and H. Yu, "Dilof: Effective and memory efficient local outlier detection in data streams," in *Proceedings of the 24th ACM SIGKDD International Conference on Knowledge Discovery & Data Mining*, 2018, pp. 1993–2002.
120. O. Anava, E. Hazan, S. Mannor, and O. Shamir, "Online learning for time series prediction," in *Conference on learning theory*, 2013, pp. 172–184.
121. C. LIU, S. C. HOI, P. ZHAO, and J. SUN, "Online learning of arima for time series prediction," 2016.
122. H. B. McMahan, G. Holt, D. Sculley, M. Young, D. Ebner, J. Grady, L. Nie, T. Phillips, E. Davydov, D. Golovin *et al.*, "Ad click prediction: a view from the trenches," in *Proceedings of the 19th ACM SIGKDD international conference on Knowledge discovery and data mining*, 2013, pp. 1222–1230.
123. B. Li and S. C. Hoi, "Online portfolio selection: A survey," *ACM Computing Surveys (CSUR)*, vol. 46, no. 3, pp. 1–36, 2014.
124. P. Das, N. Johnson, and A. Banerjee, "Online lazy updates for portfolio selection with transaction costs." in *AAAI*. Citeseer, 2013.
125. V. Nair and J. J. Clark, "An unsupervised, online learning framework for moving object detection," in *Proceedings of the 2004 IEEE Computer Society Conference on Computer Vision and Pattern Recognition, 2004. CVPR 2004.*, vol. 2. IEEE, 2004, pp. II–II.
126. A. F. Martins, N. A. Smith, M. Figueiredo, and P. Aguiar, "Structured sparsity in structured prediction," in *Proceedings of the 2011 Conference on Empirical Methods in Natural Language Processing*, 2011, pp. 1500–1511.

127. X. Gao, S. C. Hoi, Y. Zhang, J. Wan, and J. Li, "Soml: Sparse online metric learning with application to image retrieval," 2014.
128. B. Perozzi, R. Al-Rfou, and S. Skiena, "Deepwalk: Online learning of social representations," in *Proceedings of the 20th ACM SIGKDD international conference on Knowledge discovery and data mining*. ACM, 2014, pp. 701–710.
129. G. Adomavicius and A. Tuzhilin, "Toward the next generation of recommender systems: A survey of the state-of-the-art and possible extensions," *IEEE transactions on knowledge and data engineering*, vol. 17, no. 6, pp. 734–749, 2005.
130. G. Salton, *Automatic Text Processing: The Transformation, Analysis, and Retrieval of Information by Computer*. USA: Addison-Wesley Longman Publishing Co., Inc., 1989.
131. J. S. Breese, D. Heckerman, and C. Kadie, "Empirical analysis of predictive algorithms for collaborative filtering," in *Proceedings of the Fourteenth Conference on Uncertainty in Artificial Intelligence*, ser. UAI'98. San Francisco, CA, USA: Morgan Kaufmann Publishers Inc., 1998, p. 43–52.
132. P. Resnick, N. Iacovou, M. Suchak, P. Bergstrom, and J. Riedl, "Grouplens: An open architecture for collaborative filtering of netnews," in *Proceedings of the 1994 ACM Conference on Computer Supported Cooperative Work*, ser. CSCW '94, 1994, p. 175–186.
133. Y. Koren, R. Bell, and C. Volinsky, "Matrix factorization techniques for recommender systems," *Computer*, vol. 42, no. 8, pp. 30–37, 2009.
134. S. Rendle, "Factorization machines," in *2010 IEEE International Conference on Data Mining*. IEEE, 2010, pp. 995–1000.
135. S. Zhang, L. Yao, A. Sun, and Y. Tay, "Deep learning based recommender system: A survey and new perspectives," *ACM Computing Surveys (CSUR)*, vol. 52, no. 1, pp. 1–38, 2019.
136. X. He, L. Liao, H. Zhang, L. Nie, X. Hu, and T.-S. Chua, "Neural collaborative filtering," in *Proceedings of the 26th international conference on world wide web*, 2017, pp. 173–182.
137. H.-T. Cheng, L. Koc, J. Harmsen, T. Shaked, T. Chandra, H. Aradhye, G. Anderson, G. Corrado, W. Chai, M. Ispir *et al.*, "Wide & deep learning for recommender systems," in *Proceedings of the 1st workshop on deep learning for recommender systems*, 2016, pp. 7–10.
138. H. Guo, R. Tang, Y. Ye, Z. Li, and X. He, "DeepFM: a factorization-machine based neural network for CTR prediction," *arXiv preprint arXiv:1703.04247*, 2017.
139. Y. Jo and A. H. Oh, "Aspect and sentiment unification model for online review analysis," in *Proceedings of the fourth ACM international conference on Web search and data mining*, 2011, pp. 815–824.
140. S. Kim, J. Zhang, Z. Chen, A. H. Oh, and S. Liu, "A hierarchical aspect-sentiment model for online reviews." in *AAAI*. Citeseer, 2013.
141. H. Wang and M. Ester, "A sentiment-aligned topic model for product aspect rating prediction," in *Proceedings of the 2014 Conference on Empirical Methods in Natural Language Processing (EMNLP)*, 2014, pp. 1192–1202.
142. S. Xiong, K. Wang, D. Ji, and B. Wang, "A short text sentiment-topic model for product reviews," *Neurocomputing*, vol. 297, pp. 94–102, 2018.
143. Y. Tay, A. T. Luu, and S. C. Hui, "Multi-pointer co-attention networks for recommendation," in *Proceedings of the 24th ACM SIGKDD International Conference on Knowledge Discovery & Data Mining*, 2018, p. 2309–2318.
144. J. Y. Chin, K. Zhao, S. Joty, and G. Cong, "ANR: Aspect-based neural recommender," in *Proceedings of the 27th ACM International Conference on Information and Knowledge Management*, 2018, pp. 147–156.
145. X. Guan, Z. Cheng, X. He, Y. Zhang, Z. Zhu, Q. Peng, and T.-S. Chua, "Attentive aspect modeling for review-aware recommendation," *ACM Transactions on Information Systems (TOIS)*, vol. 37, no. 3, pp. 1–27, 2019.
146. G. Zhou, X. Zhu, C. Song, Y. Fan, H. Zhu, X. Ma, Y. Yan, J. Jin, H. Li, and K. Gai, "Deep interest network for click-through rate prediction," in *Proceedings of the 24th ACM SIGKDD International Conference on Knowledge Discovery & Data Mining*, 2018, pp. 1059–1068.

147. G. Zhou, N. Mou, Y. Fan, Q. Pi, W. Bian, C. Zhou, X. Zhu, and K. Gai, "Deep interest evolution network for click-through rate prediction," in *Proceedings of the AAAI conference on artificial intelligence*, vol. 33, 2019, pp. 5941–5948.
148. M. J. Pazzani and D. Billsus, "Content-based recommendation systems," in *The adaptive web*. Springer, 2007, pp. 325–341.
149. A. S. Das, M. Datar, A. Garg, and S. Rajaram, "Google news personalization: scalable online collaborative filtering," in *Proceedings of the 16th international conference on World Wide Web*, 2007, pp. 271–280.
150. M. Tavakolifard, J. A. Gulla, K. C. Almeroth, J. E. Ingvaldesn, G. Nygreen, and E. Berg, "Tailored news in the palm of your hand: a multi-perspective transparent approach to news recommendation," in *Proceedings of the 22nd International Conference on World Wide Web*, 2013, pp. 305–308.
151. J. Liu, P. Dolan, and E. R. Pedersen, "Personalized news recommendation based on click behavior," ser. IUI '10, 2010, p. 31–40.
152. C. Wang and D. M. Blei, "Collaborative topic modeling for recommending scientific articles," in *Proceedings of the 17th ACM SIGKDD international conference on Knowledge discovery and data mining*, 2011, pp. 448–456.
153. L. Li, D. Wang, T. Li, D. Knox, and B. Padmanabhan, "Scene: a scalable two-stage personalized news recommendation system," in *Proceedings of the 34th international ACM SIGIR conference on Research and development in Information Retrieval*, 2011, pp. 125–134.
154. C. Lin, R. Xie, L. Li, Z. Huang, and T. Li, "Premise: Personalized news recommendation via implicit social experts," in *Proceedings of the 21st ACM international conference on Information and knowledge management*, 2012, pp. 1607–1611.
155. J.-W. Son, A.-Y. Kim, and S.-B. Park, "A location-based news article recommendation with explicit localized semantic analysis," in *Proceedings of the 36th international ACM SIGIR conference on Research and development in information retrieval*, 2013, pp. 293–302.
156. D. Ferreira, S. Silva, A. Abelha, and J. Machado, "Recommendation system using autoencoders," *Applied Sciences*, vol. 10, no. 16, p. 5510, 2020.
157. Q. Zhu, X. Zhou, Z. Song, J. Tan, and L. Guo, "Dan: Deep attention neural network for news recommendation," in *Proceedings of the AAAI Conference on Artificial Intelligence*, vol. 33, 2019, pp. 5973–5980.
158. C. Ma, P. Kang, B. Wu, Q. Wang, and X. Liu, "Gated attentive-autoencoder for content-aware recommendation," in *Proceedings of the Twelfth ACM International Conference on Web Search and Data Mining*, 2019, p. 519–527.
159. M. Ye, P. Yin, W.-C. Lee, and D.-L. Lee, "Exploiting geographical influence for collaborative point-of-interest recommendation," in *SIGIR*, 2011.
160. Q. Yuan, G. Cong, Z. Ma, A. Sun, and N. M. Thalmann, "Time-aware point-of-interest recommendation," in *SIGIR*, 2013, pp. 363–372.
161. D. Lian, C. Zhao, X. Xie, G. Sun, E. Chen, and Y. Rui, "GeoMF: Joint geographical modeling and matrix factorization for point-of-interest recommendation," in *SIGKDD*, 2014, pp. 831–840.
162. X. Li, G. Cong, X.-L. Li, T.-A. N. Pham, and S. Krishnaswamy, "Rank-GeoFM: A ranking based geographical factorization method for point of interest recommendation," in *SIGIR*, 2015, pp. 433–442.
163. C. Cheng, H. Yang, M. R. Lyu, and I. King, "Where you like to go next: Successive point-of-interest recommendation," in *IJCAI*, 2013, pp. 2605–2611.
164. S. Feng, X. Li, Y. Zeng, G. Cong, Y. M. Chee, and Q. Yuan, "Personalized ranking metric embedding for next new POI recommendation," in *IJCAI*, 2015, pp. 2069–2075.
165. S. Feng, L. V. Tran, G. Cong, L. Chen, J. Li, and F. Li, "Hme: A hyperbolic metric embedding approach for next-poi recommendation," in *Proceedings of the 43rd International ACM SIGIR Conference on Research and Development in Information Retrieval*, 2020, pp. 1429–1438.
166. S. Feng, G. Cong, B. An, and Y. M. Chee, "Poi2vec: Geographical latent representation for predicting future visitors," in *AAAI*, 2017.

167. Q. Liu, S. Wu, L. Wang, and T. Tan, "Predicting the next location: A recurrent model with spatial and temporal contexts," in *AAAI*, 2016.
168. P. Zhao, H. Zhu, Y. Liu, Z. Li, J. Xu, and V. S. Sheng, "Where to go next: A spatio-temporal LSTM model for next POI recommendation," *AAAI*, 2019.
169. R. M. Bell and Y. Koren, "Lessons from the netflix prize challenge," *Acm Sigkdd Explorations Newsletter*, vol. 9, no. 2, pp. 75–79, 2007.
170. L. Zhao, Z. Lu, S. J. Pan, and Q. Yang, "Matrix factorization+ for movie recommendation."
171. M. Schedl, Y. Deldjoo, M. F. Dacrema, M. G. Constantin, H. Eghbal-zadeh, S. Cereda, B. Ionescu, and P. Cremonesi, "Movie genome: alleviating new item cold start in movie recommendation," *User Modeling and User-Adapted Interaction*, 2019.
172. Z. Zhao, Q. Yang, H. Lu, T. Weninger, D. Cai, X. He, and Y. Zhuang, "Social-aware movie recommendation via multimodal network learning," *IEEE Transactions on Multimedia*, vol. 20, no. 2, pp. 430–440, 2017.
173. K. B. Petersen and M. S. Pedersen, "The matrix cookbook, Nov 2012," *URL* http://www2.imm.dtu.dk/pubdb/p.php, vol. 3274, p. 14, 2012.
174. G. A. Baker, G. A. Baker Jr, G. Baker, P. Graves-Morris, and S. S. Baker, *Pade Approximants: Encyclopedia of Mathematics and It's Applications, Vol. 59 George A. Baker, Jr., Peter Graves-Morris*. Cambridge University Press, 1996, vol. 59.
175. P. P. Petrushev and V. A. Popov, *Rational approximation of real functions*. Cambridge University Press, 2011, vol. 28.
176. L. V. Ahlfors, "Complex analysis: an introduction to the theory of analytic functions of one complex variable," *New York, London*, p. 177, 1953.
177. H. Cohen, *Numerical approximation methods*. Springer, 2011.
178. M. J. D. Powell, *Approximation theory and methods*. Cambridge university press, 1981.
179. F. R. Chung, *Spectral graph theory*. American Mathematical Soc., 1997, no. 92.
180. B. Bollobás, *Extremal graph theory*. Courier Corporation, 2004.
181. T. N. Kipf and M. Welling, "Semi-supervised classification with graph convolutional networks," *ICLR*, 2017.
182. W. Hamilton, Z. Ying, and J. Leskovec, "Inductive representation learning on large graphs," in *Advances in Neural Information Processing Systems*, 2017, pp. 1024–1034.
183. K. Xu, W. Hu, J. Leskovec, and S. Jegelka, "How powerful are graph neural networks?" in *International Conference on Learning Representations*, 2019. [Online]. Available: https://openreview.net/forum?id=ryGs6iA5Km
184. D. K. Hammond, P. Vandergheynst, and R. Gribonval, "Wavelets on graphs via spectral graph theory," *Applied and Computational Harmonic Analysis*, vol. 30, no. 2, pp. 129–150, 2011.
185. J. Atwood and D. Towsley, "Diffusion-convolutional neural networks," in *Advances in Neural Information Processing Systems*, 2016, pp. 1993–2001.
186. A. Grover and J. Leskovec, "node2vec: Scalable feature learning for networks," in *Proceedings of the 22nd ACM SIGKDD international conference on Knowledge discovery and data mining*. ACM, 2016, pp. 855–864.
187. Y. Lin, Z. Liu, M. Sun, Y. Liu, and X. Zhu, "Learning entity and relation embeddings for knowledge graph completion." in *AAAI*, vol. 15, 2015, pp. 2181–2187.
188. D. Wang, P. Cui, and W. Zhu, "Structural deep network embedding," in *Proceedings of the 22nd ACM SIGKDD international conference on Knowledge discovery and data mining*, 2016, pp. 1225–1234.
189. F. Wu, A. Souza, T. Zhang, C. Fifty, T. Yu, and K. Weinberger, "Simplifying graph convolutional networks," in *International Conference on Machine Learning*, 2019, pp. 6861–6871.
190. J. Klicpera, A. Bojchevski, and S. Günnemann, "Predict then propagate: Graph neural networks meet personalized pagerank," 2018.
191. F. M. Bianchi, D. Grattarola, L. Livi, and C. Alippi, "Graph neural networks with convolutional ARMA filters," *CoRR*, 2019.

192. J. Gilmer, S. S. Schoenholz, P. F. Riley, O. Vinyals, and G. E. Dahl, "Neural message passing for quantum chemistry," in *Proceedings of the 34th International Conference on Machine Learning-Volume 70*. JMLR. org, 2017, pp. 1263–1272.
193. P. Veličković, G. Cucurull, A. Casanova, A. Romero, P. Lio, and Y. Bengio, "Graph attention networks," *arXiv preprint arXiv:1710.10903*, 2017.
194. R. Johnson and T. Zhang, "On the effectiveness of laplacian normalization for graph semi-supervised learning," *Journal of Machine Learning Research*, vol. 8, no. Jul, pp. 1489–1517, 2007.
195. N. Shervashidze, P. Schweitzer, E. J. Van Leeuwen, K. Mehlhorn, and K. M. Borgwardt, "Weisfeiler-lehman graph kernels." *Journal of Machine Learning Research*, vol. 12, no. 9, 2011.
196. M. Defferrard, X. Bresson, and P. Vandergheynst, "Convolutional neural networks on graphs with fast localized spectral filtering," in *Advances in neural information processing systems*, 2016, pp. 3844–3852.
197. J. Tang, M. Qu, M. Wang, M. Zhang, J. Yan, and Q. Mei, "Line: Large-scale information network embedding," in *Proceedings of the 24th international conference on world wide web*. International World Wide Web Conferences Steering Committee, 2015, pp. 1067–1077.
198. Q. Li, Z. Han, and X.-M. Wu, "Deeper insights into graph convolutional networks for semi-supervised learning," in *Thirty-Second AAAI Conference on Artificial Intelligence*, 2018.
199. S. Abu-El-Haija, B. Perozzi, A. Kapoor, N. Alipourfard, K. Lerman, H. Harutyunyan, G. V. Steeg, and A. Galstyan, "Mixhop: Higher-order graph convolutional architectures via sparsified neighborhood mixing," *arXiv preprint arXiv:1905.00067*, 2019.
200. Z. Chen, F. Chen, R. Lai, X. Zhang, and C.-T. Lu, "Rational neural networks for approximating jump discontinuities of graph convolution operator," *ICDM*, 2018.
201. Q. Li, X.-M. Wu, H. Liu, X. Zhang, and Z. Guan, "Label efficient semi-supervised learning via graph filtering," in *The IEEE Conference on Computer Vision and Pattern Recognition (CVPR)*, June 2019.
202. A. Loukas, A. Simonetto, and G. Leus, "Distributed autoregressive moving average graph filters," *IEEE Signal Processing Letters*, vol. 22, no. 11, pp. 1931–1935, Nov 2015.
203. E. Isufi, A. Loukas, A. Simonetto, and G. Leus, "Autoregressive moving average graph filtering," *IEEE Transactions on Signal Processing*, vol. 65, no. 2, pp. 274–288, Jan 2017.
204. R. Levie, F. Monti, X. Bresson, and M. M. Bronstein, "Cayleynets: Graph convolutional neural networks with complex rational spectral filters," *IEEE Transactions on Signal Processing*, vol. 67, no. 1, pp. 97–109, 2018.
205. X. Zhu and M. Rabbat, "Approximating signals supported on graphs," in *Acoustics, Speech and Signal Processing (ICASSP), 2012 IEEE International Conference on*. IEEE, 2012, pp. 3921–3924.
206. D. I. Shuman, S. K. Narang, P. Frossard, A. Ortega, and P. Vandergheynst, "The emerging field of signal processing on graphs: Extending high-dimensional data analysis to networks and other irregular domains," *IEEE Signal Processing Magazine*, vol. 30, no. 3, pp. 83–98, 2013.
207. D. I. Shuman, B. Ricaud, and P. Vandergheynst, "Vertex-frequency analysis on graphs," *Applied and Computational Harmonic Analysis*, vol. 40, no. 2, pp. 260–291, 2016.
208. L. N. Trefethen, *Approximation theory and approximation practice*. Siam, 2013, vol. 128.
209. R. Pachon, "Algorithms for polynomial and rational approximation," Ph.D. dissertation, University of Oxford, 2010.
210. K. Cho, B. Van Merriënboer, C. Gulcehre, D. Bahdanau, F. Bougares, H. Schwenk, and Y. Bengio, "Learning phrase representations using RNN encoder-decoder for statistical machine translation," *arXiv preprint arXiv:1406.1078*, 2014.
211. Y. Li, D. Tarlow, M. Brockschmidt, and R. Zemel, "Gated graph sequence neural networks," *arXiv preprint arXiv:1511.05493*, 2015.
212. K. S. Tai, R. Socher, and C. D. Manning, "Improved semantic representations from tree-structured long short-term memory networks," *arXiv preprint arXiv:1503.00075*, 2015.

213. J. You, R. Ying, X. Ren, W. L. Hamilton, and J. Leskovec, "GraphRNN: Generating realistic graphs with deep auto-regressive models," *arXiv preprint arXiv:1802.08773*, 2018.
214. M. Miwa and M. Bansal, "End-to-end relation extraction using LSTMs on sequences and tree structures," *arXiv preprint arXiv:1601.00770*, 2016.
215. J. You, B. Liu, Z. Ying, V. Pande, and J. Leskovec, "Graph convolutional policy network for goal-directed molecular graph generation," in *Advances in neural information processing systems*, 2018, pp. 6410–6421.
216. N. De Cao and T. Kipf, "Molgan: An implicit generative model for small molecular graphs," *arXiv preprint arXiv:1805.11973*, 2018.
217. M. Simonovsky and N. Komodakis, "Graphvae: Towards generation of small graphs using variational autoencoders," in *International Conference on Artificial Neural Networks*. Springer, 2018, pp. 412–422.
218. S. Pan, R. Hu, G. Long, J. Jiang, L. Yao, and C. Zhang, "Adversarially regularized graph autoencoder for graph embedding," *arXiv preprint arXiv:1802.04407*, 2018.
219. T. Ma, J. Chen, and C. Xiao, "Constrained generation of semantically valid graphs via regularizing variational autoencoders," in *Advances in Neural Information Processing Systems*, 2018, pp. 7113–7124.
220. A. Bojchevski, O. Shchur, D. Zügner, and S. Günnemann, "Netgan: Generating graphs via random walks," *arXiv preprint arXiv:1803.00816*, 2018.
221. H. Wang, J. Wang, J. Wang, M. Zhao, W. Zhang, F. Zhang, X. Xie, and M. Guo, "Graphgan: Graph representation learning with generative adversarial nets," in *Thirty-second AAAI conference on artificial intelligence*, 2018.
222. X. Guo and L. Zhao, "A systematic survey on deep generative models for graph generation," 2020.
223. J. Zhou, G. Cui, Z. Zhang, C. Yang, Z. Liu, and M. Sun, "Graph neural networks: A review of methods and applications," *arXiv preprint arXiv:1812.08434*, 2018.
224. Z. Chen, F. Chen, L. Zhang, T. Ji, K. Fu, L. Zhao, F. Chen, and C.-T. Lu, "Bridging the gap between spatial and spectral domains: A survey on graph neural networks," 2020.
225. Z. Wu, S. Pan, F. Chen, G. Long, C. Zhang, and S. Y. Philip, "A comprehensive survey on graph neural networks," *IEEE Transactions on Neural Networks and Learning Systems*, 2020.
226. Z. Zhang, P. Cui, and W. Zhu, "Deep learning on graphs: A survey," *IEEE Transactions on Knowledge and Data Engineering*, 2020.
227. P. Cui, X. Wang, J. Pei, and W. Zhu, "A survey on network embedding," *IEEE Transactions on Knowledge and Data Engineering*, vol. 31, no. 5, pp. 833–852, 2018.
228. D. Zhang, J. Yin, X. Zhu, and C. Zhang, "Network representation learning: A survey," *IEEE transactions on Big Data*, 2018.
229. P. W. Battaglia, J. B. Hamrick, V. Bapst, A. Sanchez-Gonzalez, V. Zambaldi, M. Malinowski, A. Tacchetti, D. Raposo, A. Santoro, R. Faulkner, C. Gulcehre, F. Song, A. Ballard, J. Gilmer, G. Dahl, A. Vaswani, K. Allen, C. Nash, V. Langston, C. Dyer, N. Heess, D. Wierstra, P. Kohli, M. Botvinick, O. Vinyals, Y. Li, and R. Pascanu, "Relational inductive biases, deep learning, and graph networks," 2018.
230. A. Sanchez-Gonzalez, J. Godwin, T. Pfaff, R. Ying, J. Leskovec, and P. W. Battaglia, "Learning to simulate complex physics with graph networks," 2020.
231. X. Ju, S. Farrell, P. Calafiura, D. Murnane, L. Gray, T. Klijnsma, K. Pedro, G. Cerati, J. Kowalkowski, G. Perdue *et al.*, "Graph neural networks for particle reconstruction in high energy physics detectors," *arXiv preprint arXiv:2003.11603*, 2020.
232. S. Seo, C. Meng, and Y. Liu, "Physics-aware difference graph networks for sparsely-observed dynamics," in *International Conference on Learning Representations*, 2019.
233. F. Alet, A. K. Jeewajee, M. Bauza, A. Rodriguez, T. Lozano-Perez, and L. P. Kaelbling, "Graph element networks: adaptive, structured computation and memory," *arXiv preprint arXiv:1904.09019*, 2019.
234. D. K. Duvenaud, D. Maclaurin, J. Iparraguirre, R. Bombarell, T. Hirzel, A. Aspuru-Guzik, and R. P. Adams, "Convolutional networks on graphs for learning molecular fingerprints," in *Advances in neural information processing systems*, 2015, pp. 2224–2232.

235. S. Kearnes, K. McCloskey, M. Berndl, V. Pande, and P. Riley, "Molecular graph convolutions: moving beyond fingerprints," *Journal of computer-aided molecular design*, vol. 30, no. 8, pp. 595–608, 2016.
236. M. Zitnik, M. Agrawal, and J. Leskovec, "Modeling polypharmacy side effects with graph convolutional networks," *Bioinformatics*, vol. 34, no. 13, pp. i457–i466, 2018.
237. A. Fout, J. Byrd, B. Shariat, and A. Ben-Hur, "Protein interface prediction using graph convolutional networks," in *Advances in neural information processing systems*, 2017, pp. 6530–6539.
238. K. Do, T. Tran, and S. Venkatesh, "Graph transformation policy network for chemical reaction prediction," in *Proceedings of the 25th ACM SIGKDD International Conference on Knowledge Discovery & Data Mining*, 2019, pp. 750–760.
239. H. Dai, C. Li, C. Coley, B. Dai, and L. Song, "Retrosynthesis prediction with conditional graph logic network," in *Advances in Neural Information Processing Systems*, 2019, pp. 8872–8882.
240. C. R. Qi, H. Su, K. Mo, and L. J. Guibas, "Pointnet: Deep learning on point sets for 3d classification and segmentation," in *Proceedings of the IEEE conference on computer vision and pattern recognition*, 2017, pp. 652–660.
241. Y. Wang, Y. Sun, Z. Liu, S. E. Sarma, M. M. Bronstein, and J. M. Solomon, "Dynamic graph CNN for learning on point clouds," *Acm Transactions On Graphics (tog)*, vol. 38, no. 5, pp. 1–12, 2019.
242. L. Wang, Y. Huang, Y. Hou, S. Zhang, and J. Shan, "Graph attention convolution for point cloud semantic segmentation," in *Proceedings of the IEEE Conference on Computer Vision and Pattern Recognition*, 2019, pp. 10 296–10 305.
243. W. Norcliffe-Brown, S. Vafeias, and S. Parisot, "Learning conditioned graph structures for interpretable visual question answering," in *Advances in neural information processing systems*, 2018, pp. 8334–8343.
244. M. Narasimhan, S. Lazebnik, and A. Schwing, "Out of the box: Reasoning with graph convolution nets for factual visual question answering," in *Advances in neural information processing systems*, 2018, pp. 2654–2665.
245. S. Qi, W. Wang, B. Jia, J. Shen, and S.-C. Zhu, "Learning human-object interactions by graph parsing neural networks," in *Proceedings of the European Conference on Computer Vision (ECCV)*, 2018, pp. 401–417.
246. H. Xu, C. Jiang, X. Liang, and Z. Li, "Spatial-aware graph relation network for large-scale object detection," in *Proceedings of the IEEE Conference on Computer Vision and Pattern Recognition*, 2019, pp. 9298–9307.
247. H. Hu, J. Gu, Z. Zhang, J. Dai, and Y. Wei, "Relation networks for object detection," in *Proceedings of the IEEE Conference on Computer Vision and Pattern Recognition*, 2018, pp. 3588–3597.
248. J. Gu, H. Hu, L. Wang, Y. Wei, and J. Dai, "Learning region features for object detection," in *Proceedings of the European Conference on Computer Vision (ECCV)*, 2018, pp. 381–395.
249. L. Yao, C. Mao, and Y. Luo, "Graph convolutional networks for text classification," in *Proceedings of the AAAI Conference on Artificial Intelligence*, vol. 33, 2019, pp. 7370–7377.
250. Y. Zhang, P. Qi, and C. D. Manning, "Graph convolution over pruned dependency trees improves relation extraction," *arXiv preprint arXiv:1809.10185*, 2018.
251. S. Vashishth, M. Bhandari, P. Yadav, P. Rai, C. Bhattacharyya, and P. Talukdar, "Incorporating syntactic and semantic information in word embeddings using graph convolutional networks," *arXiv preprint arXiv:1809.04283*, 2018.
252. D. Marcheggiani, J. Bastings, and I. Titov, "Exploiting semantics in neural machine translation with graph convolutional networks," *arXiv preprint arXiv:1804.08313*, 2018.
253. J. Bastings, I. Titov, W. Aziz, D. Marcheggiani, and K. Sima'an, "Graph convolutional encoders for syntax-aware neural machine translation," *arXiv preprint arXiv:1704.04675*, 2017.
254. R. v. d. Berg, T. N. Kipf, and M. Welling, "Graph convolutional matrix completion," *arXiv preprint arXiv:1706.02263*, 2017.

255. R. Ying, R. He, K. Chen, P. Eksombatchai, W. L. Hamilton, and J. Leskovec, "Graph convolutional neural networks for web-scale recommender systems," in *Proceedings of the 24th ACM SIGKDD International Conference on Knowledge Discovery & Data Mining*, 2018, pp. 974–983.
256. F. Monti, D. Boscaini, J. Masci, E. Rodola, J. Svoboda, and M. M. Bronstein, "Geometric deep learning on graphs and manifolds using mixture model CNNs," in *Proceedings of the IEEE Conference on Computer Vision and Pattern Recognition*, 2017, pp. 5115–5124.
257. H. Yao, F. Wu, J. Ke, X. Tang, Y. Jia, S. Lu, P. Gong, J. Ye, and Z. Li, "Deep multi-view spatial-temporal network for taxi demand prediction," in *Thirty-Second AAAI Conference on Artificial Intelligence*, 2018.
258. J. Zhang, X. Shi, J. Xie, H. Ma, I. King, and D.-Y. Yeung, "GaAN: Gated attention networks for learning on large and spatiotemporal graphs," *arXiv preprint arXiv:1803.07294*, 2018.
259. Y. Li, R. Yu, C. Shahabi, and Y. Liu, "Diffusion convolutional recurrent neural network: Data-driven traffic forecasting," *arXiv preprint arXiv:1707.01926*, 2017.
260. B. Yu, H. Yin, and Z. Zhu, "Spatio-temporal graph convolutional networks: A deep learning framework for traffic forecasting," *arXiv preprint arXiv:1709.04875*, 2017.
261. M. Schlichtkrull, T. N. Kipf, P. Bloem, R. Van Den Berg, I. Titov, and M. Welling, "Modeling relational data with graph convolutional networks," in *European Semantic Web Conference*. Springer, 2018, pp. 593–607.
262. C. Shang, Y. Tang, J. Huang, J. Bi, X. He, and B. Zhou, "End-to-end structure-aware convolutional networks for knowledge base completion," in *Proceedings of the AAAI Conference on Artificial Intelligence*, vol. 33, 2019, pp. 3060–3067.
263. D. Nathani, J. Chauhan, C. Sharma, and M. Kaul, "Learning attention-based embeddings for relation prediction in knowledge graphs," *arXiv preprint arXiv:1906.01195*, 2019.
264. Z. Wang, Q. Lv, X. Lan, and Y. Zhang, "Cross-lingual knowledge graph alignment via graph convolutional networks," in *Proceedings of the 2018 Conference on Empirical Methods in Natural Language Processing*, 2018, pp. 349–357.
265. K. Xu, L. Wang, M. Yu, Y. Feng, Y. Song, Z. Wang, and D. Yu, "Cross-lingual knowledge graph alignment via graph matching neural network," *arXiv preprint arXiv:1905.11605*, 2019.
266. I. Balazevic, C. Allen, and T. Hospedales, "Multi-relational poincaré graph embeddings," in *Advances in Neural Information Processing Systems*, 2019, pp. 4463–4473.
267. J. Tang, J. Zhang, L. Yao, J. Li, L. Zhang, and Z. Su, "Arnetminer: extraction and mining of academic social networks," in *Proceedings of the 14th ACM SIGKDD international conference on Knowledge discovery and data mining*, 2008, pp. 990–998.
268. L. Tang and H. Liu, "Relational learning via latent social dimensions," in *Proceedings of the 15th ACM SIGKDD international conference on Knowledge discovery and data mining*, 2009, pp. 817–826.
269. M. Zitnik and J. Leskovec, "Predicting multicellular function through multi-layer tissue networks," *Bioinformatics*, vol. 33, no. 14, pp. i190–i198, 2017.
270. N. Wale, I. A. Watson, and G. Karypis, "Comparison of descriptor spaces for chemical compound retrieval and classification," *Knowledge and Information Systems*, vol. 14, no. 3, pp. 347–375, 2008.
271. A. K. Debnath, R. L. Lopez de Compadre, G. Debnath, A. J. Shusterman, and C. Hansch, "Structure-activity relationship of mutagenic aromatic and heteroaromatic nitro compounds. correlation with molecular orbital energies and hydrophobicity," *Journal of medicinal chemistry*, vol. 34, no. 2, pp. 786–797, 1991.
272. P. D. Dobson and A. J. Doig, "Distinguishing enzyme structures from non-enzymes without alignments," *Journal of molecular biology*, vol. 330, no. 4, pp. 771–783, 2003.
273. K. M. Borgwardt, C. S. Ong, S. Schönauer, S. Vishwanathan, A. J. Smola, and H.-P. Kriegel, "Protein function prediction via graph kernels," *Bioinformatics*, vol. 21, no. suppl_1, pp. i47–i56, 2005.
274. H. Toivonen, A. Srinivasan, R. D. King, S. Kramer, and C. Helma, "Statistical evaluation of the predictive toxicology challenge 2000–2001," *Bioinformatics*, vol. 19, no. 10, pp. 1183–1193, 2003.

275. H. V. Jagadish, J. Gehrke, A. Labrinidis, Y. Papakonstantinou, J. M. Patel, R. Ramakrishnan, and C. Shahabi, "Big data and its technical challenges," *Communications of the ACM*, vol. 57, no. 7, pp. 86–94, 2014.
276. Y. LeCun, L. Bottou, Y. Bengio, and P. Haffner, "Gradient-based learning applied to document recognition," *Proceedings of the IEEE*, vol. 86, no. 11, pp. 2278–2324, 1998.
277. M. Fey and J. E. Lenssen, "Fast graph representation learning with PyTorch Geometric," in *ICLR Workshop on Representation Learning on Graphs and Manifolds*, 2019.
278. M. Wang, L. Yu, D. Zheng, Q. Gan, Y. Gai, Z. Ye, M. Li, J. Zhou, Q. Huang, C. Ma, Z. Huang, Q. Guo, H. Zhang, H. Lin, J. Zhao, J. Li, A. J. Smola, and Z. Zhang, "Deep graph library: Towards efficient and scalable deep learning on graphs," *ICLR Workshop on Representation Learning on Graphs and Manifolds*, 2019. [Online]. Available: https://arxiv.org/abs/1909.01315
279. Z. Yang, W. Cohen, and R. Salakhudinov, "Revisiting semi-supervised learning with graph embeddings," in *International Conference on Machine Learning*, 2016, pp. 40–48.
280. X. Zhu, Z. Ghahramani, and J. D. Lafferty, "Semi-supervised learning using gaussian fields and harmonic functions," in *Proceedings of the 20th International conference on Machine learning (ICML-03)*, 2003, pp. 912–919.
281. D. Zhou, O. Bousquet, T. N. Lal, J. Weston, and B. Schölkopf, "Learning with local and global consistency," in *Advances in neural information processing systems*, 2004, pp. 321–328.
282. Y. Bengio, O. Delalleau, and N. Le Roux, "11 label propagation and quadratic criterion," 2006.
283. M. Wess, S. M. P. Dinakarrao, and A. Jantsch, "Weighted quantization-regularization in DNNs for weight memory minimization towards HW implementation," *IEEE Transactions on Computer Aided Systems of Integrated Circuits and Systems*, 2018.
284. E. Ackerman, "How drive.ai is mastering autonomous driving with deep learning," accessed August 2018. [Online]. Available: https://spectrum.ieee.org/cars-that-think/transportation/self-driving/how-driveai-is-mastering-autonomous-driving-with-deep-learning
285. A. Krizhevsky, I. Sutskever, and G. E. Hinton, "ImageNet classification with deep convolutional neural networks," in *International Conference on Neural Information Processing Systems*, 2012.
286. J. Demme, M. Maycock, J. Schmitz, A. Tang, A. Waksman, S. Sethumadhavan, and S. Stolfo, "On the feasibility of online malware detection with performance counters," in *International Symposium on Computer Architecture*, 2013.
287. M. Chiappetta, E. Savas, and C. Yilmaz, "Real time detection of cache-based side-channel attacks using hardware performance counters," *Appl. Soft Comput.*, vol. 49, no. C, Dec 2016.
288. K. N. Khasawneh, M. Ozsoy, C. Donovick, N. Abu-Ghazaleh, and D. Ponomarev, "EnsembleHMD: Accurate hardware malware detectors with specialized ensemble classifiers," 2018.
289. F. Brasser and et al, "Hardware-assisted security: Understanding security vulnerabilities and emerging attacks for better defenses," in *International Conference on Compilers, Architecture, and Synthesis for Embedded Systems (CASES)*, 2018.
290. H. Sayadi, N. Patel, P. D. S. Manoj, A. Sasan, S. Rafatirad, and H. Homayoun, "Ensemble learning for hardware-based malware detection: A comprehensive analysis and classification," in *ACM/EDAA/IEEE Design Automation Conference*, 2018.
291. C. Szegedy, W. Zaremba, I. Sutskever, J. Bruna, D. Erhan, I. J. Goodfellow, and R. Fergus, "Intriguing properties of neural networks," in *International Conference on Learning Representations (ICLR)*, 2014.
292. I. J. Goodfellow, J. Shlens, and C. Szegedy, "Explaining and harnessing adversarial examples," in *International Conference on Learning Representations (ICLR)*, 2015.
293. N. Papernot, P. McDaniel, S. Jha, M. Fredrikson, Z. B. Celik, and A. Swami, "The limitations of deep learning in adversarial settings," in *IEEE European Symposium on Security and Privacy (Euro S&P)*, 2016.
294. Y. Liu, X. Chen, C. Liu, and D. Song, "Delving into transferable adversarial examples and black-box attacks," in *International Conference on Learning Representations (ICLR)*, 2017.

295. Y. LeCun, C. Cortes, and C. J. Burges, "Mnist digit dataset," accessed August 2018. [Online]. Available: http://yann.lecun.com/exdb/mnist/
296. N. Dalvi, P. Domingos, Mausam, S. Sanghai, and D. Verma, "Adversarial classification," in *ACM SIGKDD International Conference on Knowledge Discovery and Data Mining*, 2004.
297. D. Lowd and C. Meek, "Adversarial learning," in *ACM SIGKDD International Conference on Knowledge Discovery in Data Mining*, 2005.
298. T. Matsumoto, H. Matsumoto, K. Yamada, and S. Hoshino, "Impact of artificial "gummy" fingers on fingerprint systems." vol. 26, 04 2002.
299. B. Nelson, M. Barreno, F. J. Chi, A. D. Joseph, B. I. P. Rubinstein, U. Saini, C. Sutton, J. D. Tygar, and K. Xia, "Exploiting machine learning to subvert your spam filter," in *Usenix Workshop on Large-Scale Exploits and Emergent Threats*, 2008.
300. B. I. Rubinstein, B. Nelson, L. Huang, A. D. Joseph, S.-h. Lau, S. Rao, N. Taft, and J. D. Tygar, "ANTIDOTE: Understanding and defending against poisoning of anomaly detectors," in *ACM SIGCOMM Conference on Internet Measurement*, 2009.
301. B. Biggio, B. Nelson, and P. Laskov, "Poisoning attacks against support vector machines," in *International Conference on Machine Learning*, 2012.
302. H. Xiao, B. Biggio, G. Brown, G. Fumera, C. Eckert, and F. Roli, "Is feature selection secure against training data poisoning?" in *International Conference on Machine Learning*, 2015.
303. L. Muñoz-González, B. Biggio, A. Demontis, A. Paudice, V. Wongrassamee, E. Lupu, and F. Roli, "Towards poisoning of deep learning algorithms with back-gradient optimization," in *ACM Workshop on Artificial Intelligence and Security*, 2017.
304. U. Shaham, Y. Yamada, and S. Negahban, "Understanding adversarial training: increasing local stability of neural nets through robust optimization," *ArXiv e-prints*, 2015.
305. A. Kurakin, I. Goodfellow, and S. Bengio, "Adversarial examples in the physical world," in *International Conference on Learning Representations*, 2017.
306. Y. Dong, F. Liao, T. Pang, H. Su, J. Zhu, X. Hu, and J. Li, "Boosting adversarial attacks with momentum," in *Neural Information Processing Systems Conference*, 2017.
307. ——, "Boosting adversarial attacks with momentum," in *IEEE Conf. on Computer Vision and Pattern Recognition*, 2018.
308. A. Madry *et al.*, "Towards deep learning models resistant to adversarial attacks," *arXiv preprint arXiv:1706.06083*, 2017.
309. N. Papernot, P. McDaniel, X. Wu, S. Jha, and A. Swami, "Distillation as a defense to adversarial perturbations against deep neural networks," in *IEEE Symposium on Security and Privacy (S&P)*, 2016.
310. S. Moosavi-Dezfooli, A. Fawzi, and P. Frossard, "Deepfool: a simple and accurate method to fool deep neural networks," in *IEEE Conf. on computer vision and pattern recognition*, 2016.
311. N. Carlini and D. Wagner, "Towards evaluating the robustness of neural networks," in *IEEE Symposium on Security and Privacy (SP)*, 2017.
312. J. Su, D. V. Vargas, and K. Sakurai, "One pixel attack for fooling deep neural networks," *IEEE Transactions on Evolutionary Computation*, vol. 23, no. 5, pp. 828–841, 2019.
313. S.-M. Moosavi-Dezfooli, A. Fawzi, O. Fawzi, and P. Frossard, "Universal adversarial perturbations," in *IEEE conference on computer vision and pattern recognition*, 2017, pp. 1765–1773.
314. I.-T. Chen and B. Sirkeci-Mergen, "A comparative study of autoencoders against adversarial attacks," in *Int. Conf. on Image Processing, Computer Vision, and Pattern Recognition (IPCV)*, 2018.
315. D. Meng and H. Chen, "Magnet: a two-pronged defense against adversarial examples," in *Proceedings of the 2017 ACM SIGSAC Conf. on Computer and Communications Security*. ACM, 2017, pp. 135–147.
316. A. S. Ross and F. Doshi-Velez, "Improving the adversarial robustness and interpretability of deep neural networks by regularizing their input gradients," in *Thirty-second AAAI Conf. on artificial intelligence*, 2018.
317. T. Pang, C. Du, Y. Dong, and J. Zhu, "Towards robust detection of adversarial examples," in *Advances in Neural Information Processing Systems*, 2018, pp. 4579–4589.

318. X. Wang, S. Wang, P.-Y. Chen, Y. Wang, B. Kulis, X. Lin, and P. Chin, "Protecting neural networks with hierarchical random switching: Towards better robustness-accuracy trade-off for stochastic defenses," in *Int. Joint Conf. on Artificial Intelligence (IJCAI)*, 2019.
319. Z. Liu, Q. Liu, T. Liu, N. Xu, X. Lin, Y. Wang, and W. Wen, "Feature distillation: Dnn-oriented jpeg compression against adversarial examples," *Arxiv*, 2019.
320. X. Jia, X. Wei, X. Cao, and H. Foroosh, "Comdefend: An efficient image compression model to defend adversarial examples," *CoRR*, 2018.
321. H. Drucker and Y. Le Cun, "Improving generalization performance using double backpropagation," *IEEE Trans. on Neural Networks*, vol. 3, no. 6, 1992.
322. K. Grosse, P. Manoharan, N. Papernot, M. Backes, and P. D. McDaniel, "On the (statistical) detection of adversarial examples," *CoRR*, vol. abs/1702.06280, 2017.
323. J. H. Metzen, T. Genewein, V. Fischer, and B. Bischoff, "On detecting adversarial perturbations." in *International Conference on Learning Representations*, 2017.
324. M. Sabokrou, M. Khalooei, and E. Adeli, "Self-supervised representation learning via neighborhood-relational encoding," in *ICCV*, 2019.
325. J. Liu, W. Zhang, Y. Zhang, D. Hou, Y. Liu, H. Zha, and N. Yu, "Detection based defense against adversarial examples from the steganalysis point of view," in *IEEE Conf. on Computer Vision and Pattern Recognition*, 2019.
326. zalandoresearch, "Mnist fashion dataset," accessed August 2018. [Online]. Available: https://github.com/zalandoresearch/fashion-mnist
327. N. Papernot and et al, "Technical report on the cleverhans v2.1.0 adversarial examples library," *arXiv preprint arXiv:1610.00768*, 2018.
328. keras, "Mnist model," accessed August 2018. [Online]. Available: https://github.com/keras-team/keras/blob/master/examples/mnist_mlp.py
329. Z. Zhong, L. Zheng, G. Kang, S. Li, and Y. Yang, "Random erasing data augmentation," *arXiv preprint arXiv:1708.04896*, 2017.
330. J. Rauber, W. Brendel, and M. Bethge, "Foolbox: A python toolbox to benchmark the robustness of machine learning models," *arXiv preprint arXiv:1707.04131*, 2017.
331. G. W. Ding, L. Wang, and X. Jin, "Advertorch v0. 1: An adversarial robustness toolbox based on pytorch," *arXiv preprint arXiv:1902.07623*, 2019.
332. A. Reiss and D. Stricker, "Introducing a new benchmarked dataset for activity monitoring," in *Wearable Computers (ISWC), 2012 16th International Symposium on*. IEEE, 2012, pp. 108–109.
333. A. Jafari *et al.*, "A low-power wearable stand-alone tongue drive system for people with severe disabilities," *IEEE transactions on biomedical circuits and systems*, vol. 12, no. 1, pp. 58–67, 2018.
334. J. Birjandtalab *et al.*, "A non-EEG biosignals dataset for assessment and visualization of neurological status," in *Signal Processing Systems (SiPS), IEEE International Workshop on*. IEEE, 2016, pp. 110–114.
335. S. van der Walt, S. C. Colbert, and G. Varoquaux, "The numpy array: a structure for efficient numerical computation," *CoRR*, vol. abs/1102.1523, 2011. [Online]. Available: http://arxiv.org/abs/1102.1523
336. W. Mckinney, "pandas: a foundational python library for data analysis and statistics," *Python High Performance Science Computer*, 01 2011.
337. M. Abadi, P. Barham, J. Chen, Z. Chen, A. Davis, J. Dean, M. Devin, S. Ghemawat, G. Irving, M. Isard, M. Kudlur, J. Levenberg, R. Monga, S. Moore, D. G. Murray, B. Steiner, P. Tucker, V. Vasudevan, P. Warden, M. Wicke, Y. Yu, and X. Zheng, "Tensorflow: A system for large-scale machine learning," in *12th USENIX Symposium on Operating Systems Design and Implementation (OSDI 16)*, 2016, pp. 265–283. [Online]. Available: https://www.usenix.org/system/files/conference/osdi16/osdi16-abadi.pdf
338. F. Pedregosa, G. Varoquaux, A. Gramfort, V. Michel, B. Thirion, O. Grisel, M. Blondel, P. Prettenhofer, R. Weiss, V. Dubourg, J. Vanderplas, A. Passos, D. Cournapeau, M. Brucher,

M. Perrot, and Édouard Duchesnay, "Scikit-learn: Machine learning in python," *Journal of Machine Learning Research*, vol. 12, no. 85, pp. 2825–2830, 2011. [Online]. Available: http://jmlr.org/papers/v12/pedregosa11a.html
339. F. Pérez and B. E. Granger, "IPython: a system for interactive scientific computing," *Computing in Science and Engineering*, vol. 9, no. 3, pp. 21–29, May 2007. [Online]. Available: https://ipython.org
340. D. Gafurov, K. Helkala, and T. Søndrol, "Gait recognition using acceleration from mems," in *The First International Conference on Availability, Reliability and Security*, Apr. 2006.
341. M. Muaaz and R. Mayrhofer, "An analysis of different approaches to gait recognition using cell phone based accelerometers," in *International Conference on Advances in Mobile Computing and Multimedia*, Vienna, Austria, 2013.
342. G. Yang, L. Xie, M. Mantysalo, X. Zhou, Z. Pang, L. D. Xu, S. Kao-Walter, Q. Chen, and L.-R. Zheng, "A health-IoT platform based on the integration of intelligent packaging, unobtrusive bio-sensor, and intelligent medicine box," *IEEE Transactions on Industrial Informatics*, vol. 10, no. 4, 2014.
343. B. Xu, L. D. Xu, H. Cai, C. Xie, J. Hu, and F. Bu, "Ubiquitous data accessing method in IoT-based information system for emergency medical services," *IEEE Transactions on Industrial Informatics*, vol. 10, no. 2, May 2014.
344. H. B. Menz, S. R. Lord, and R. C. Fitzpatrick, "Acceleration patterns of the head and pelvis when walking are associated with risk of falling in community-dwelling older people," *Journal of Gerontology: Medical Sciences*, vol. 58, 2003.
345. D. Anguita, A. Ghio, L. Oneto, X. Parra, and J. L. Reyes-Ortiz, "Human activity recognition on smartphones using a multiclass hardware-friendly support vector machine," in *International Workshop of Ambient Assisted Living (IWAAL)*, Vitoria-Gasteiz, Spain, Dec. 2012.
346. "Data collection as a barrier to personalized medicine," *Trends in Pharmacological Sciences*, vol. 36, no. 2, pp. 68–71, 2015.
347. A. Stisen, H. Blunck, S. Bhattacharya, T. S. Prentow, M. B. Kjærgaard, A. Dey, T. Sonne, and M. M. Jensen, "Smart devices are different: Assessing and mitigatingmobile sensing heterogeneities for activity recognition," in *Proceedings of the 13th ACM Conference on Embedded Networked Sensor Systems*, ser. SenSys '15. New York, NY, USA: ACM, 2015, pp. 127–140.
348. S. A. Rokni and H. Ghasemzadeh, "Synchronous dynamic view learning: A framework for autonomous training of activity recognition models using wearable sensors," in *2017 16th ACM/IEEE International Conference on Information Processing in Sensor Networks (IPSN)*, April 2017, pp. 79–90.
349. D. Cook, K. D. Feuz, and N. C. Krishnan, "Transfer learning for activity recognition: a survey," *Knowledge and Information Systems*, vol. 36, no. 3, pp. 537–556, Sep 2013.
350. A. Calatroni, D. Roggen, and G. Tröster, "Automatic transfer of activity recognition capabilities between body-worn motion sensors: Training newcomers to recognize locomotion," in *Eighth International Conference on Networked Sensing Systems (INSS'11)*, Penghu, Taiwan, Jun. 2011.
351. Z. Zhao, Y. Chen, J. Liu, and M. Liu, "Cross-mobile elm based activity recognition," *International Journal of Engineering and Industries*, vol. 1, 01 2011.
352. R. Fallahzadeh and H. Ghasemzadeh, "Personalization without user interruption: Boosting activity recognition in new subjects using unlabeled data," in *2017 ACM/IEEE 8th International Conference on Cyber-Physical Systems (ICCPS)*, April 2017, pp. 293–302.
353. H. Hachiya, M. Sugiyama, and N. Ueda, "Importance-weighted least-squares probabilistic classifier for covariate shift adaptation with application to human activity recognition," *Neurocomputing*, vol. 80, pp. 93–101, 2012, special Issue on Machine Learning for Signal Processing 2010.
354. J. Quionero-Candela, M. Sugiyama, A. Schwaighofer, and N. D. Lawrence, *Dataset Shift in Machine Learning*. The MIT Press, 2009.
355. B. W. Silverman, "Density ratios, empirical likelihood and cot death," *Journal of the Royal Statistical Society*, vol. 27, no. 1, pp. 26–33, 1978.

356. J. Ćwik and J. Mielniczuk, "Estimating density ratio with application to discriminant analysis," *Communications in Statistics–Theory and Methods*, vol. 18, no. 8, pp. 3057–3069, 1989.
357. H. Daume Iii, "From zero to reproducing kernel hilbert spaces in twelve pages or less," Mar 2004.
358. J. C. Spall, *Introduction to Stochastic Search and Optimization*, 1st ed. New York, NY, USA: John Wiley & Sons, Inc., 2003.
359. P. Bartlett and H. Lei, "Representer theorem and kernel examples," 2008.
360. D. Anguita, A. Ghio, L. Oneto, X. Parra, and J. L. Reyes-Ortiz, "A public domain dataset for human activity recognition using smartphones," in *ESANN*, 2013.
361. M. Hall, E. Frank, G. Holmes, B. Pfahringer, P. Reutemann, and I. H. Witten, "The WEKA data mining software: An update," *SIGKDD Explor. Newsl.*, vol. 11, no. 1, pp. 10–18, Nov. 2009.
362. A. Tang, S. Sethumadhavan, and S. J. Stolfo, "Unsupervised anomaly-based malware detection using hardware features," in *International Workshop on Recent Advances in Intrusion Detection (RAID'14)*. Springer, 2014, pp. 109–129.
363. M. Ozsoy, C. Donovick, I. Gorelik, N. Abu-Ghazaleh, and D. Ponomarev, "Malware-aware processors: A framework for efficient online malware detection," in *HPCA'15*, Feb 2015, pp. 651–661.
364. B. Singh, D. Evtyushkin, J. Elwell, R. Riley, and I. Cervesato, "On the detection of kernel-level rootkits using hardware performance counters," in *ASIACCS'17*, 2017.
365. Y. Yarom and et.all, "Flush+ reload: A high resolution, low noise, l3 cache side-channel attack." in *USENIX Security Symposium*, 2014.
366. D. Gruss, C. Maurice, K. Wagner, and S. Mangard, "Flush+ flush: a fast and stealthy cache attack," in *International Conference on Detection of Intrusions and Malware, and Vulnerability Assessment*. Springer, 2016, pp. 279–299.
367. C. Disselkoen, D. Kohlbrenner, L. Porter, and D. Tullsen, "Prime+ abort: A timer-free high-precision L3 cache attack using intel TSX," in *26th USENIX Security Symposium (USENIX Security 17),(Vancouver, BC)*, 2017, pp. 51–67.
368. P. Kocher and et.all, "Spectre attacks: Exploiting speculative execution," *arXiv preprint arXiv:1801.01203*, 2018.
369. M. Lipp, M. Schwarz, D. Gruss, T. Prescher, W. Haas, S. Mangard, P. Kocher, D. Genkin, Y. Yarom, and M. Hamburg, "Meltdown," *arXiv preprint arXiv:1801.01207*, 2018.
370. N. Patel, A. Sasan, and H. Homayoun, "Analyzing hardware based malware detectors," in *DAC'17*. ACM, 2017, pp. 25:1–25:6.
371. M. R. Guthaus, J. S. Ringenberg, D. Ernst, T. M. Austin, T. Mudge, and R. B. Brown, "Mibench: A free, commercially representative embedded benchmark suite," in *Fourth Annual IEEE International Workshop on Workload Characterization (IISWC'01)*, Dec 2001, pp. 3–14.
372. J. L. Henning, "Spec cpu2006 benchmark descriptions," *SIGARCH Comput. Archit. News*, vol. 34, no. 4, pp. 1–17, Sep. 2006.
373. V. intelligence service, "virusshare.com," in http://www.virustotal.com/intelligence/, April 2018.
374. M. Helsely, "Lxc: Linux container tools," in *IBM developer works technical library*, 2009.
375. Intel, "Intel 64 and ia-32 architectures software developer manual, volume 3b: System programming guide," 2016.
376. Z. Allaf, M. Adda, and A. E. Gegov, "A comparison study on flush+ reload and prime+ probe attacks on AES using machine learning approaches," in *UK Workshop on Computational Intelligence*, 2017.
377. I. Prada, F. D. Igual, and K. Olcoz, "Detecting time-fragmented cache attacks against AES using performance monitoring counters," *arXiv preprint arXiv:1904.11268*, 2019.
378. F. Yao, H. Fang, M. Doroslovacki, and G. Venkataramani, "Towards a better indicator for cache timing channels," *arXiv preprint arXiv:1902.04711*, 2019.

379. T. Zhang and et.all, "Cloudradar: A real-time side-channel attack detection system in clouds," in *RAID*. Springer, 2016.
380. C. Li and J.-L. Gaudiot, "Online detection of spectre attacks using microarchitectural traces from performance counters," in *2018 30th SBAC-PAD*. IEEE.
381. C. Delimitrou and C. Kozyrakis, "Quasar: resource-efficient and QoS-aware cluster management," in *ACM SIGARCH Computer Architecture News*, vol. 42, no. 1. ACM, 2014, pp. 127–144.
382. O. Alipourfard, H. H. Liu, J. Chen, S. Venkataraman, M. Yu, and M. Zhang, "Cherrypick: Adaptively unearthing the best cloud configurations for big data analytics," in *14th {USENIX} Symposium on Networked Systems Design and Implementation ({NSDI} 17)*, 2017, pp. 469–482.
383. S. Venkataraman, Z. Yang, M. Franklin, B. Recht, and I. Stoica, "Ernest: efficient performance prediction for large-scale advanced analytics," in *13th {USENIX} Symposium on Networked Systems Design and Implementation ({NSDI} 16)*, 2016, pp. 363–378.
384. N. J. Yadwadkar, B. Hariharan, J. E. Gonzalez, B. Smith, and R. H. Katz, "Selecting the best VM across multiple public clouds: A data-driven performance modeling approach," in *ACM SoCC*, 2017.
385. Q. Zhang *et al.*, "Dynamic energy-aware capacity provisioning for cloud computing environments," in *ACM ICAC*, 2012, pp. 145–154.
386. M. Guevara, B. Lubin, and B. C. Lee, "Navigating heterogeneous processors with market mechanisms," in *High Performance Computer Architecture (HPCA2013), 2013 IEEE 19th International Symposium on*. IEEE, 2013, pp. 95–106.
387. B. Guenter, N. Jain, and C. Williams, "Managing cost, performance, and reliability tradeoffs for energy-aware server provisioning," in *IEEE INFOCOM*, 2011.
388. A. Altomare, E. Cesario, and A. Vinci, "Data analytics for energy-efficient clouds: design, implementation and evaluation," *Journal of Parallel, Emergent and Distributed Systems*, 2018.
389. C. Delimitrou and C. Kozyrakis, "Paragon: Qos-aware scheduling for heterogeneous datacenters," in *ACM SIGPLAN Notices*, vol. 48, no. 4. ACM, 2013, pp. 77–88.
390. G. Kousiouris *et al.*, "Dynamic, behavioral-based estimation of resource provisioning based on high-level application terms in cloud platforms," *Elsevier Future Generation Computer Systems*, 2014.
391. C. Delimitrou and C. Kozyrakis, "Hcloud: Resource-efficient provisioning in shared cloud systems," *ACM SIGOPS Operating Systems Review*, vol. 50, no. 2, pp. 473–488, 2016.
392. S. M. Zahedi and B. C. Lee, "Ref: Resource elasticity fairness with sharing incentives for multiprocessors," *ACM SIGARCH Computer Architecture News*, vol. 42, no. 1, pp. 145–160, 2014.
393. N. Kulkarni, F. Qi, and C. Delimitrou, "Leveraging approximation to improve datacenter resource efficiency," *IEEE Computer Architecture Letters*, vol. 17, no. 2, pp. 171–174, 2018.
394. C. Delimitrou and C. Kozyrakis, "Bolt: I know what you did last summer... in the cloud," in *ACM SIGARCH Computer Architecture News*, vol. 45, no. 1. ACM, 2017, pp. 599–613.
395. *Rightscale Inc. 2017. Amazon EC2: Rightscale.* http://www.rightscale.com/, 2017.
396. Y. Apache Hadoop, "Yet another resource negotiator," *Proceedings of ACM SoCC*, 2013.
397. P. Bodik, R. Griffith, C. Sutton, A. Fox, M. I. Jordan, and D. A. Patterson, "Automatic exploration of datacenter performance regimes," in *Proceedings of the 1st workshop on Automated control for datacenters and clouds*. ACM, 2009, pp. 1–6.
398. N. J. Yadwadkar, G. Ananthanarayanan, and R. Katz, "Wrangler: Predictable and faster jobs using fewer resources," in *Proceedings of the ACM Symposium on Cloud Computing*. ACM, 2014, pp. 1–14.
399. A. K. Maji, S. Mitra, B. Zhou, S. Bagchi, and A. Verma, "Mitigating interference in cloud services by middleware reconfiguration," in *Proceedings of the 15th International Middleware Conference*. ACM, 2014, pp. 277–288.
400. F. Romero and C. Delimitrou, "Mage: Online interference-aware scheduling in multi-scale heterogeneous systems," *arXiv preprint arXiv:1804.06462*, 2018.

Index

S. Rafatirad et al., *Machine Learning for Computer Scientists and Data Analysts*,
https://doi.org/10.1007/978-3-030-96756-7

Q

R

S

MIX
Papier aus verantwortungsvollen Quellen
Paper from responsible sources
FSC® C105338
FSC
www.fsc.org

If you have any concerns about our products,
you can contact us on
ProductSafety@springernature.com

In case Publisher is established outside the EU,
the EU authorized representative is:
Springer Nature Customer Service Center GmbH
Europaplatz 3, 69115 Heidelberg, Germany

Printed by Libri Plureos GmbH
in Hamburg, Germany